Statistical Methods for SPC and TQM

CHAPMAN & HALL STATISTICS TEXTBOOK SERIES

OTHER TITLES IN THE SERIES INCLUDE

Practical Statistics for Medical Research
D. G. Altman
The Analysis of Time Series an Introduction
Fourth edition
C. Chatfield
Problem Solving A Statistician's Guide
C. Chatfield
Applied Statistics Principles and Examples
D. R. Cox and E. J. Snell
An Introduction to Generalized Linear Models
Second Edition
A. J. Dobson
Multivariate Statistics A Practical Approach
B. Flury and H. Riedwyl
Readings in Decision Analysis
S. French
Multivariate Statistical Methods A Primer
B. F. J. Manly
Applied Non-parametric Statistical Methods
Second Edition
P. Sprent
Probability Methods and Measurement
A. O'Hagan
Decision Analysis A Bayesian Approach
J. Q. Smith
Applied Statistics A Handbook of GENSTAT Analyses
E. J. Snell and H. R. Simpson
Elementary Applications of Probability Theory
H. C. Tuckwell
Statistical Process Control Theory and Practice
Third Edition
G. B. Wetherill and D. W. Brown
Statistics in Research and Development
Second Edition
R. Caulcutt
Modelling Binary Data
D. Collett
Statistical Analysis of Reliability Data
M. J. Crowder, A. C. Kimber, T. J. Sweeting and R. L. Smith
Statistical Methods in Agriculture and Experimental Biology
Second Edition
R. Mead, R. N. Curnow and A. M. Hasted
Statistical Methods for SPC and TQM
Derek Bissell

Full information on the complete range of Chapman & Hall statistics books is available from the publishers.

Statistical Methods for SPC and TQM

Derek Bissell

Consultant and lecturer in
Statistics for Industrial Applications and consultant to
Department of Trade and Industry on Statistics in Weights and Measures.
Honorary Professor in Statistics for SPC and TQM,
University of Wales, Swansea, UK.

CHAPMAN & HALL
London · Glasgow · New York · Tokyo · Melbourne · Madras

Published by Chapman and Hall, 2–6 Boundary Row, London SE1 8HN, UK

Chapman & Hall, 2–6 Boundary Row, London SE1 8HN, UK

Blackie Academic & Professional, Wester Cleddens Road, Bishopbriggs, Glasgow G64 2NZ, UK

Chapman & Hall Inc., One Penn Plaza, 41st Floor, New York NY 10119, USA

Chapman & Hall Japan, Thomson Publishing Japan, Hirakawacho Nemoto Building, 6F, 1-7-11 Hirakawa-cho, Chiyoda-ku, Tokyo 102, Japan

Chapman & Hall Australia, Thomas Nelson Australia, 102 Dodds Street, South Melbourne, Victoria 3205, Australia

Chapman & Hall India, R. Seshadri, 32 Second Main Road, CIT East, Madras 600 035, India

First edition 1994

Typeset in 10/12 Times by Thomson Press (India) Ltd, New Delhi
Printed in Great Britain by T. J. Press Padstow (Cornwall) Ltd

ISBN 0 412 39440 5

A catalogue record for this book is available from the British Library

Library of Congress Cataloging-in-Publication data available

Printed on permanent acid-free text paper, manufactured in accordance with the proposed ANSI/NISO Z 30.48–199X and ANSI Z 39.48–1984

Contents

Preface

The 1980's saw a re-awakening of interest in statistical methods in industry. Several factors contributed to this, notably the adoption of 'statistical process control (SPC) by some major multinational firms, and their insistence on its use by their suppliers and sub-suppliers. Other influences included the 1979 Weights and Measures legislation with its requirements for sampling and monitoring, the flood of firms seeking BS 5750 or ISO 9000 accreditation, an increasing consumer demand for high quality combined with value for money and the recognition that, to meet this demand, quality improvement, productivity and business survival are closely related. Acceptance of this principle, not only in manufacturing industries but in distribution, transport, retailing, catering and public service, has led to the philosophy of 'total quality' (in several guises such as TQ-management, TQ-involvement, TQ-excellence – hence TQM, TQI, TQE, etc.), requiring performance improvement in all aspects of any operation.

Based on the maxim that progress means 1000 people taking one step forward rather than one person taking 1000 steps, attention in SPC has focussed on a limited range of simple techniques that can be effectively used by many individuals. The 'seven statistical tools' for 'quality circles' provide an example, although there are several different versions of these tools!

However, when these methods are effectively used, some of their limitations also become apparent. For example, the normal probability plot is a useful device, and in many cases leads to drawing a best-fit line by eye to represent a distribution and its relation to a specification. In less straightforward situations – with appreciable scatter, or a sprinkling of 'rogue' values, for example – a line may still be useful but less easy to draw. Alternatively, a curve may appear to be more appropriate than a straight line. In these applications, a different approach may be required, necessitating the use of slightly more advanced statistical methods.

One might cite many other examples: the need to adapt basic types of control chart ($\bar{x}$, R etc.) to unusual circumstances; the handling of non-normal distributions in capability analysis, and the relevance of simple experimental designs to the investigation of problems or trying out ideas for improvement.

When SPC, linked to a TQ-philosophy, is applied in process support activities and general business areas, the statistical base needs to be widened to take in useful methods beyond any particular kit of seven tools. This book sets out to cover these extensions. It is not an SPC primer, nor an evangelical message to executives on a new approach to management. It assumes a commitment to TQ, or continuous improvement, along with a reasonable working knowledge of the basic quantitative methods of SPC. An ability to use a scientific calculator (or basic statistical software for a computer) is required, along with some familiarity with algebraic principles of representing numbers by symbols in formulae. It will not be necessary for the reader to understand matters like differentiation, integration, the summing of series, or trigonometry.

Thus the techniques covered here are explored in greater depth than is required by most individuals engaged in SPC, quality circles, TQ teams, etc. It is intended for those who will advise, co-ordinate and lead these activities, or analyse the data arising from them. Many firms are finding it useful, even essential, for one or two 'local experts' to acquire this deeper understanding, and it is to the SPC co-ordinator or quality improvement engineer that this book is addressed.

The text is based on successful courses run by the author under the title 'SPC+'. I am grateful to the many who have encouraged (even nagged) me to adapt my course notes for publication. They include clients who have allowed me to use examples of applications, colleagues who have assisted in developing simplified versions of established methods, and course participants who have offered suggestions on presentation.

Finally, I dedicate the book to my family – to my first wife, Yvonne, with whom I was planning my first SPC+ course just prior to her death; to my second wife, Ena, who has encouraged me to continue with writing the final draft, and to my daughter and son-in-law for their support and very practical assistance in word-processing both a tentative first draft and this final version.

Derek Bissell*
Abergavenny
Gwent, UK

*Derek Bissell is the adopted business name of Professor A. F. Bissell.

1

Introduction: statistics, SPC and total quality

1.1 Statistics and quality

1.1.1 Much of this book is concerned with statistical methods relevant to **statistical process control** (SPC). As the reader will discover, the author takes a wide view of the scope of SPC, and therefore of the methods which are useful in its implementation. This introduction emphasizes that the activities embraced herein are all of proven value in quality assurance, quality improvement and the management of industrial production and technology. Many of the methods covered are also applicable to the non-technological aspects of **total quality management** (TQM). The emphasis in these areas is on the improvement of efficiency in the use of time and materials, smoother flow of documents and materials through administration, warehousing, maintenance, etc., and monitoring of occurrences such as absence, accidents and equipment faults, while applying the SPC principles of problem solving and continuous improvement.

1.1.2 The function of SPC is often (erroneously) narrowed to that of using simple control charts to detect assignable (or special) causes of variation, and taking prompt corrective action whenever 'out-of-control' is signalled. In many processes, there may be no simple corrective action: detailed investigation may be necessary to identify causes. In others, the purpose of a control technique may be to indicate changes that do not require correction but which result in management decisions of various kinds. In some cases, tests such as tensile, electrical, etc. may be carried out as part of overall **quality assurance** activity. While instant corrections may be totally impracticable, charting and monitoring of such data can contribute to a long term understanding of the process or system. Similar considerations apply to such data as absence or accidents, where trends, relationships to environmental conditions or the effects of changes of policy may be the object of study.

1.1.3 In follow-up analysis, problem-solving and technical development, methods beyond the scope of this book may be required. Topics such as regression analysis, model fitting and the design/analysis of experiments come to mind. We cannot in one volume cover all of these techniques, and therefore we concentrate on those which support or expand the scope of SPC and quality assurance, at the expense of those concerned with development and experimentation. The interface between SPC and what is often termed **quality engineering** is vague, and there may be readers who deplore the omission of some methods but find the inclusion of others surprising. We hope the rationale for the selection will emerge in the succeeding chapters.

1.2 What is SPC?

1.2.1 A glimpse of the real meaning of SPC is afforded by reversing the order of the acronym:

Control: effective management
Process: the activity of converting inputs into output
Statistics: the collection, presentation and effective analysis of data.

The objective is not, even in manufacturing applications, simply that of monitoring to achieve conformity to a specification, but of *continuous improvement* of the whole system. Furthermore, the regular discipline of data collection and presentation provides a basis for objective, rational decision-making.

1.2.2 It is essential that the data collection does not become an end in itself. Data costs time, effort and money. It must therefore be

Collected wisely
Analysed effectively
Used efficiently (i.e. generating 'action').

We now examine the first two elements, control and the process, in more detail before proceeding to statistics.

1.3 Control: management of the process

1.3.1 Modern ideas of management are increasingly focussed on 'quality' as a strategy for survival and growth. Much of the credit for this must be attributed to W. Edwards Deming and J.K. Juran. It is unfortunate that, although their ideas were available to the Western nations in the 1950s, the effectiveness of their approaches was proved by the Japanese before they gained wide acceptance in the United States, Britain and Europe.

1.3.2 The ideas underlying most **quality management** philosophies may be summarized as follows, although different 'gurus' vary the emphasis on different aspects.

Strategy: plan and organize for long term goals. Quality improvement must be led by senior management.
Logistics: provide methods, resources to implement the strategy. The resources will involve training and the commitment of senior management time to directing and supporting the programme.
Development: be prepared for change, solve problems, improve systems in all areas of activity.
Human relations: people are the most valuable resource in the organization-treat them accordingly. Participative management, job satisfaction, giving the right to pride of workmanship, must be implemented, not merely talked about.

1.3.3 Only in an environment where each member of the organization recognizes that he or she forms part of many supplier–customer relationships can SPC and associated methods fully succeed. Suppliers (and receivers) of service, facilities, documents, information, instructions, as well as materials and components, need this collaborative approach to promote real improvements in all systems. One may identify three levels of motivation for SPC.

1. We do it (albeit reluctantly) with respect to products for a major customer because he insists.
2. We use it to improve the quality of our products and the efficiency of our manufacturing operations.
3. SPC forms part of our (total quality) philosophy, which extends to all aspects of our business. It is applied to quality and productivity of operations, to services such as maintenance, transport and material supply, and to administrative areas like safety, personnel, computer management and sales activities.

The first will (with luck) avoid the loss of the major customer, but will achieve little in terms of system improvement. The second can realize real advantages in both productivity and quality, but may result in conflict or misunderstanding with supporting services who do not fully understand or appreciate the potential of the methods available. Only the third approach, energetically promoted and maintained from the top, can realize this potential for improving not only quality, productivity and cost-effectiveness, but human relations, removal of obstacles, balancing workloads and smoother running of all operations.

1.4 The process

1.4.1 The word **process**, in an SPC context, tends to be associated with some form of manufacture – a machining process, steelmaking or chemical distillation, perhaps. We shall take a much wider view, and regard a process as any manufacturing, clerical, administrative, service or other system that constitutes the sequence

$$\text{Input} \rightarrow \text{Activity} \rightarrow \text{Output}$$

Table 1.1 Examples of processes

Input	Activity	Output
Manuscript, audio tape →	Word processing →	Letters, documents
Clock cards, time records →	Computing →	Pay slips, credit transfers
Raw data →	Data processing →	Reports, statements, invoices, VAT returns
Enquiries →	Sales activity →	Orders
Orders →	Manufacture or selection from stock →	Delivery
Seeds, fertilizers →	Cultivation →	Crop
Materials or components →	Manufacturing →	Finished goods or 'value-added' materials

For example we have

$$\text{Food} \rightarrow \text{Digestion} \rightarrow \text{Energy}$$

(Is this not, in any case, known as the digestive **process?**)

Here, food provides the input, digestion the activity. Both need to be controlled to ensure the right output. One might even consider monitoring energy input and charting weight gain or loss as examples of the use of SPC techniques!

Some less flippant examples include agricultural, clerical, service and manufacturing processes, such as those in Table 1.1, where the inputs may comprise materials, documents or verbal statements, and the activities may be carried out by individual action or the use of equipment.

1.4.2 The traditional role of quality control has been to check the output of the process, identifying substandard output for rectification, repair, re-work, downgrading (for sale to undiscerning customers as 'seconds'?) or outright rejection. Such control (which is certainly not the activity intended by the statistical quality control pioneers like Shewhart in the USA and Pearson and Dudding in Britain) is analogous to steering a ship by looking at the wake. The identification of half-a-dozen aspects common to all processes and systems can focus attention on problem prevention rather than trouble detection. Various authors list four, five or all six of these aspects under different names, e.g. men or people, machinery or equipment, methods or procedures. We adopt the acronym IMPOSE, i.e. a means of applying the system strategy so that the intended outputs are realised. Figure 1.1 places these aspects in the system context, and they should be regarded as sources of data, areas for control and improvement and for the application of problem-solving techniques.

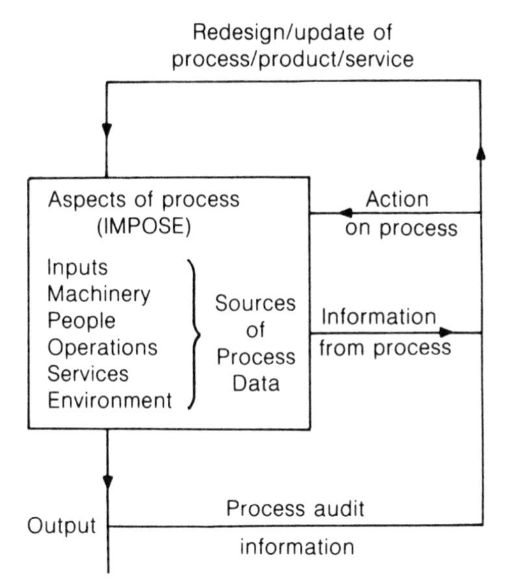

Fig. 1.1 Process control.

Whether in production, design, administration, service, the results of control (rather than detection of faults) are smoother operations, reduced cost, higher outputs, improved quality and customer satisfaction and better human relations. The latter result from the sense of a job done well, and from the satisfaction gained by people with a day-to-day knowledge of problems in the system actually having that knowledge tapped and acted upon.

1.5 Statistics

1.5.1 Having briefly discussed 'control' and 'process', we must now define **Statistics**. In this particular context it is any quantitative technique which contributes to

monitoring and control
understanding the system
problem identification
problem diagnosis
problem solving
improving the system
rational and objective decisions.

These techniques may be grouped as follows.

Logical data collection: aimed at solving problems and understanding systems.

Well-designed record sheets: making it easy to collect the required data, using tally sheets, control charts, fault location diagrams, questionnaires, etc.
Statistical summaries: both numerical and graphical; tables, charts and diagrams, measures of magnitude and variation, rates and percentages.
Establishing priorities: pareto analysis, ranking, consensus judgement.
Generating and organizing ideas: including brainstorming, cause and effect analysis, establishing quality circles or problem-solving techniques. **Failure mode** and **effects analysis** (FMEA) could be usefully employed here in a broad sense.
Looking at relationships: two-way tables and scatter diagrams for investigating numerical cause and effect patterns.
Monitoring and control: observing patterns over time, using run and control charts.

These activities are served by the basic statistical tools of SPC/TQM, and some are briefly reviewed in early chapters of this book. The bibliography lists sources of more detailed information.

The main purpose, however, is to provide a grounding in the statistical concepts on which these and other tehchniques depend, with a minimum of mathematical theory. These concepts will then be applied to SPC topics, dealing especially with the following areas.

First principles of control charts: with applications to individual values, moving averages, non-normal data and processes with special problems or features.
Cusum methods: not only as an alternative to control charts, but as a general tool for data presentation and analysis.
Capability analysis: evaluating the ability of a process, machine or measurement system to produce results acceptable to the user or customer. Included here is 'gauge capability', or the evaluation of **repeatability** and **reproducibility**. Methods based on **normal** and **non-normal distributions** are presented.

1.5.2 We bear in mind, however, that SPC, in either product, process or total quality applications, must not be dominated by techniques. To use an often-quoted aphorism, SPC is 20% statistics, 10% common sense and 70% engineering. For non-engineering applications, one would substitute a detailed expertise in the area of interest – medicine, computers, plastics technology. Or, as W. Edwards Deming puts it, 'statistics is a basic but comprehensive way of thinking and acting which combines information, opinion and common-sense for the better design of products, systems, strategies and decision-making'. The accent is on communication, teamwork and problem-solving.

1.5.3 Successful implementation of SPC, or the effective use of statistical techniques for effective system management, thus depends on a blend of management skills, technology, human relations and statistics. Figure 1.2 summarizes the content of this chapter with regard to the role of the statistical methods,

Statistical Process Control = Efficient Management (*Control*) of any System (*Process*) using quantitative information effectively presented and analysed (*statistics*)

Continuous improvement	Areas of application	Examples of statistical tools
Measure performance	Data gathering	sampling, checksheets, tallies, fault diagrams, control and run charts
↓	Data presentation	stem/leaf, box/whisker, histogram, bar chart, run chart, scatter/blob diagram, other graphs and tables, control charts.
Highlight problems	Setting priorities	Pareto analysis, ranking, risk analysis, FMEA
↓	Finding causes	Cause-effect diagrams, brainstorming, collation/correction, FMEA, flow sheets
Identify causes	Performance measures	Averages, measures of variation, rates, proportions, capability indices
↓	Monitoring	Control charts (various), cusums
Develop solutions	Relating to process	Stratification, collation, annotation, planned observation, experiments
↓		
Measure improvement (→ back to Measure performance)		

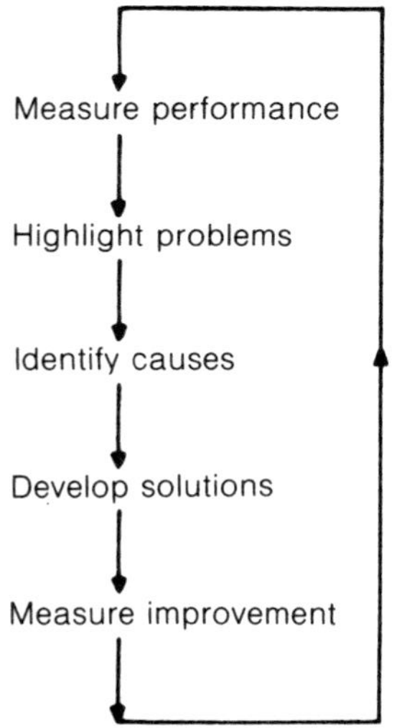

(FMEA = Failure mode and effects analysis)

Fig. 1.2 Statistical methods in system and process management.

irrespective of the nature of the process, the product or the size or type of organization. Often the most difficult step is that of finding a suitable measure of performance, especially in applications that are not industrial nor driven by a customer specification.

The statistical methods are often limited to a so-called set of 'seven statistical tools'. Any set of seven is necessarily restrictive, so Fig. 1.2 uses seven potential areas of application, with illustrations of *some* of the appropriate tools. There is considerable overlap between these areas of application and the tools – control charts provide convenient data collection forms, effective means of monitoring, and a useful form of data presentation. They also provide a ready format for further analysis, including the calculation of performance measures.

The rest of this book is concerned with developing an understanding of statistical principles so that the various tools can be properly and usefully applied to process management and improvement.

2

Data collection and graphical summaries

2.1 Introduction

2.1.1 This Chapter provides a review of some basic methods of presenting raw data in a more digestible form, and some less well known but very useful techniques. None of these tools should be considered in isolation: they will almost always be used in a progression or combination in the course of a real application.

It is worth reviewing some of the essential principles of data collection and recording. Often data is collected as a habit: perhaps someone started it long ago for a purpose now obscured, and the routine has continued ever since. No one dares to question its relevance or suitability in changing conditions. Some important points include:

Objective: Is the data required for legal purposes (accounts, VAT, customer documentation), control, solving a specific problem or management information?

Type: Is the data obtained by counting or measurement? Samples or 100% checks?

Frequency: Is it required once only, or for regular checks (once per hour, every batch or shift, etc.)?

Recording format: Will this be automatic data capture, record sheets, tallies, control charts, questionnaires? Document design should be considered in terms of relevance, simplicity, ease of transfer of information to other media (e.g. computer), possible multiple use of the same data for various purposes, and facilitating calculation of summaries.

Communication and training: Ensuring all concerned know the relevance and importance, and are familiar with the correct procedure, methods of calculation and avoiding errors.

Integrity: Data is used for control, making decisions, diagnosing problems and other important functions. It needs to be objective, honest, legible and (especially where sampling is concerned) properly representative of the system or process.

The collection of data does not, of itself, solve problems. It is sometimes thought that SPC in particular is concerned with displaying control charts on machines or workstations and carrying out various routine calculations (e.g. capability indices). These are only the outward signs; nearer to the true meaning of SPC is **solving problems collectively**. This places the emphasis on solving problems rather than just collecting information about them, and on teamwork rather than buck-passing.

Data collection must therefore be accompanied by such activities as flow-charting, brainstorming, cause and effect analysis and Pareto analysis. Texts covering 'statistical process control' and 'total quality management' deal with these topics – they are not the subject-matter of the present book, but their importance must not be under-rated.

2.2 Organizing raw data

2.2.1 In its raw form, data is rarely useful. Care is needed in presentation and extracting useful summary statistics. Consider the following set of data:

238.9, 238.3, 240.4, 241.0, 239.0, 239.3, 239.6, 236.9, 239.3, 240.1, 238.8, 240.7, 241.0, 240.1, 239.3, 239.1, 239.5, 239.9, 238.9, 237.4, 238.2, 238.4, 239.2, 239.7, 239.4, 240.1, 240.3, 239.5, 239.0, 238.6, 239.8, 240.3, 239.7, 238.7, 240.7, 240.0, 240.1, 241.6, 239.8, 240.6, 239.7, 239.7, 240.4, 239.5, 238.6, 239.2, 237.6, 238.9, 239.2, 240.3, 239.4, 240.8.

In this form, the figures convey little except that they are all fairly close to 240. What are they? And have we any background information?

The data make rather more sense when we learn that they are weights (in grams) of valve liners moulded in a corrosion resistant plastic material. They are weighed as a quality check (if they weigh too little, they may contain voids or be undersize; if too much, they may be of poor shape due to improper mould closure). There is a design specification of 240 ± 5 g.

While one can now examine whether the weights (about equal to a half-pound of butter!) satisfy the specification, it is impossible to discern any pattern. Some organization is needed.

2.2.2 The stem-and-leaf table

The 'stem and leaf' table is a useful first step. It is based on identifying which digits do not vary at all (the hundreds, in this case), which vary little (the tens) and hence where the real variation begins(units.) Taking a set of units which covers the range of the data (say 235–242 to be on the generous side) as 'stems', the remaining digits (the decimal part in this example) are allocated as leaves on the appropriate stems. In other cases the stems might be hundreds or tens, the leaves tenths or hundredths, etc.

Table 2.1 Stem-and-leaf table: valve liner data

Stem	Leaves
242	
241	0 0 6
240	4 1 7 1 1 3 3 7 0 1 6 4 3 8
239	0 3 6 3 3 1 5 9 2 7 4 5 0 8 7 8 7 7 5 2 2 4
238	9 3 8 9 2 4 6 7 6 9
237	4 6
236	9
235	

Table 2.2 Final (ordered) stem-and-leaf table

Stem	Leaves
241	0 0 6
240	0 1 1 1 1 3 3 3 4 4 6 7 7 8
239	0 0 1 2 2 2 3 3 3 4 4 5 5 5 6 7 7 7 7 8 8 9
238	2 3 4 6 6 7 8 9 9 9
237	4 6
236	9

In the example, the first value is 238.9, so the leaf .9 is allocated to the stem 238, then .3 also to stem 238, .4 to 240, etc., as in Table 2.1. Immediately a pattern emerges – the values cluster around a centre (stem 239), tapering off in each direction. Even at this stage, it is apparent that although all the values lie within the specification, there are more values in the lower half (239.9 and below) than in the upper half (240.0 and above). Of course, this may be a deliberate saving on materials!

If required, the leaves within each stem can now be rearranged in ascending order to give a complete ranking order which is useful for identifying quantiles such as the median and first and third quartiles, as described in section 2.3.3. This yields (deleting the unused stems 242 and 235) the presentation in Table 2.2.

2.2.3 *The frequency table*

Useful though it is in initially sorting data, the stem-and-leaf table does not necessarily give a satisfactory final presentation. The frequency table, in which data are collected into a reasonable number of classes, is a useful device, and leads naturally to the diagrammatic form of the histogram.

For the most effective presentation, a table with about ten or twelve classes of equal width is usually preferred. The stem-and-leaf table suggests that a set of classes based on half-gram intervals should achieve this. Care is necessary to avoid ambiguity at the class boundaries, e.g. groups comprising 236.5–237,

Table 2.3 Frequency table: valve liner data

Class limits	Mid-value (x)	Tallies	Frequency (f)
236.75–237.25	237.0	|	1
237.25–237.75	237.5	||	2
237.75–238.25	238.0	|	1
238.25–238.75	238.5	𝍸	5
238.75–239.25	239.0	𝍸 𝍸	10
239.25–239.75	239.5	𝍸 𝍸 |||	13
239.75–240.25	240.0	𝍸 |||	8
240.25–240.75	240.5	𝍸 |||	8
240.75–241.25	241.0	|||	3
241.25–241.75	241.5	|	1
			52

237–237.5, 237.5–238, etc., could lead to inconsistency, omission or even double counting. Possible unambiguous class limits include:

236.5–236.9 237.0–237.4 (defining the values to be included in any class)	or	236.45–236.95 236.95–237.45 (the extra decimal avoids ambiguity)

Another possible approach is to make the class **mid-values** as simple as possible. This is useful where the mid-values may be used to represent the whole class for subsequent calculations. Often, by calculating class limits half-way between the mid-values, ambiguity looks after itself, as in Table 2.3.

When the classes have been defined, a tally chart is compiled, using the 'five-bar gate' method. For the present example, half-gram mid-values are used in Table 2.3. With this number of classes, the pattern remains clear, but a little extra detail emerges. In this case, no major anomalies exist, but in other cases lopsidedness (skewness) or a tendency to multiple peaks (bimodality or multimodality) may be observed, leading to diagnosis of possible problems.

2.3 Graphical presentation

2.3.1 Histogram

While the tally chart is often adequate for displaying the pattern, the presentation may be improved by drawing a scaled bar chart or histogram. It is here that the advantage of using equal class widths becomes apparent, as the histogram

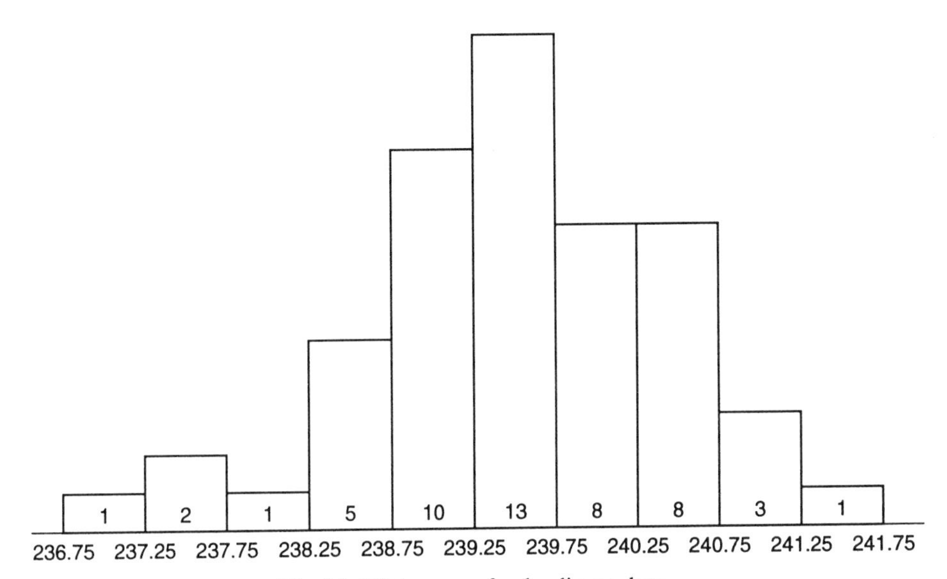

Fig. 2.1 Histogram of valve linear data.

columns have heights proportional to frequencies. If class widths vary, it is necessary to calculate areas to give the correct representation. Figure 2.1 shows a histogram for the valve liner data.

At this point it is appropriate to consider the relationship between a **sample** and the system, population or process from which it is drawn and which it purports to represent. All samples will contain minor irregularities, and the fact that two values in the above sample occurred in the 237.5 class, but only one each in the 237 and 238 classes, would not be taken to imply that the batch from which the sample was drawn contains, say, 200 valve liners in the 237.25–237.75 weight range, but only 100 in each of the neighbouring classes. To obtain an impression of the whole distribution, only the broad features are required. In this case, we have a fairly central hump and a pair of roughly equal tails.

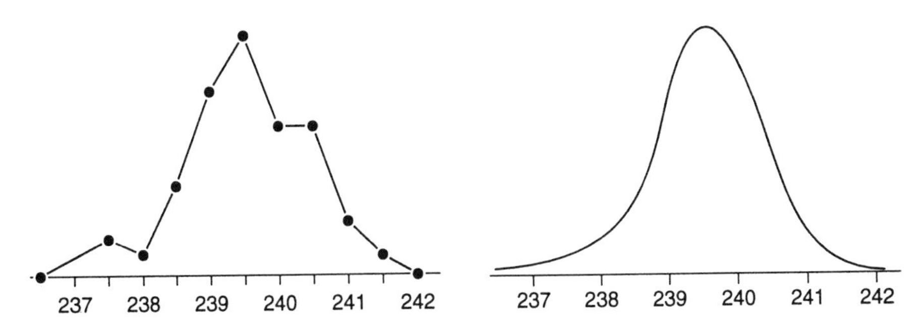

Fig. 2.2 (a) Frequency polygon; (b) frequency curve.

These indications are better shown by a frequency curve (Fig. 2.2(b)) than a polygon (Fig. 2.2(a)) which merely joins the tops of the histogram columns (and in fact does not even properly portray the **sample**, as the areas are incorrectly distributed over the classes).

2.3.2 *Blob diagrams and box-and-whisker plots*

A further graphical technique, which in some respects combines features of both the stem-and-leaf table and the frequency table, is the 'blob' diagram. A suitable scale is drawn and divided to represent the range of the values, and a blob is placed on the scale to represent each measurement. Repeated values are shown by piling up the blobs, so that frequencies are readily counted up. If desired, values can be rounded whilst compiling the blobs, thus yielding an alternative to the grouped frequency table. The final appearance is then as for the 'digidot' of the next section. Figure 2.3 shows an unrounded blob diagram for the valve-liner data.

Back-to-back blob diagrams are also useful for presenting data from two samples to highlight differences in average or variation. In Fig. 2.4, it is clear that sample B exhibits more variation than sample A, and in general the whole pattern for B is displaced to the right of that for A, i.e. there is a difference in 'average' (note that we are not yet defining precisely which average – mean, median, mode, etc. The discrepancy is one of general location of the A and B data sets).

2.3.3 *Box-and-whisker plots*

Where a sample or data set is numerous, the blob diagram may preserve too much detail, possibly masking the main features of location and dispersion. Again, if several or many (rather than two) samples are to be compared, multiple blob diagrams are not easy to interpret. In these cases, the box-and-whisker plot may be advantageous.

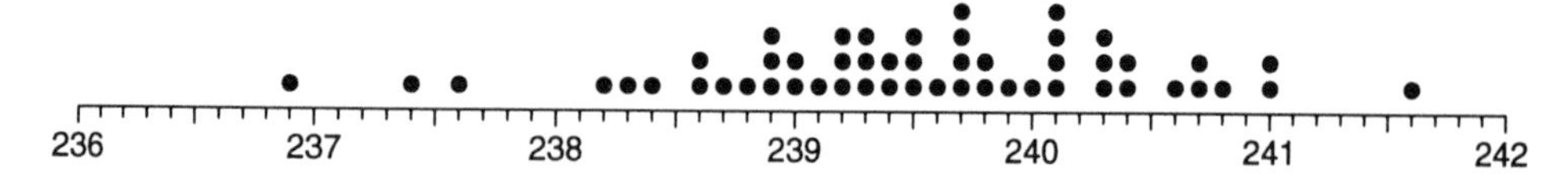

Fig. 2.3 Blob diagram for valve liner data.

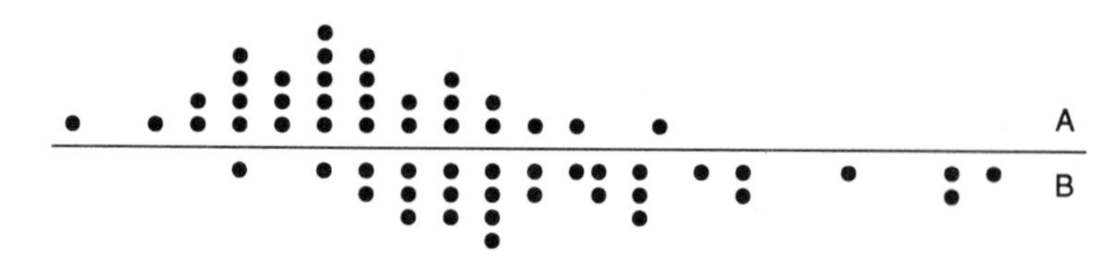

Fig. 2.4 Back-to-back blob diagrams.

We here assume that the reader is familiar with the **median** as a measure of location. When all individual values are available, it is defined as the middle value, or the average of the middle two, when the values are arranged in order. We also introduce the first and third **quartiles**, the values which separate the lower and upper 25% of the data from the central 50%. Note that the median is also the second quartile.

To locate the median and quartiles in the ordered data, for example in a stem-and-leaf table or a blob diagram, use the following expressions. The rationale will be explained in connection with probability plotting in Chapter 13.

$$Q_1 = x_i, \quad \text{where} \quad i = \tfrac{1}{4}(n+1);$$
$$M = x_i, \quad i = \tfrac{1}{2}(n+1);$$
$$Q_3 = x_i, \quad i = \tfrac{3}{4}(n+1).$$

Thus, for the 52 items in the data set forming the main example in this chapter,

For Q_1, $i = \frac{1}{4}(52+1) = 13.25$; so Q_1 lies between the thirteenth and fourteenth values from the lower end. These are in fact 238.9 and 239.0, so we may take Q_1 as 238.9 approximately (second decimal place accuracy is not justified). For M, $i = \frac{1}{2}(52+1) = 26.5$; with x_{26} and x_{27} both 239.5, the median is also 239.5.

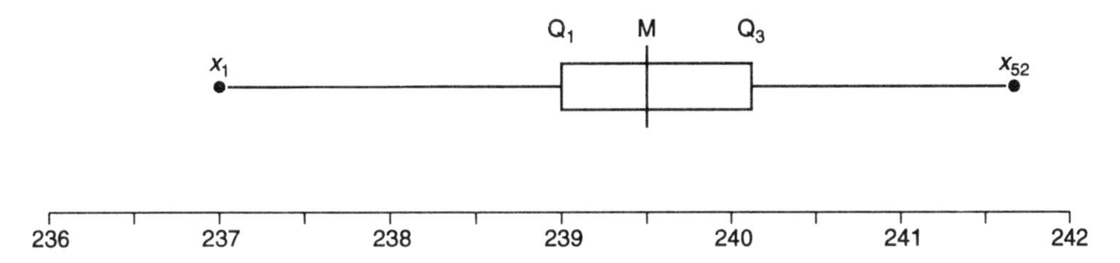

Fig. 2.5 Box-and-whisker plot for valve liner data.

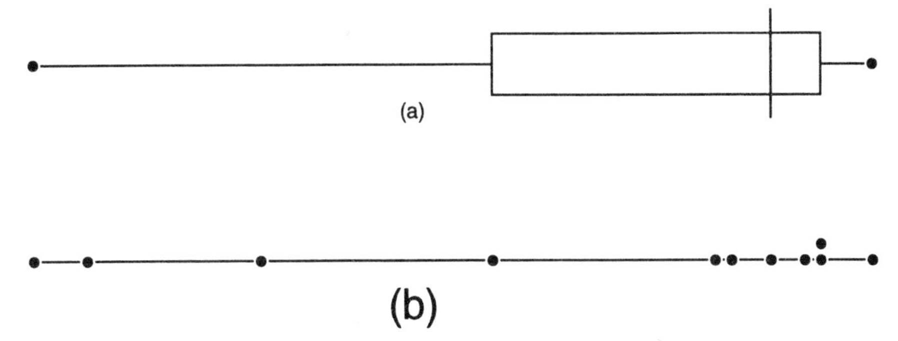

Fig. 2.6 Box-and-whisker plot expanded to blob diagram. A box-and-whisker plot (a) would suggest some peculiarity in the original data. A follow-up diagram (b) reveals values clustered at the upper end with wide gaps between values below the median.

For Q_3, $i = 39.75$; x_{39} and x_{40} are both 240.1. So we have $Q_1 = 238.9$, $M = 239.5$, $Q_3 = 240.1$.

The box-and-whisker plot is shown in Fig. 2.5. The median is identified by a line across the box, which covers the middle 50% range from Q_1 to Q_3. The whiskers extend from the box to the extreme values at each end of the data.

Blob diagrams contain more information than box-and-whisker plots, and a choice may need to be made according to the amount of detail required. Blobs as a follow-up to unusual appearance in the box-and-whisker plots may be a useful tactic, as illustrated in Fig. 2.6.

2.3.4 *Run chart and 'digidot'*

The time sequence in which data occur may often be significant, and plotting data in time (or other sequence) order will then be useful. A plot with time as the horizontal (left-to-right) axis and a vertical scale for the variable of interest is known variously as a line graph, run chart or time plot. In effect it is a control chart without statistically based limits, and serves as a simple but effective data summary permitting the detection of trends, cyclic behaviour, outlying values and other features. It may be used with guidelines or zones based on practical considerations, such as temperatures outside which extra heating or cooling equipment may need to be switched on, limits to process or material characteristics beyond which special action needs to be taken, and so on. Strict statistical control is not required for all cases, though it is always worth recognising and exploiting possibilities for preventing problems and reducing variation.

Figure 2.7 shows a run-chart (with specification limits and nominal value) and digidot for the valve-liner data. It emphasizes that many more results are below the nominal than above, and there would seem to be more risk of violating the lower specification limit than the upper. This point will recur later in the book for this set of data.

2.4 Two-dimensional data

2.4.1 *Scatter diagram*

Often two or more characteristics or properties are measured on the same items or on the same occasions. It is then useful to assess whether there are relationships between the variables, either in pairs or multiples. Numerical methods involve regression and correlation techniques which lie beyond the scope of the present text, and the presentation of three or more variables in graphical form also involves some complications. The most effective way of presenting two variables is the scatter diagram (or scatter plot).

The horizontal axis of the diagram is usually labelled x, the vertical y. If it is suspected that one variable may, in some sense, be the 'cause' of the other,

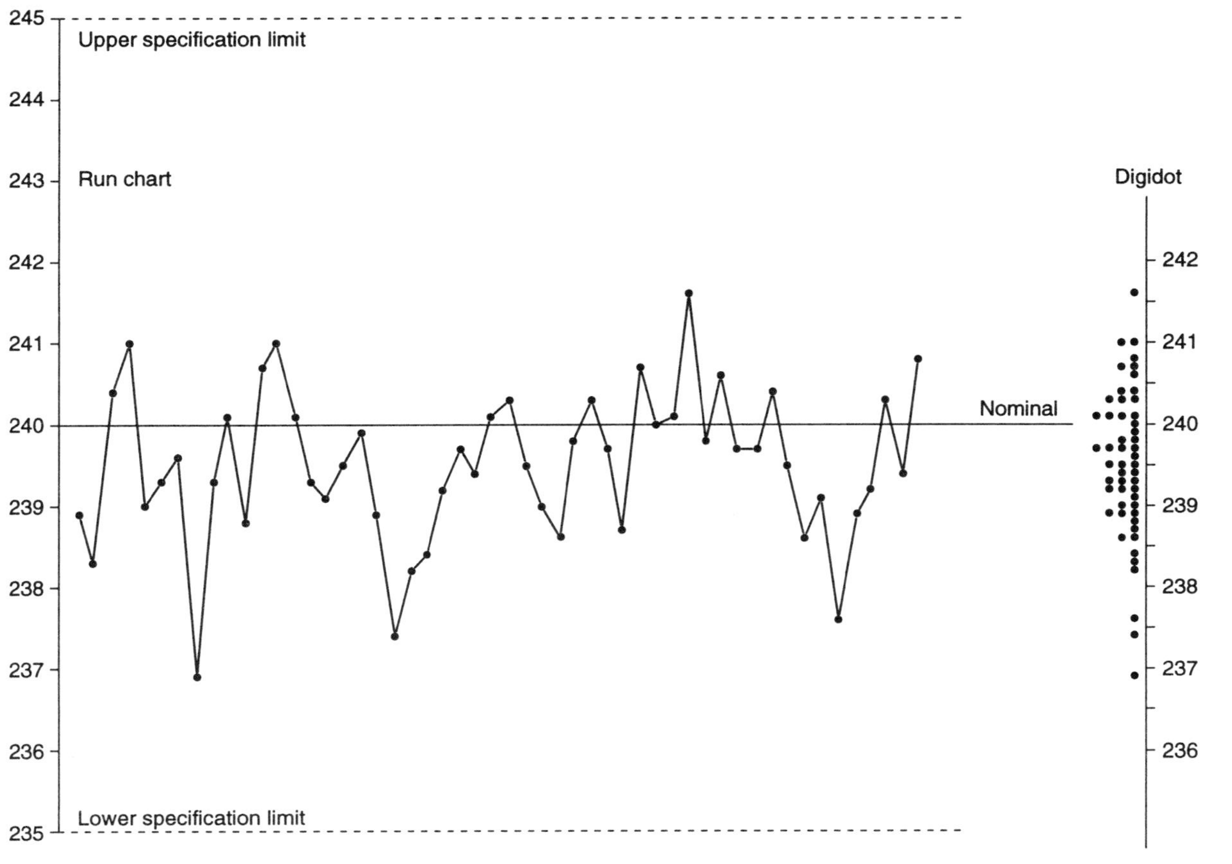

Fig. 2.7 Run chart and digidot for valve liner data.

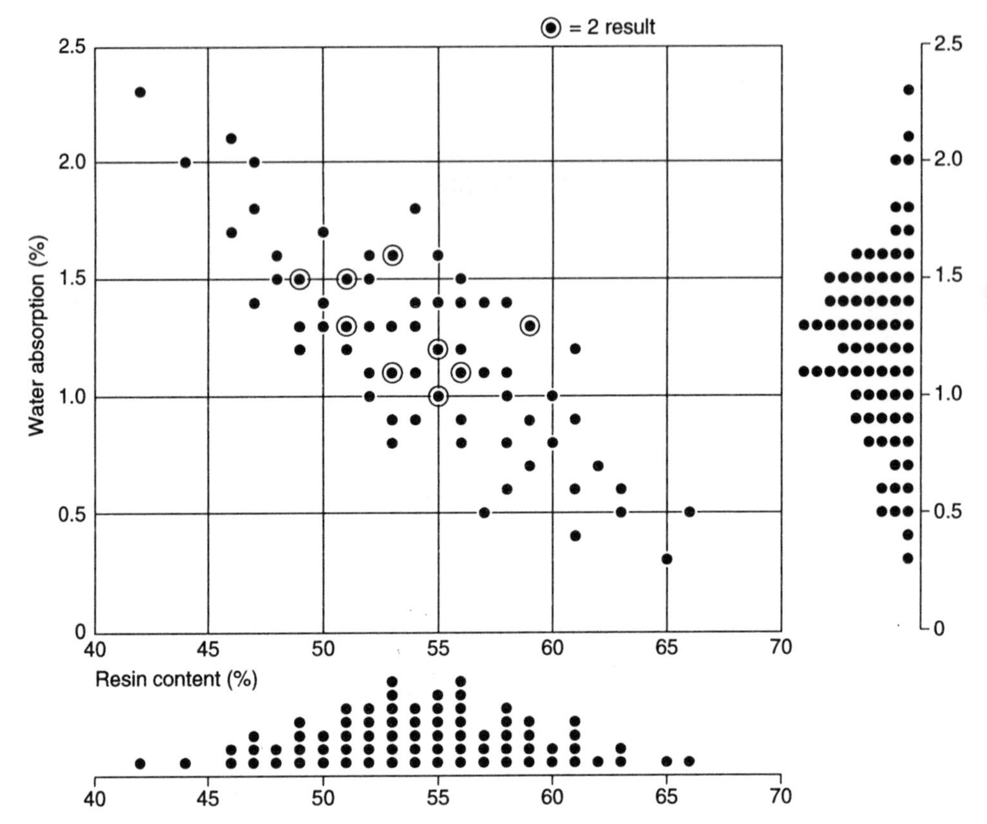

Fig. 2.8 Scatter diagram with marginal blob diagrams.

the cause variable is assigned to x and the effect variable to y. In fact the run chart uses this convention in that time is often regarded as a cause of changes in a system.

Each pair of x, y values is plotted at the appropriate coordinates, and any relationship will appear as a pattern. If the data arise from different sources (e.g. machines, operators, days, material batches) symbols or colour coding for these subgroupings may reveal further relationships. It may also be useful to compile blob diagrams for x and y on their respective axes, as shown in Fig. 2.8.

2.5 Tabular presentation of two-way classifications

2.5.1 Guidelines for tabulation

Observations often need to be classified in two or more dimensions, as for example sales value by both product and destination, or absence records by department, age and sex. We confine our attention here to two-way tables. Such tables deserve thought in their construction and presentation, and some important points are:

avoid unnecessary detail;
– it is easier to scan rows than columns;
clearly label rows and columns;
arrange the table so that the reader is guided to the important features;
provide row, column and overall summaries.

The number of significant digits in table entries should also be as small as possible consistent with clarity. While for percentages in the 10–90 region whole percentage points, or perhaps one decimal place, will suffice, for percentage values near 100 or 0 it may be important to distinguish between 0.18 and 0.23 (which would not be differentiated if rounded to the nearest 0.1). Again, if a two-way table were set up for the familiar valve liner data (say by machine and date classification), at least four significant figures (e.g. 238.6) will be required unless re-presentation as deviations from nominal is possible. Here another problem arises: negative values are less easily understood than positive.

Ehrenberg (1975) gives excellent advice on tabular and other aspects of data presentation.

2.5.2 *Example of tabular presentation*

The 'message' in a data set can be obscured by indifferent or poor presentation, but will be emphasized by attention to the preceding guidelines. The data in Table 2.4 concerns rejections for mis-shapes on items preformed on one machine and finished on another. The collection of the data involved painstaking identification of boxes of components over a period of months; a little attention to presentation is justified after so much effort in collecting the data!

Table 2.4 tends to confuse by its detail. The varying quantities produced and rejected convey little of the real information. Of course, the very low rejects

Table 2.4 Rejects for mis-shapes, January–June

	Preform position							
Finishing	1		2		3		4	
position	Made	Rejects	Made	Rejects	Made	Rejects	Made	Rejects
1	625	19	705	29	570	51	685	35
2	530	9	535	12	515	33	610	20
3	390	20	350	13	470	56	495	29
4	820	16	730	15	460	35	510	20
5	590	13	495	9	565	33	645	21
6	455	4	645	3	650	28	575	19

Table 2.5 Percentage mis-shapes by preform and finishing positions

Finishing position	Preform Position 3	4	1	2	Total made	Total rejected	% rejected
3	11.9	5.9	5.1	3.7	1705	118	6.9
1	8.9	5.1	3.0	4.1	2585	134	5.2
4	7.6	3.9	2.0	2.1	2520	86	3.4
2	6.4	3.3	1.7	2.2	2190	74	3.4
5	5.8	3.3	2.2	1.8	2295	76	3.3
6	4.3	3.3	0.9	0.5	2325	54	2.3
Total made	3230	3520	3410	3460	13620	–	–
Total rejected	236	144	81	81	–	542	–
% rejected	7.3	4.1	2.4	2.3	–	–	4.0

like ‘preform 2, finishing 6, and the very high, such as ‘preform 3, finishing 3’, stand out, but no overall pattern emerges.

If we first express the rejects purely as percentages, some confusing detail is removed. Next, the rows and columns can be re-arranged to bring the largest reject rates towards the upper rows and left-hand columns. Row and column summaries provide production totals for each preform and finishing position. The resulting revision appears as Table 2.5.

It is now evident that some positions on both machines are consistently better (or worse) than others. The combinations of finishing positions 2, 4, 5, 6 with preform positions 1 and 2 are particularly good, and the grouping of these values in the lower right of the table draws attention to this. Similarly, in the upper left corner, preform 3 is generally bad, particularly in conjunction with finishing positions 3, 1, 4. Such indications from the data can now lead to diagnosis (*why* do these differences occur?) and problem solving (can the *causes* be dealt with?).

2.6 Conclusion

There are many other ways of presenting numerical data in eye-catching form. Pictograms use repeated symbols (men, cars, houses) to compare different groups; pie charts use ‘slices’ to indicate contributions to a total; bar charts, if necessary subdivided, compare output, wealth, expenditure in different areas or organizations. All these may occasionally help in presenting data in an SPC or TQM context.

The availability of copiers and transparent materials widens the range of media, using overlays to show changes over time or other variables.

We have deferred mention of the cumulative frequency curve, or ogive, until

Chapter 13, where it will be related to probability plotting for capability analysis and the study of distribution forms.

Above all, it should be recognized that the use of graphical methods exploits the ability of the human brain to deal with visual input. Drawing graphs is not mere pandering to the simplistic, but a valuable means of recognizing patterns and interpreting important features of numerical data.

3

Numerical data summaries: location and dispersion

3.1 Introduction

Chapter 2 introduced several graphical methods for presenting data and identifying important features of distribution shape. We now consider numerical methods for describing these features, especially of location (i.e. the general magnitude of the data values) and dispersion (the extent of any variation among the values). These **summary statistics** permit concise statements to be made which nevertheless convey the essential information content of the data. For example, one might write, in full, that on day 1 a total of 516 items were made and 26 rejected, while on day 2 the figures were 354 and 21, respectively. It is both more concise *and* more informative to express this in the form:

Day 1 516 made, 26 rejected; rejects 5.04%
Day 2 354 made, 21 rejected; rejects 5.93%.

We now see at a glance that the performance on day 2 was rather worse than on day 1 despite the lower output.

The main summary statistics are averages and measures of variation. Some other statistics—rates and percentages, for example, are specialized forms of average. Measures of variation may be either absolute or relative, i.e. expressing variation as some proportion of an average. Some other shape features, such as skewness, are deferred to Chapter 14 in the context of determination of capability.

3.2 Measures of location: average

3.2.1 Arithmetic mean ($\bar{x}$)

(a) Definition and notation

Averages are intended to convey an impression of the general magnitude of numerical values. The word 'average' originally carried an implication of a fair share (as in compensation for losses, i.e. average adjusting in modern insurance terminology, or in sharing out the spoils of battle). This feature is most clearly seen in the arithmetic mean, where a total is divided by a number of contributions.

Gradually the term has come to indicate something akin to typical, so that 'average' now embraces various statistics which may bring out somewhat different features of the data. It is therefore advisable to use more precise terms to indicate which particular measure of central tendency is intended or implied.

The simplest, and therefore the most widely used, average is the arithmetic mean, usually denoted by the symbol $\bar{x}$ (although, in general, any alphabetic symbol having some useful association, like w for weight, R for range, etc., becomes an arithmetic mean by the addition of the bar symbol). In SPC applications, the bar does not always indicate a true arithmetic mean, but a weighted mean, e.g. $\bar{p}$, $\bar{c}$. This point will be considered in section 3.2.4.

In words, the arithmetic mean is obtained as the total of the data values (e.g. the x's) divided by their number (usually denoted by n). In a convenient algebraic form, this operation is represented by

$$\bar{x} = \frac{1}{n}\sum x \quad \text{or} \quad \frac{\sum x}{n} \tag{3.1}$$

where $\sum$ represents summation of the x-values. In full, this notation is

$$\bar{x} = \frac{1}{n}\sum_{i=1}^{n} x_i \tag{3.2}$$

indicating that the x values numbered (conceptually) from $i = 1$ to n are summed and divided by the number of elements included. Where all values in a set are to be used, subscript i may be omitted, giving the familiar format of formula (3.1). The full notation is sometimes useful when some values are to be omitted, for example a 'censored' mean

$$\dot{x} = \frac{1}{n-10}\sum_{i=6}^{n-5} x_i$$

indicates that the first 5 and last 5 items are to be omitted. Full subscripts can also be used to merge subsets of data; for example

$$\sum_{j=1}^{k}\sum_{i=1}^{n_j} x_{ij}$$

is an instruction to add all the x values in k data sets which may contain various numbers (n_j) of elements. Thus the subtotal for the n_1 items in group 1 is added to subtotal 2 for the n_2 items in the next group and so on. While multiple ('several layered') summation can also be useful, this book will avoid it beyond two levels.

(b) Use, misuse and interpretation of $\bar{x}$

Although well-known, the arithmetic mean is often misused. For example, it requires all data values to be equally important and carrying the same kind of numerical information. The value of $\bar{x}$ then indicates the quantity that each of n participants would receive if the total, $\sum x$, were equally shared among them, i.e. the 'fair-share' principle. In a graphical sense, $\bar{x}$ also corresponds to the point along the x-axis of a histogram or frequency curve at which the system will balance. This feature illustrates one of the drawbacks of the arithmetic mean; it is heavily influenced by extreme or outlying values, in the same way that one small child a long way from the fulcrum of a see-saw can distort the overall balance.

In some applications, it is also useful to note that where the deviations of the x_i are measured from some constant, say c, then the sum of squared deviations $\sum(x_i - c)^2$ has a minimum when $c = \bar{x}$; thus $\bar{x}$ may be considered the **least-squares** estimator of central tendency. It is this facet of $\bar{x}$ that often justifies its use in estimating the parameters of distribution models, such as normal or Poisson distributions etc., which are described in Chapter 4.

We now illustrate some of these points by further reference to the valve liner weight data of section 2.1. Using the 52 individual values, we have

$$\sum_{i=1}^{52} x_i = 12\,454.5, \; n = 52$$

giving $\bar{x} = 239.509\,615\,384$ etc. How many of these digits should be retained?

A useful general rule is that the mean, $\bar{x}$, should be recorded to at least one more significant figure than the raw values; with large data sets (say 50 or more) a further digit may be justified. In the present case, we have $\bar{x} = 239.51$ in either case (i.e. to *three* decimal places we have 239.510).

Suppose that one value had been misrecorded, say 23.98 instead of 239.8. Then $\sum x$ would become 12 238.68, and the apparent value of $\bar{x}$ falls to 235.36. Quite simple data validity checks can either avoid such accidental distortion or draw attention to anomalous values requiring correction or investigation.

(c) Calculation of $\bar{x}$ from a frequency table

Quite often data will have been organized into a tally or frequency table. Even if the original values have been rounded or approximated by grouping, use of the tabulated data can speed up calculation with little loss of accuracy. It is important that, where data are grouped, the central value of each class is used to represent all the members of the class. Thus for the valve liner data of Table 2.3 and Fig. 2.1, where half-gram class widths are used, the weights are effectively rounded to the nearest 0.5 g. Thus the mid-class values 237.0, 237.5, 238.0 $\cdots$ 241.5 are used for calculation.

For each class, a subtotal is calculated by multiplying the class mid-value by its corresponding frequency, and the subtotals are then added to form the grand total. The total number of data values is simply the sum of the class frequencies. Symbolically, these operations are represented by the expression

$$\bar{x} = \frac{\sum fx}{\sum f} \tag{3.3}$$

A procedure for carrying out these operations is available on most scientific calculators having statistical functions. We defer a detailed example until the corresponding routine for standard deviation has been covered in section 3.3.3.

3.2.2 Median (M or $\tilde{x}$)

(a) Definition and notation

The median has been noted in section 2.3.3 in connection with box and whisker plots. When data are arranged in order of magnitude, as in the stem-and-leaf table, the median is the middle value in the ordered array. For an even number of items, the median lies between the middle two values and is usually taken to be the arithmetic mean of these two values. A formal definition is that the median is the ith ordered value where $i = \frac{1}{2}(n+1)$. A useful convention for indicating ordered values is to bracket the subscript, viz. $x_{(i)}$, so that we have

$$M \text{ (or } \tilde{x}) = x_{(i)}, \quad i = \tfrac{1}{2}(n+1) \tag{3.4}$$

For the 52 valve liners, $\frac{1}{2}(n+1) = 26.5$, so that $M = x_{(26.5)}$, i.e. it lies between the 26th and 27th ordered values. Because

$$x_{(26)} = x_{(27)} = 239.5, \ M \text{ is also } 239.5$$

Unfortunately the notation for the median is less standardized than that for the mean. Many statistical texts use M, but motor industry documents prefer $\tilde{x}$. In the remainder of the book, to avoid possible confusion with $\bar{x}$, we shall denote the median by M.

(b) Use and interpretation of the median

Because the median lies at the centre of the ordered values, it immediately conveys the idea of something that is typical – the 'average man' idea. For data exhibiting a fairly symmetrical pattern of distribution, the mean and median will be very similar. For asymmetrical (skewed) patterns, the median and mean will differ: the greater the asymmetry, the larger the discrepancy between $\bar{x}$ and M.

The median is appreciably less affected by irregularities in the data, such as rogue or outlying values. It is therefore said to be **robust**. To take the example from section 3.21(b), if the value 239.8 were misrecorded as 23.98, then this erroneous value would become $x_{(1)}$, but $x_{(26)}$ and $x_{(27)}$ would still be values of 239.5, so that the median is unaffected by the error, whereas $\bar{x}$ was distorted from 239.51 to 235.36. However, this robustness is two-edged, as it may conceal changes in distribution shape when medians are used in monitoring applications, as in the case of control charts.

In the absence of irregularities, the median varies more from sample to sample than does the mean. Thus $\bar{x}$ is generally a more efficient estimator than M, i.e. it uses the information content of the sample more effectively. A further factor

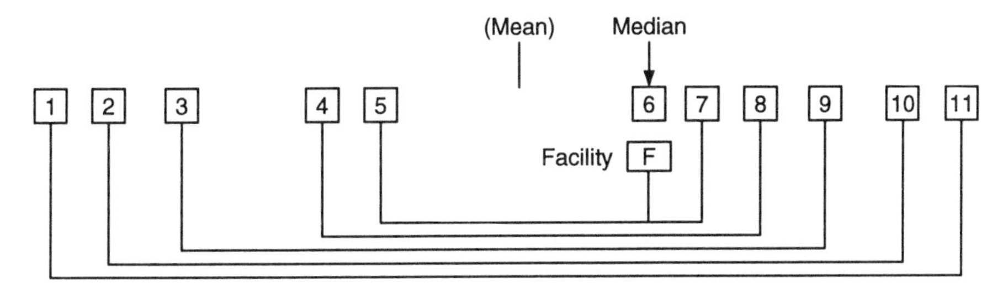

Fig. 3.1 The median and minimum sum of absolute deviations. Note: total of distances from machines to facility is a minimum when the facility is adjacent to median (not mean) machine position.

in favour of $\bar{x}$ is its ease of calculation except for very small samples (say less than 10), containing an odd number of values so that the central item stands alone.

In section 3.2.1(b), the 'least-squares' property of $\bar{x}$ was noted, i.e. it minimizes the sum of **squared** deviations. The corresponding property of the median is that it minimizes the sum of the absolute deviations, so that $\sum|x_i - c|$ is a minimum when $c = M$. There is an interesting application of this property which occasionally has a practical use when siting a facility. If, say, a facility (such as a grinder) needs to be located along a row of irregularly spaced machines, the minimum total distance from all of the machines to the common facility occurs when the latter is placed close to the median machine in the row, as in Fig. 3.1.

(c) Locating the median from a tally or frequency table

Where data are grouped into a tally or frequency table, it is usually sufficient to identify the class in which the median item falls. If greater precision is required, the following method is appropriate. Identify the class within which the $\frac{1}{2}(n+1)$th value, i.e. the median, falls. Note the upper boundary of this class, and its cumulative frequency, say x_j and F_j, respectively. Note also the upper boundary and cumulative frequency for the next lower class, say x_{j-1} and F_{j-1}. Evidently F_j will be at least $\frac{1}{2}(n+1)$ and $F_{(j-1)}$ will not exceed $\frac{1}{2}(n+1)$.

Then the estimated median is given by

$$\hat{M} = x_{j-1} + \left(\frac{F_j - \frac{1}{2}(n+1)}{F_j - F_{j-1}}\right)(x_j - x_{j-1}) \tag{3.5}$$

The procedure is illustrated in Table 3.1 for the data of section 2.2.1.

3.2.3 Mode

(a) Definition and estimation

A common, but sometimes misleading, definition of the mode is as the most frequently occurring value in the data set. This definition may be adequate

Table 3.1 Estimation of median for valve liner data

Upper class boundary	Class frequency	Cumulative frequency
237.25	1	1
237.75	2	3
238.25	1	4
238.75	5	9
239.25	10	19 ← x_{j-1}, F_{j-1}
239.75	13	32 ← x_j, F_j
240.25	8	40
240.75	8	48
241.25	3	51
241.75	1	52

$\frac{1}{2}(n+1) = 26.5$; median is in class with upper boundary 239.75.

$$\hat{M} = 239.25 + \left(\frac{32 - 26.5}{32 - 19}\right)(239.75 - 239.25) = 239.46.$$

(Compare the median of 239.5 obtained from the individual, i.e. ungrouped, data values).

when numerous data values are available, but where only limited data exist, the vagaries of sampling may produce anomalies. Consider the following (artificial) data set, comprising values rounded to the nearest integer:

8, 10, 11, 13, 14, 15, 16, 18, 19, 21, 24, 29, 37, 50, 68, 68, 85, 102

A tally or histogram would indicate a skewed distribution, with a long upper tail, short lower tail, and a concentration of values in the 'teens. A blob diagram, Fig. 3.2, gives a similar impression.

However, according to the above definition, the mode is 68 (the only value occurring more than once). This hardly offers an indication of central tendency!

A better definition is along the lines of 'the region in which the data values have the greatest concentration'. This concurs with estimation of the mode from a tally, frequency table or histogram as the most numerous **class**, though even

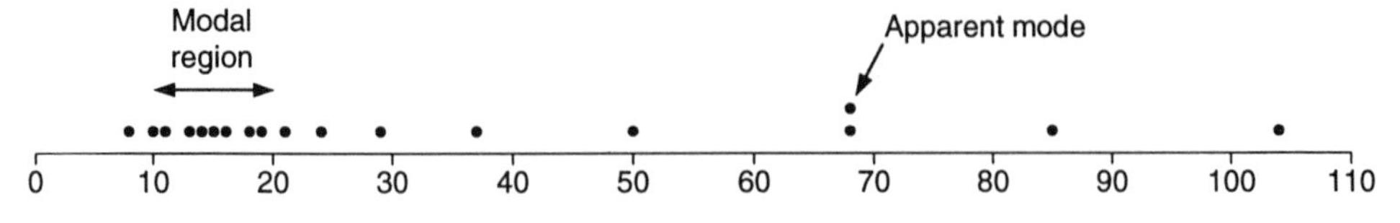

Fig. 3.2 Blob diagram and mode.

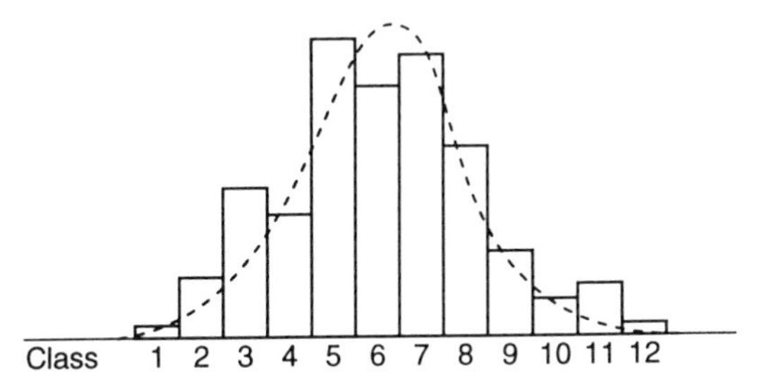

Fig. 3.3 Estimating the mode: an awkward case. Here the most numerous class is 5 although the frequency curve suggests a mode in the upper part of class 6.

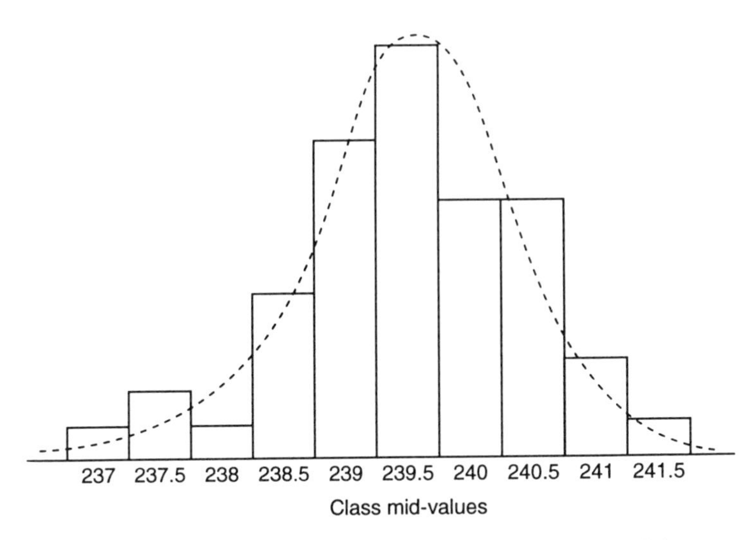

Fig. 3.4 Estimation of mode for valve liner data. Both histogram and frequency curve suggest the mode at approximately 239.5.

here some care is necessary, as illustrated in Fig. 3.3. The peak of a frequency curve drawn through the data is perhaps the best guide, as shown in Fig. 3.3 and also Fig. 3.4 for the valve liner data. Blob diagrams provide a further possible means of identifying or estimating the mode, as can be seen in Fig. 2.3, where the greatest concentration appears to be around 239.5.

Sometimes data exhibit two or more modes, and minor peaks may occur even towards the tails of an empirical distribution. One must be careful not to over-interpret, as random samples from a perfectly regular distribution can give untypical and apparently anomalous patterns.

There is no generally agreed symbol to represent the mode, and we shall therefore refer to it subsequently by name rather than symbol.

(b) Interpretation and use of the mode

Unlike $\bar{x}$ and M, the mode rarely provides an estimate of a parameter, but is useful in interpreting the implications of a set of data, and in detecting problems or anomalies.

It is well known that, for the normal distribution (see Chapter 4), the mode, median and mean are identical. It follows that for a **sample** from a normal distribution, they should at least be similar. A strongly displaced mode, or any clear bimodality (even multimodality), may indicate a contaminated or otherwise disturbed parent distribution – mixing of data from two or more subdistributions, problems in rounding, recording or sampling, etc.

It is less well known that for non-normal but unimodal distributions, the mode, median and mean are approximately related by the following expression:

$$(\text{Mean} - \text{Mode}) \doteqdot 3 \times (\text{Mean} - \text{Median})$$

where the symbol $\doteqdot$ indicates 'approximately equals'.

In words, this implies that the mean, median and mode appear, in alphabetical order, with the mean nearest to the longer tail of the distribution, and the distance from median to mode (in the direction of the shorter tail) is about twice that from median to mean (in the direction of the longer tail). Figure 3.5 illustrates the relationship, which can be useful for estimating the mode, provided that the data are smoothly distributed and unimodal. Then

$$\text{Mode} \doteqdot (3 \times \text{Median} - 2 \times \text{Mean}) \tag{3.6}$$

3.2.4 Other measures of location

Some other 'averages', although of more limited application, are worth noting.

(a) Weighted averages

Occasionally, some data points are more important, in some sense, than others. This can be reflected in the averaging process by giving varying weight factors

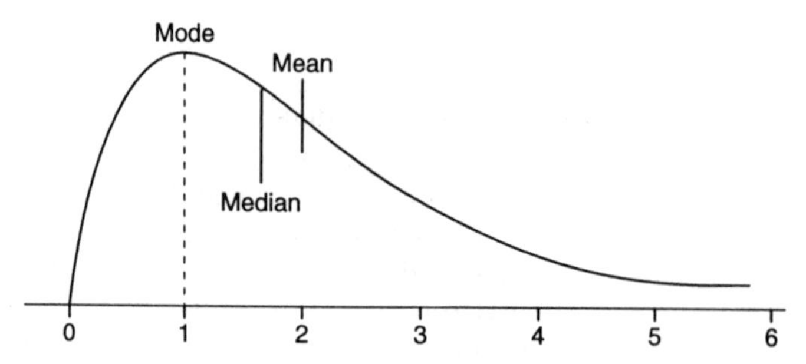

Fig. 3.5 Relation between $\bar{x}, M$ and mode. Mean – Mode = 3 (Mean – Median) or Median – Mode = 2 (Median – Mean).

to the x-values. A well known example is the retail prices index, where obviously the price of a kilogram of bread is more important than that of a kilogram of, say, salt. Weighted averages may be applicable in constructing specialized quality indices, where certain dimensions or properties are more critical than others.

The calculation of $\bar{x}$ from a frequency table, section 3.2.1(c), is a particular application of the weighting principle. Here, some x-values are more 'important' simply because there are more of them.

The general expression for a weighted average is similar to (3.3). Where w_i is the weight appropriate to x_i, we have

$$\bar{x}_w = \sum w_i x_i / \sum w_i \tag{3.7}$$

the summation extending over the appropriate range of x-values, usually $i = 1$ to n.

(b) The geometric mean

The formal definition of the geometric mean of n values is as the nth root of their product, i.e.

$$GM(x) = \left(\prod_{i=1}^{n} \right)^{1/n} \tag{3.8}$$

where $\prod$ indicates 'product of' in a manner analogous to $\sum$ for 'sum of'.

The x-values must necessarily be positive, otherwise an nth root may not exist. Calculation via (3.8) is generally impracticable because, for example, the product of 50 values each around 200 in magnitude will cause overflow of a calculator or computer. The solution is to use logarithms, and GM then becomes the antilog of the mean of log (x). Thus the geometric mean is especially useful where data have been logarithmically transformed. Natural (base e) logarithms are preferable for reasons outlined in section 3.3.4, and then

$$GM = \exp\left\{ \frac{1}{n} \sum_{i=1}^{n} \ln(x_i) \right\} \tag{3.9}$$

As will be further discussed in Chapter 14, the log transformation is a useful means of handling positively skewed data values, and it also has applications in dealing with analysis of experimental designs.

(c) Harmonic mean

The harmonic mean is the reciprocal of the arithmetic mean of the reciprocals of the data values, i.e.

$$HM = \left(\frac{1}{n} \sum_{i=1}^{n} \frac{1}{x_i} \right)^{-1} \tag{3.10}$$

Like the geometric mean, its principal application is in conjunction with transformations, in this case the reciprocal or $1/x$ transformation.

The reader will probably known of the 'catch' question concerning a car performing two circuits of a race track. If the first lap is covered at 30 m.p.h., at what speed must the second lap be driven to average 60 m.p.h. Of course, the 60 m.p.h. average is impossible; the point is that the question really concerns times to complete a lap, i.e. hours per mile, rather than miles per hour. The harmonic mean is really more relevant in such a case. Suppose that we have three laps covered at 20, 30 and 40 m.p.h. for a one mile circuit. Then the times taken for the three laps are 3, 2 and 1.5 minutes, totalling 6.5 minutes. The average lap time is $2.1\dot{6}$ minutes, and the average speed is $60 \div 2.16 = 27.69$ m.p.h.

The same result is obtained via (3.10), thus

$$HM = \left\{\frac{1}{3}\left(\frac{1}{20}+\frac{1}{30}+\frac{1}{40}\right)\right\}^{-1} = 27.69 \text{ m.p.h.}$$

(d) Midrange

The midrange can sometimes provide a quick estimate of central tendency. It is simply the average of the largest and smallest values in the data set. Representing ordered values by $x_{(i)}$,

$$\text{Midrange} = \tfrac{1}{2}\{x_{(1)} + x_{(n)}\} \tag{3.11}$$

In graphical terms, the midrange is halfway between the extreme points on a blob diagram or box and whisker plot.

An obvious possibility is to use the midrange in association with the range (section 3.3.1), but it suffers from its inefficiency (in that all intermediate values are discounted), and the fact that it may be grossly affected by any rogue value in the data.

3.2.5 *Proportions and rates of occurrence*

(a) Attributes and $\bar{p}$

As will be well known to SPC practitioners, when x non-conforming items are observed in a sample of size n, the proportion $p = x/n$ is a useful way of summarizing the information and for making comparisons.

In fact if each conforming item is scored as zero and each non-conforming item as 1, p corresponds to the mean score per sampled item, as the sum of the 1's is identical to the number non-conforming, x.

Because $p = x/n$, one may also (by cross-multiplying) write $x = np$, so the dual symbol np is often used to represent the number non-conforming.

When combining attribute data sets of unequal sizes, the p values for the various sets need to be weighted by their matching sample sizes. We then have, for k data sets,

$$\bar{p} = \sum_{j=1}^{k} n_j p_j \Bigg/ \sum_{j=1}^{k} n_j$$

but since each $p_j = n_j p_j / n_j$, this becomes

$$\bar{p} = \sum_{j=1}^{k} n_j \frac{np_j}{n_j} \Bigg/ \sum_{j=1}^{k} n_j = \sum_{j=1}^{k} np_j \Bigg/ \sum_{j=1}^{k} n_j \tag{3.12}$$

familiar to users of p-charts.

(b) Events and $\bar{c}$, $\bar{u}$

For non-conformities observed in samples or areas of opportunity of size n_j, similar considerations apply. If each non-conformity is scored as 1, then $\bar{c}$ simply measures the average score provided all the n_j are identical. If the n_j vary, it is necessary to weight the scores from combined data sets, and this leads to an average rate of occurrence per sampling unit,

$$\bar{u} = \sum_{j=1}^{k} c_j \Bigg/ \sum_{j=1}^{k} n_j \tag{3.13}$$

as used in conjunction with the u-chart.

(c) Other rates, %, p.p.m.

Most of the averages we have considered yield a measure of typical magnitude per item or unit. Often one prefers a measure related to hundreds or millions of units, or to rates of trimming scrap per tonne of output, material used per litre of a distillate produced, or even miles per litre of fuel used by a vehicle. In the latter example, one might note that European users prefer litres consumed per 100 kilometres, rather than kilometres per litre.

All these rates, and other indices, are specialized averages (percentages hark back to Roman times, when provisions were ordered for groups of 100 men under a centurion). Any of these special measures may provide input to statistical process control as aim values for specifications, centre lines (targets) for control charts, or summaries of performance.

3.3 Measures of dispersion or variation

3.3.1 Range, R

The range is appealing as a measure of variation by its simplicity. It is the (unsigned) difference between the largest and smallest values. For observations arranged in order from $x_{(1)}$, the smallest, to $x_{(n)}$, the largest, the symbolic form is

$$R = x_{(n)} - x_{(1)}, \quad \text{or equivalently } x_{\max} - x_{\min} \tag{3.14}$$

Partly because of its very simplicity, it suffers from some disadvantages. Some of these are as follows:

(a) Especially for moderate or large sample sizes ($n > 10$) it is rather inefficient, in that it takes no direct account of $n-2$ of the data values, except that they lie between x_{max} and x_{min}. Thus the following two sets of five observations have the same range, although blob diagrams would reveal large differences in the **nature** of the variation, as would some of the other statistics that measure dispersion, like those described in sections 3.3.2–3.3.4.

Set 1	90, 99, 100, 101, 110	$R = 20$
Set 2	90, 90, 91, 110, 110	$R = 20$

(b) Even for samples drawn from a particular batch, lot or population, samples of different sizes will yield different ranges, with a strong tendency for larger samples to exhibit larger ranges. A large sample size provides more opportunity for some of the more extreme (though not necessarily anomalous) values to turn up.

For the data of section 2.2.1, it can be seen from the stem-and-leaf table (Table 2.2) or the blob diagram (Fig. 2.3) that the range of the 52 values is

$$R = 241.6 - 236.9 = 4.6$$

Had the sample size been only 5, it is fairly unlikely that these two extreme values would occur. The range for the next pair inwards from the extremes would have been $241.0 - 237.4 = 3.6$, a large discrepancy, and even these tail values might not have occurred. Five randomly chosen subsets from the 52 are (in rank order number):

5,	9,	37,	49,	60	$R = 2.6$
10,	24,	41,	42,	52	$R = 2.8$
3,	24,	38,	44,	46	$R = 3.0$
5,	14,	23,	28,	48	$R = 2.5$
8,	11,	14,	25,	38	$R = 1.5$

Conversely, had a much larger sample (200, 2500, etc.) been taken, even wider variation might have been noted. This dependence of range on sample size makes R unsuitable as a sole measure of variation, though the average range over several or many samples of the same size can be both convenient and useful, as noted in Chapters 8, 13 and 16.

(c) Perhaps surprisingly in view of its simplicity to the SPC practitioner, the range does not lend itself so readily to use with small calculators with $\bar{x}$, s facilities. In terms of computer logic, each of the n values has to be tested to see if it is either the largest or smallest, and registers or memory units are required to store the current maximum and minimum. Small calculators do not have the logic circuitry or storage for these operations, and the calculation of standard deviation (section 3.3.3) is in fact much more straight forward.

(d) The relationship of range to other measures of variation depends to some extent on the shape of the distribution of individual values. These relationships are well documented for the normal distribution but less so for most other distribution forms. As a result, the use of range to standard deviation conversion factors (Table 13.1) is limited to those applications where 'approximate normality' is demonstrated or can reasonably be assumed.
(e) Finally, the range (to a greater extent than other measures) can be distorted by anomalous data values. Interestingly, in terms of detecting disturbances to a process this can sometimes be a useful feature, but for purposes of estimating the underlying system variation it becomes a disadvantage.

The range finds its main application in connection with manually managed (rather than computerized) control charts such as $\bar{x}$, R, using repeated samples of the same size. Provided the process is stable, the mean range, $\bar{R}$, over a suitable period does provide a convenient measure subject to its interpretation in terms of the sample sizes used for monitoring purposes.

3.3.2 *Variance*

(a) Definitions
Particularly for SPC, the standard deviation, covered in section 3.3.3, has become the definitive measure of variation. The reasons for this include both its status, especially for the normal distribution, as a distribution parameter indicating an interpretable aspect of distribution shape (as discussed in section 4.6), and its ease of calculation via electronic calculators and computers using 'one-pass' routines. This means that summations ($\sum x, \sum x^2$) can be compiled in one pass through the data and then used for calculating the required statistics, without either storage of the individual values or further reference to them.

The standard deviation is the square root of the variance, which itself has important properties and applications. In its simplest form, the variance is the mean of the squared deviations from the arithmetic mean, and is thus the 'mean-square deviation'. The standard deviation is therefore the root-mean-square measure of variation. Unfortunately these verbal definitions are slightly inaccurate in practice, as it is necessary to distinguish population parameters from sample estimates of these parameters.

We shall first define the parameters and estimates in symbols, and go on to consider their relationship.

$$\text{Population variance, } \sigma^2 = \frac{1}{N} \sum_{i=1}^{N} (x_i - \mu)^2 \qquad (3.14)$$

where N is the number of members in the population, and μ is the population mean, i.e.

$$\mu = \frac{1}{N}\sum_{i=1}^{N} x_i$$

$$\text{Population standard deviation, } \sigma = \sqrt{\sigma^2}$$

$$\text{Sample estimate of variance, } s^2 = \frac{1}{n-1}\sum_{i=1}^{n}(x_i - \bar{x})^2 \tag{3.15}$$

(where n is the sample size, the sample contains only a small proportion, say not more than 10%, of the population, and $\bar{x}$ is the sample mean as in section 3.2.1).

(b) Calculation of s^2

In the form of (3.14) or (3.15) the variance requires two passes of the data, one to obtain $\sum x$ and thence $\bar{x}$, and a second to calculate, square and add the deviations from $\bar{x}$. For computational purposes, the algebraic identity (3.16) is useful, permitting the summations and counting of data values to be completed in one pass.

$$\sum(x - \bar{x})^2 \equiv \sum x^2 - \frac{1}{n}\left(\sum x\right)^2 \tag{3.16}$$

The $\sum x^2$ total is often referred to as the 'crude' sum of squares, and the term $\frac{1}{n}(\sum x)^2$ (i.e. the square of the grand total divided by the sample size) as the correction factor. The resultant $\sum(x - \bar{x})^2$ is the 'corrected' sum of squares, and its incorporation into the expression for s^2 yields:

$$s^2 = \frac{1}{n-1}\left\{\sum x^2 - \frac{1}{n}\left(\sum x\right)^2\right\} \tag{3.17}$$

(note that the correction factor may be written in various ways, including $(\sum x)^2/n$, $n\bar{x}^2$ and $\bar{x}\sum x$).

Whichever form is used, it is essential to carry all possible significant figures in the intermediate calculation steps, rounding only at the final step. Indeed, while (3.16) is an algebraic identity it can fail to yield correct arithmetic results in cases where the data contain several initial digits which do not vary. Consider, for instance,

$$\left.\begin{aligned} x_1 &= 1\,000\,000 \\ x_2 &= 1\,000\,001 \\ x_3 &= 1\,000\,002 \end{aligned}\right\} \qquad \left\{\begin{aligned} &\textstyle\sum x = 3\,000\,003 \\ &\textstyle\sum x^2 = 3\,000\,006\,000\,005 \\ &\textstyle\frac{1}{3}(\sum x)^2 = 3\,000\,006\,000\,003 \end{aligned}\right.$$

Then $\sum(x - \bar{x}^2) = \sum x^2 - \frac{1}{3}(\sum x)^2 = 2$. However, most calculators, and many computers, cannot handle more than a dozen or so significant digits, so that the difference between $\sum x^2$ and $1/n(\sum x)^2$ is 'lost' and will yield a variance, and therefore a standard deviation, of (apparently) zero.

This problem can be overcome, for what is known as 'stiff' data, by suitable coding. This means omitting the unvarying digits, operating with those that do vary, and adjusting the mean afterwards. In the above example, one would subtract a million from each value, leaving residues of 0, 1, 2; then $\sum x$ becomes 3, $\sum x^2 = 5$, $\sum(x - \bar{x})^2 = 5 - 3 = 2$. This agrees with the sum of the squared deviations from the mean of 1 000 001, i.e. $1^2 + 1^2 = 2$. The true mean of the original values is recovered by adding one million to the mean of the coded values, viz

$$1\,000\,000 + \tfrac{3}{3} = 1\,000\,001$$

Other forms of coding may involve re-scaling by division or multiplication, as when fractional values contain several zeros following the decimal point. This is analogous to working in millimetres rather than metres, or in milligrams rather than minute fractions of a kilogram.

(c) Use and interpretation of variance

The main difficulty with the variance is that of interpretation. Its units are the squares of the original units of measurement, and at best this may be confusing; at worst such squared units may not exist. If original measurements are, say, in centimetres, then the units of variance are cm^2 – yet these do *not* refer to area! For the data of section 2.2.1 the units are grams, but square grams are meaningless.

Nevertheless, the variance has important properties so that it, rather than standard deviation, is regarded as the key measure of variation by statisticians. It can be broken down, using the 'analysis of variance', into contributions to the overall variation, each of which may be recognizable as a feature of a process. The overall variance can also, in many cases, by synthesized by combining separate estimates of these contributions. Simplified techniques based on the analysis of variance form the basis of the methods described in Chapter 16. As an indicative example, suppose that it has been established that the delivery performance of four dispensers, independently contributing to a formulation, have been studied. The following statistics are obtained.

Dispenser	A (filler)	$\bar{x} = 500$ kg,	$s = 8.0$;
	B (reinforcement)	$\bar{x} = 350$ kg,	$s = 5.0$;
	C (resin)	$\bar{x} = 600$ kg,	$s = 11.0$;
	D (reagent)	$\bar{x} = 80$ kg,	$s = 2.5$.

We require the variances for A, B, C, D which are, respectively, 64, 25, 121 and 6.25. The average total delivery for batches using one shot from each dispenser is obviously 1530 kg. Less obviously, the overall variance is the sum of the contributions, giving 216.25. The standard deviation is then $\sqrt{216.25} = 14.7$, and from these synthesized estimates of mean and standard deviation the performance of the system, as a whole, can be readily evaluated.

Because of the difficulty in interpreting σ^2 or s^2, the standard deviation is used as a practical and informative measure of variation, and we shall now consider it before proceeding to extend the analysis of the valve liner data.

3.3.3 The standard deviation, σ or s

(a) Definition and calculation

For present purposes, we may define the standard deviation as the square root of the variance, i.e.

$$\sigma = \sqrt{\left[\frac{1}{N}\sum(x-\mu)^2\right]} \tag{3.18}$$

the population standard deviation, and

$$s = \sqrt{\left[\frac{1}{n-1}\sum(x-\bar{x})^2\right]} \tag{3.19}$$

the sample estimate of σ.

As for the variance, there are more suitable forms for actual computation. The form most often used is based on the count of data values (n), the sample total ($\sum x$) and the sum of squares ($\sum x^2$). Then

$$s = \sqrt{\left[\frac{1}{n-1}\left(\sum x^2 - \frac{(\sum x)^2}{n}\right)\right]} \tag{3.20}$$

the summations extending from $i = 1$ to n. The form for the population parameter (rarely required in SPC applications, though examples do occur in measuring non-normality) is correspondingly

$$\sigma = \sqrt{\left[\frac{1}{n}\left(\sum x^2 - \frac{(\sum x)^2}{N}\right)\right]} \tag{3.21}$$

The discrepancy between the formulae for σ and s, i.e. the reason for the divisor $n-1$ rather than n, will be resolved in Chapter 5 when dealing with sampling and estimation.

For dealing with theoretical distributions, other forms using probabilities of occurrence of the various x-values (instead of actual frequencies in data sets) are adopted, but we shall not venture into the theory in this text.

(b) Interpretation of σ and s

The standard deviation has particular relevance as a measure of variation in connection with some theoretical distributions, or **distribution models**, especially the Normal distribution. This will be considered in Chapter 4. Otherwise, σ or s gather into one numerical summary the variation of the data values. All values contribute (unlike the range, where 'inner' values have no effect), but the standard deviation can be grossly inflated by untypically extreme or rogue observations.

Section 3.4 offers a comparison between the standard deviation and other measures of variation – again generally in terms of values that are, at least approximately, Normally distributed.

While s is the most used measure of variation in SPC applications, its importance should not blind one to its limitations nor to the occasional usefulness of alternative measures that may be less prone to distortion in some circumstances.

3.3.4 *Coefficient of variation*

Sometimes a measure of **relative variation** is required. When manufacturing items over a range of sizes, it may be reasonable to expect that those which are larger, or heavier, or stronger, may also exhibit greater absolute variation but that this variation is proportional to the average size (or weight, or strength).

The coefficient of variation is strictly defined as the standard deviation divided by the mean, and is therefore generally a fraction. For this reason, it is often multiplied by 100 and expressed as a percentage. Thus for a population,

$$C(\text{or } Cv) = \frac{\sigma}{\mu} \quad \text{or} \quad 100\frac{\sigma}{\mu}\% \tag{3.22}$$

and for a sample,

$$\hat{C} = \frac{s}{\bar{x}} \quad \text{or} \quad 100\frac{s}{\bar{x}}\% \tag{3.23}$$

Again referring to the valve liner data, with $\bar{x} = 239.51$ and $s = 0.9335$, $\hat{C} = 0.9335/239.51 = 0.0039$ or 0.39%, implying that the standard deviation is 0.39% of the mean.

A useful feature of the coefficient of variation occurs in connection with logarithmic transformations, mentioned in section 3.2.4(b). When natural (base e) logarithms are used, the standard deviation of $\ln x$ is usually close to the coefficient of variation of the original x-values. For the valve liner data, we would find that for $y = \ln x$, $\bar{y} = 5.4786$ and $s_y = 0.3902$, the latter almost identical to $\hat{C}_x$ above. Incidentally, the exponential function of $\bar{y}$, 239.508, is the geometric mean of the x-values, in this case very similar to the arithmetic mean because of the small coefficient of variation.

3.3.5 *Other measures of variation*

Three further statistics are worth noting as measures of variation.

(a) Mean absolute deviation

The mean absolute deviation (MAD) is the arithmetic mean of the deviations of the x_i from $\bar{x}$. We have noted that $\bar{x}$ has the property that the sum of the deviations of x_i from it is zero, because the negative deviations exactly balance the positive. If the negative signs are ignored, and we use the absolute or unsigned deviations, MAD is then obtained as

$$MAD = \frac{1}{n}\sum |x - \bar{x}|, \tag{3.24}$$

the vertical bars indicating 'absolute value of'.

Unfortunately MAD cannot be calculated in one pass of the data. It is necessary to find $\bar{x}$, and then the deviations can be summed in a second pass. For some programmable calculators, it is worth noting that the absolute value of the deviations can be obtained by squaring $x - \bar{x}$ and then square-rooting. Most computers, and some scientific calculators, have the $|x|$ function.

MAD is not widely used as a general measure of variation. In the SPC area, it forms part of a useful test of normality; for the normal distribution, MAD is approximately 0.8σ and appreciable divergence from this relationship is an indicator of non-normality. This point is taken up in Chapter 14.

The reader may recall that one of the properties of the median is that it mimimizes the sum of absolute deviations (see section 3.2.2(b) and Fig. 3.1). It might therefore appear logical to use the median as the central parameter for MAD, but the general practice is to use $\bar{x}$ as defined in (3.24).

(b) Interquartile and Semi-interquartile Range

In fact the interquartile range (IQR) has already been noted, as it is the length of the 'box' of the box-and-whisker plot of section 2.3.3. Denoting the lower and upper quartiles of the data by Q_1 and Q_3,

$$IQR = Q_3 - Q_1 \tag{3.25}$$

and we note also that convenient means of finding Q_1, Q_3 are via the stem-and-leaf table or the blob diagram. Formally, the quartiles are derived from the ordered $x_{(i)}$ values, with

$$Q_1 = x_{(i)}, \qquad i = \tfrac{1}{4}(n-1)$$
$$Q_3 = x_{(i)}, \qquad i = \tfrac{3}{4}(n+1)$$

The 'missing' second quartile is, of course, the median. This suggests that the median is the natural measure of location for use in association with IQR, and the box-and-whisker plot is a natural form of graphical presentation. IQR can also be identified on a blob diagram, and the quartiles highlighted on the stem-and-leaf table.

For the normal distribution, IQR corresponds to approximately 1.35σ (i.e. 0.675σ on each side of the mean). However, IQR is generally used purely as part of a data summary or for comparisons between data sets.

Finally, the semi-interquartile range $SIQR$ is one-half of the IQR. We mention it here simply because it occurs in some statistics texts and occasionally in computer software. Where Q_1 and Q_3 are not fairly evenly spaced on either side of the median, the *SIQR* is of little value and may even be misleading. In such case, it is better to quote Q_1 , M and Q_3.

3.4 Summary and comparisons: measures of variation

Of the several measures discussed in section 3.3, the variance has the disadvantage of being expressed in units which are not those of the original measurements; σ^2 (or s^2) is thus difficult to interpret, although of great theoretical and practical importance. The problem of interpretation is overcome by using σ (or s), the square root of the variance.

We now consider some advantages and disadvantages of these statistics, and look at their interrelationships when the base variable, x, is normally distributed.

Table 3.2 lists the pros and cons of R, s, *MAD* and *IQR* (because of the obvious relationship between *IQR* and *SIQR*, the latter is omitted from the Table).

The relationships between the statistics measuring variation depend on the form of the underlying distribution. The following approximate indications are those for the normal distribution, and the ratios may be very different for non-normal forms. Indeed, as noted for *MAD* in section 3.3.5(a), such ratios may

Table 3.2 Advantages and disadvantages of measures of dispersion: R, s, *MAD* and *IQR*

Measure	Advantages	Disadvantages
Range, R	Simplicity Use in $\bar{x}, R$ charts Use in box-and-whisker plots	Dependence on sample size Uses only extreme values, so less efficient
Standard deviation, s	Efficient use of all sample data Ease of calculation Relationship to Normal distribution Additive property of s^2	Less intuitive appeal than other measures
Mean absolute deviation, *MAD*	Conceptually simple Uses all data values Useful in testing Normality (MAD/σ)	Two-pass calculation needed Few uses other than as data summary or Normality test
Inter-quartile range, *IQR*	Simple relation to box-and-whisker plots	Needs ordering of data values to define quartiles

Table 3.3 Relationships between R, s, MAD and IQR (Normally distributed variable)

Because R uses only the extreme values in the data, its relationship to the other statistics is much more variable than the remaining ratios. All values are approximate and for guidance only.

Measure 2	Measure 1			
	R	s	MAD	IQR
R	(1)	$\sqrt{n}$*	$\frac{4}{5}\sqrt{n}$*	$\frac{4}{3}\sqrt{n}$*
s	$\frac{1}{\sqrt{n}}$*	(1)	$\frac{4}{5}$	$\frac{4}{3}$
MAD	$\frac{5}{4\sqrt{n}}$*	$\frac{5}{4}$	(1)	$\frac{3}{5}$
IQR	$\frac{3}{4\sqrt{n}}$*	$\frac{3}{4}$	$\frac{5}{3}$	(1)

Upper right triangle: Measure 1 ÷ Measure 2
Lower left triangle: Measure 2 ÷ Measure 1
*Very approximate ratios for $n = 2$ to 12. For larger samples, R/s increases to about 3.7 at $n = 20$, 4.5 at $n = 50$, 5.0 to $n = 100$, 6.1 at $n = 500$.

even be used as tests for normality or non-normality – the ratio between R and s is a further example.

In the case of the range, the relationship to s or σ depends heavily on the sample size, and full details of conversion factors will be deferred to Chapter 13. Table 3.3 relates the measures described in section 3.3.

As well as giving a rough indication of the relationships, the ratios in Table 3.3 may be useful for checking as a safeguard against gross errors or severe non-normality. Thus, if a sample of (say) 25 items yields a range of 17.6, a standard deviation of 2.2 and an IQR of 3.0, there would appear to be an anomaly in the range, viz:

$$R/s = 8 \text{ (much larger than is typical for } n = 25)$$

but

$$s/IQR = 0.73, \text{ not far from } \tfrac{3}{4} \text{ as listed in the Table.}$$

3.5 Further examples

3.5.1 Using the 52 raw values in section 2.2.1, we find:

$\sum x = 12454.50; \quad \sum x^2 = 2983016.95; \quad n = 52$

$$s^2 = \frac{1}{51}\left(2\,983\,016.95 - \frac{12\,454.50^2}{52}\right) = 0.871\,474\,5;$$

$$s = 0.933\,528.$$

We already have (from section 3.2.1(a)) $\bar{x} = 239.51$, and following the guideline that one should retain one further significant figure in s than in $\bar{x}$, we may round s to 0.934.

We note also that $R = 4.7$ (from section 3.3.1(b)), and from the calculations carried out for the box-and-whisker plot in section 2.3.3, $IQR = 240.1 - 238.9 = 1.2$.

It remains to calculate MAD, and this requires the 52 deviations of x from $\bar{x}$ to be obtained and averaged (without regard to + or − sign). This is a fairly tedious operation, but results in 37.319 23 whence $MAD = 0.7177$.

As a check, we find the ratios

$R/s = 5.04$ (reasonable for $n = 52$),
$MAD/s = 0.768$ (close to 4/5),
and $IQR/s = 1.285$ (not far from 4/3).

Table 3.4 Calculation of $\bar{x}$ and s from frequency table

Class mid-value x	Frequency f	fx	fx^2
237.0	1	237.0	56 169.00
237.5	2	475.0	112 812.50
238.0	1	238.0	56 644.00
238.5	5	1 192.5	284 411.25
239.0	10	2 390.0	571 210.00
239.5	13	3 113.5	745 683.25
240.0	8	1 920.0	460 800.00
240.5	8	1 924.0	462 722.00
241.0	3	723.0	174 243.00
241.5	1	241.5	58 322.25
	52 ($\sum f$)	12 454.5 ($\sum fx$)	2 983 017.25 ($\sum fx^2$)

$\bar{x} = 12\,454.5 \div 52 = 239.51$

$$s^2 = \frac{1}{51}\left(2\,983\,017.25 - \frac{12\,454.5^2}{52}\right) = 0.877\,36$$

$s = \sqrt{0.877\,36} = 0.937$

3.5.2 It is often useful to calculate $\bar{x}$ and s from a grouped frequency table when data are sufficiently numerous. For this purpose, the values within any class are represented by the class mid-value, and the frequency formula (3.3) is then applied. The corresponding expression for s is

$$s\text{ (grouped data)} = \sqrt{\left[\frac{1}{\sum f}\left(\sum f x^2 - \frac{(\sum f x)^2}{\sum f}\right)\right]} \tag{3.26}$$

For convenience, part of the frequency table (Table 2.3) for the valve liner data is reproduced in Table 3.4, along with other intermediate results in the calculation of $\sum fx$ and $\sum fx^2$.

The values of $\bar{x}$, s obtained in Table 3.4 are in good agreement with those calculated from the complete raw data. Note that the same basic data with a different grouping would not give identical results. For example, groups covering 236.5–236.9, 237.0–237.4, 237.5–237.9, etc. would give class mid-values 236.7, 237.2, 237.7 etc. The resulting frequency table and $\bar{x}$, s are shown in Table 3.5, along with a table based on 1 gram class widths.

Table 3.5 Alternative frequency tables for value liner data

(a)

Class contents	Mid-value (x)	Frequency (f)
236.5–236.9	236.7	1
237.0–237.4	237.2	1
237.5–237.9	237.7	1
238.0–238.4	238.2	3
238.5–238.9	238.7	7
239.0–239.4	239.2	14
239.5–239.9	239.7	8
240.0–240.4	240.2	10
240.5–240.9	240.7	4
241.0–241.4	241.2	2
241.5–241.9	241.7	1

$\sum f = 52$
$\sum fx = 12\,452.4$
$\sum fx^2 = 2\,982\,015.38$
$\bar{x} = 239.47$
$s = 0.977\,5$

(b)

Class contents	Mid-value (x)	Frequency (f)
236.0–236.9	236.45	1
237.0–237.9	237.45	2
238.0–238.9	238.45	10
239.0–239.9	239.45	22
240.0–240.9	240.45	14
241.0–241.9	241.45	3

$\sum f = 52$
$\sum fx = 12\,454.4$
$\sum fx^2 = 2\,982\,977.43$
$\bar{x} = 239.51$
$s = 1.017\,8$

Compare with $\bar{x} = 239.51$, $s = 0.9335$ for ungrouped data,
Compare with $\bar{x} = 239.51$, $s = 0.937$ for grouping of Table 3.4

Note the appreciably larger value of s obtained with the coarser grouping. In general, the wider the classes, the greater the estimate s will become, and this is one of the reasons for advising a minimum of 8–10 classes, or alternatively a width of not more than half of the standard deviation. Adjustment can be made for grouping via Sheppard's correction:

$$s^2(\text{adjusted}) = s^2(\text{unadjusted}) - \tfrac{1}{12}(\text{class width})^2 \tag{3.27}$$

For Table 3.4, this adjustment would reduce s to 0.9255, while for Table 3.5(a) and (b) the adjusted standard deviations become 0.9668 and 0.9760, respectively. The best practice, whenever possible, is to calculate $\bar{x}$ and s directly from raw data, and to use frequency tables as useful data summaries and as the basis for histograms and other forms of graphical presentation.

3.6 Conclusion

In this Chapter we have examined measures of location and variation in some detail. These measures, especially $\bar{x}$, s and (in the more advanced techniques of later chapters) s^2 provide the principal tools for analysis and diagnosis, in addition to their role in data description.

A thorough working knowledge of calculators with $\bar{x}$, s facility, including recovery of the summations n, $\sum x$, $\sum x^2$, will enable the reader to deal with later developments. In some respects, a good calculator may be useful even when computer software is available, because of flexibility considerations. Where a personal computer is available, a spreadsheet system with means of operating on rows and columns will facilitate many forms of statistical analysis.

We now proceed to examine the relationships between the descriptive statistics (of this and the preceding Chapter) and the parameters of theoretical distribution models. The linking of observation, usually via sampling, with modelling, provides the basis of effective application of statistical techniques in SPC and TQM.

4

Probability and distribution

4.1 Introduction

This Chapter gives a brief account of the statistical basis of the methods widely used in statistical process control, especially those concerned with control charts and capability. The reader requiring a fuller treatment should refer to one of the textbooks listed in the bibliography.

4.2 Types of data, levels of measurement

The data encountered in most applications of statistics are of three basic types:

counted data (attributes and events);
rankings (ordered data);
measurements (each observation is a value on a numerical scale).

The second of these types, though important in many applications in market research, sensory tests, etc., rarely occurs in SPC. The third type is capable of subdivision into interval and ratio levels of measurement, and a complete classification along with some of the properties, relevant statistics and examples, is given in Table 4.1. Sometimes intermediate types occur, e.g. there may be ordered categories (good, moderate, bad); or data from a number of groups where attributes have been counted may yield proportions which are subsequently treated as measurements.

4.3 Distribution

In most natural situations and manufacturing processes, it is found that observations made on similar items or under similar conditions vary to some extent. The pattern of occurrence of the different values forms a **distribution**. While a great variety of distributions occur in practice, many may be approximated by a few theoretical models. These models provide a convenient means of processing or summarizing numerical data, subject to certain assumptions underlying a particular model.

Table 4.1 Levels of measurement

	Events	Nominal	Ordinal	Interval and ratio
Nature	Occurrences	Classification	Comparison	Individual characterization
Procedure	Counting	Counting	Ranking	Measurement on linear scale Ratio level must have a 'true' zero
Measures of:-				
Location	Rate of occurrence	Proportion	Median	Mode, median, various means (arithmetic, harmonic, geometric, weighted, etc.)
Dispersion	*	*	Quantiles	Range, standard deviation, variance Coefficient of variation for ratio level
Examples	Faults Breakdowns Accidents Particles	Go/no go Quality grades Present/absent Yes/no/don't know	Any ranked data	Most data from technology, production, financial

*Over several or many groups, counts from event or nominal levels can be treated as for ratio level for location and dispersion
Notes: (i) Measurement can be downgraded, e.g. Ratio → Interval → Ordinal → Nominal, but not up-graded;
(ii) Care is necessary with 'zeros of convenience', e.g. °C, °F and vacuum (where 0 lb/in^2. corresponds to 1 atmosphere) etc.

The model is expressed as a mathematical function, permitting calculation of the proportion of values lying in various regions along the scale of measurement. In real life, complete recording of all the values in a **population** is impracticable, but under the assumption of random sampling, the proportions estimated from the model give a guide to the true proportions (or probabilities of occurrence) in the population. When occurrences of low probability are observed, this should lead to questioning the basis of the model, its assumptions or the parameters, and hence to some action such as rejection of a batch, investigation or correction of a process, etc.

In SPC and TQM, two main classes of distribution occur. One class (which includes the Normal distribution) deals with observations that may take any value along a continuous numerical scale: lengths, weights, material properties, etc. These arise from interval or ratio measurement. The other class is concerned with data obtained by counting, and thus deals with events or nominal measurement.

Before describing the three simplest models, we consider the underlying topic of probability.

4.4 Probability

Many SPC actions and decisions are based on probabilistic arguments, and some understanding of the basic axioms and rules is useful.

The frequency definition of probability states that where an 'experiment of chance' may result in different outcomes, of which some yield an event A, the probability of occurrence of A is given by:

$$P(\mathrm{A}) = \frac{\text{Number of outcomes yielding A}}{\text{Total number of possible outcomes}} \tag{4.1}$$

In empirical work, $P(\mathrm{A})$ has to be estimated as

$$P(\mathrm{A}) = \frac{\text{Number of observations of A}}{\text{Total number of observations}} \tag{4.2}$$

this estimate becoming increasingly reliable for larger numbers of observations.

As an example, an **experiment** may comprise drawing a sample of one item from a batch and measuring its length. If the specification requires, say, 22.2 ± 0.1 mm, then any of the **outcomes** 22.31, 22.32, 22.33 ... etc. lead to the **event** 'oversize to specification'. Short of 100% measurement of the batch, we might estimate the probability of such an outcome by sampling a number of items and noting those that are undersize. Then

$$P(\text{oversize}) = \frac{\text{Number of oversize items}}{\text{Total number of items in sample}}$$

Thus if two out of 100 are oversize,

$$P(\mathrm{A}) = \frac{2}{100} = 0.02$$

Evidently this would be a much less reliable estimate than, say, 200 oversize in a sample of 10 000, still giving a proportion of 0.02.

Often we need to combine probabilities to deal with more complex events. There are three basic rules:

(i) The complement rule for an event $\bar{\mathrm{A}}$, defined as the complement of A (i.e. any outcome *not* classified as A); then

$$P(\bar{\mathrm{A}}) = 1 - P(\mathrm{A}) \tag{4.3}$$

(ii) The union rule where $\mathrm{A} \cup \mathrm{B}$ indicates the occurrence of A or B or both, and $\mathrm{A} \cap \mathrm{B}$ indicates the occurrence of both A and B; then

$$P(\mathrm{A} \cup \mathrm{B}) = P(\mathrm{A}) + P(\mathrm{B}) - P(\mathrm{A} \cap \mathrm{B}) \tag{4.4}$$

For events which cannot occur together (mutually exclusive events), $P(\mathrm{A} \cap \mathrm{B}) = 0$, and we then have the simple addition rule as a special case,

$$P(\mathrm{A} \cup \mathrm{B}) = P(\mathrm{A}) + P(\mathrm{B}) \tag{4.5}$$

(iii) The intersection rule where $P(\mathrm{B}|\mathrm{A})$ is the probability of B given that A has occurred (the probability of B conditional upon A); then

$$P(\mathrm{A} \cap \mathrm{B}) = P(\mathrm{A}) \times \mathrm{P}(\mathrm{B}|\mathrm{A}) \tag{4.6}$$

Figures 4.1 and 4.2 illustrate exclusive, non exclusive, independent and conditional events.

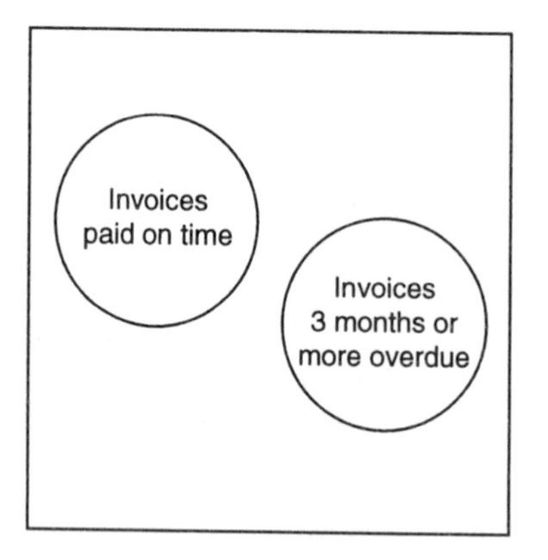

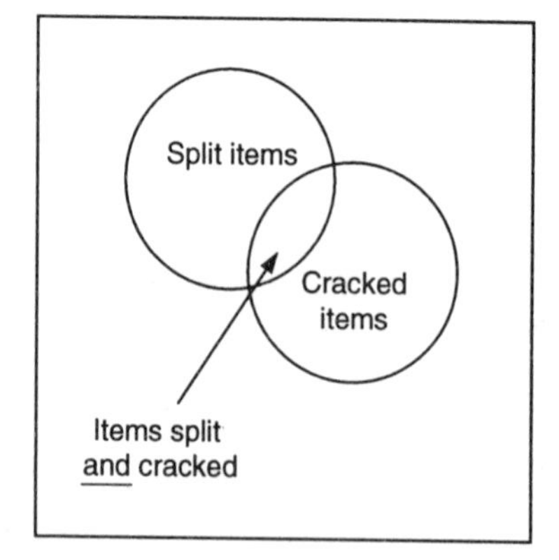

Fig. 4.1 Independent and dependent events.

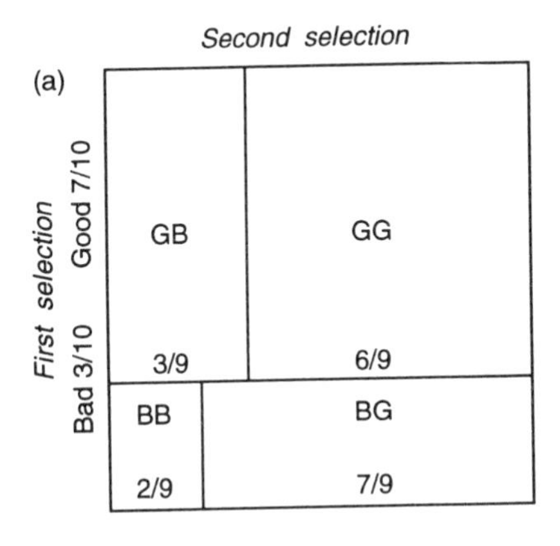

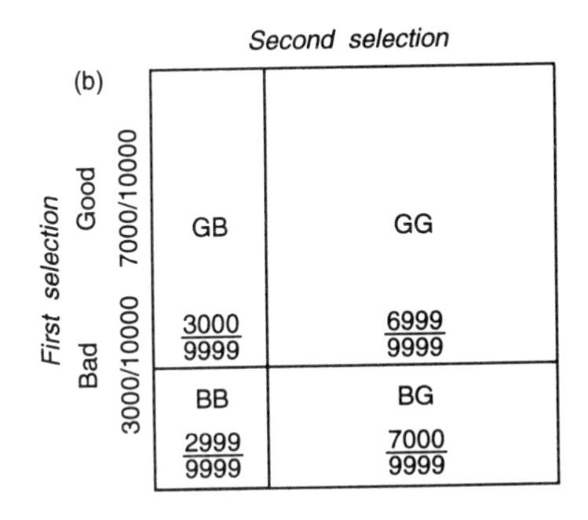

Second selection for (a): the result depends on the first selection.

$$GB = \frac{7}{10} \times \frac{3}{9} = 0.233$$

$$BG = \frac{3}{10} \times \frac{7}{9} = 0.233$$

$$GG = \frac{7}{10} \times \frac{6}{9} = 0.467$$

$$BB = \frac{3}{10} \times \frac{2}{9} = \underline{0.067}$$

$$\underline{1.000}$$

Second selection for (b): the result is almost independent of the first selection.

$$GB = \frac{7000}{10\,000} \times \frac{3000}{9999} = 0.210$$

$$BG = \frac{3000}{10\,000} \times \frac{7000}{9999} = 0.210$$

$$GG = \frac{7000}{10\,000} \times \frac{6999}{9999} = 0.490$$

$$BB = \frac{3000}{10\,000} \times \frac{3999}{9999} = \underline{0.090}$$

$$\underline{1.000}$$

Fig. 4.2 Exclusive and non-exclusive events. (a) Sample two items from 3 bad, 7 good, in a box of 10. (b) Sample two items from 3000 bad, 7000 good, in a consignment of 10 000.

If A and B are **independent** events, then $P(B|A) = P(B)$ irrespective of the occurrence of A, and the multiplication rule follows as the special case

$$P(A \cap B) = P(A) \cdot P(B) \tag{4.7}$$

Note that $A \cap B$ may imply either the occurrence of A followed by B in successive trials, or the occurrence of an outcome in one trial that simultaneously satisfies the definitions of A and B.

As an example of the interplay of these rules, consider the use of a proof-load procedure which states:

(i) Test two items. If both withstand the proof load, pass the batch; if both fail, reject the batch. If one fails, test two more.
(ii) If either of the further items fails, reject the batch.

In stage 1, suppose the proportion of defectives in the batch in 1%. Denote the selection of a defective as event A. Then $P(A) = 0.01$. Selection of a satisfactory item is event $\bar{A}$, with

$$P(\bar{A}) = 1 - P(A) = 0.99$$

Now selection of *two* faulty items is given by

$$P(A) \times P(A) = 0.0001$$

since we can regard the second selection as being virtually independent of the first in a large batch.

Also, selection of two good items is given by

$$P(\bar{A}) \times P(\bar{A}) = 0.9801$$

and for one good, one defective,

$P(A) \times P(\bar{A})$, bad followed by good, $= 0.0099$

and

$P(\bar{A}) \times P(A)$, good followed by bad, $= 0.0099$

giving 0.0198 probability for one of each (via the union rule, with $A \cap \bar{A}$ mutually exclusive to $\bar{A} \cap A$). Thus,

P(rejection on 1st two items) $= 0.0001$,
P(acceptance on 1st two items) $= 0.9801$,
P(second sample required) $= 0.0198$.

Note that these three (mutually exclusive) probabilities sum to 1.0.

Now if (i.e. a **condition**) the second sample of two items is taken,
P(both satisfactory) $= 0.9801$
P(one or two fail) $= 0.0001 + 0.0198 = 0.0199$, (or $1 - 0.9801 = 0.0199$).

Using the intersection rule, a further batch rejection probability arises from P (one failure on 1st pair) $\times$ P (one or both fail in second pair given that one failed in the first pair)

$$= 0.0198 \times 0.0199$$
$$= 0.000\,394\,02$$

Also, a further acceptance probability arises from

$$0.0198 \times 0.9801 = 0.019\,405\,98$$

The overall rejection and acceptance probabilities, again via the 'union rule' with mutually exclusive events, are

Acceptance: $0.9801 + 0.019\,405\,98 = 0.999\,505\,98$
Rejection: $0.0001 + 0.000\,394\,02 = 0.000\,494\,02$

With two **exhaustive** events, these necessarily add to exactly 1.0.

Often a probability tree diagram is useful in unraveling tricky questions. For example, if a box contains 9 components, 2 from one supplier, 3 from a second and 4 from a third, what is the probability that when three items are drawn from the box (without replacement) they are all from the same source? Or all from

Select three items from 2 of supplier A, 3 of supplier B, 4 of supplier C, without replacement.

1st selection	2nd selection	3rd selection	Overall probability
		$P(A \mid A\cap A)=0$	$P(A\cap A\cap A)=\frac{2}{9}\times\frac{1}{8}\times 0=0$‡
	$P(A \mid A)=\frac{1}{8}$	$P(B \mid A\cap A)=\frac{3}{7}$	
		$P(C \mid A\cap A)=\frac{4}{7}$	
		$P(A \mid A\cap B)=\frac{1}{7}$	
$P(A)=\frac{2}{9}$	$P(B \mid A)=\frac{3}{8}$	$P(B \mid A\cap B)=\frac{2}{7}$	
		$P(C \mid A\cap B)=\frac{4}{7}$	$P(A\cap B\cap C)=\frac{2}{9}\times\frac{3}{8}\times\frac{4}{7}=\frac{24}{504}$*
		$P(A \mid A\cap C)=\frac{1}{7}$	
	$P(C \mid A)=\frac{4}{8}$	$P(B \mid A\cap C)=\frac{3}{7}$	$P(A\cap C\cap B)=\frac{2}{9}\times\frac{4}{8}\times\frac{3}{7}=\frac{24}{504}$*
		$P(C \mid A\cap C)=\frac{3}{7}$	
		$P(A \mid B\cap A)=\frac{1}{7}$	
	$P(A \mid B)=\frac{2}{8}$	$P(B \mid B\cap A)=\frac{2}{7}$	
		$P(C \mid B\cap A)=\frac{4}{7}$	$P(B\cap A\cap C)=\frac{3}{9}\times\frac{2}{8}\times\frac{4}{7}=\frac{24}{504}$*
		$P(A \mid B\cap B)=\frac{2}{7}$	
$P(B)=\frac{3}{9}$	$P(B \mid B)=\frac{2}{8}$	$P(B \mid B\cap B)=\frac{1}{7}$	$P(B\cap B\cap B)=\frac{3}{9}\times\frac{2}{8}\times\frac{1}{7}=\frac{6}{504}$‡
		$P(C \mid B\cap B)=\frac{4}{7}$	
		$P(A \mid B\cap C)=\frac{2}{7}$	$P(B\cap C\cap A)=\frac{3}{9}\times\frac{4}{8}\times\frac{2}{7}=\frac{24}{504}$*
	$P(C \mid B)=\frac{4}{8}$	$P(B \mid B\cap C)=\frac{2}{7}$	
		$P(C \mid B\cap C)=\frac{3}{7}$	
		$P(A \mid C\cap A)=\frac{1}{7}$	
	$P(A \mid C)=\frac{2}{8}$	$P(B \mid C\cap A)=\frac{3}{7}$	$P(C\cap A\cap B)=\frac{4}{9}\times\frac{2}{8}\times\frac{3}{7}=\frac{24}{504}$*
		$P(C \mid C\cap A)=\frac{3}{7}$	
		$P(A \mid C\cap B)=\frac{2}{7}$	$P(C\cap B\cap A)=\frac{4}{9}\times\frac{3}{8}\times\frac{2}{7}=\frac{24}{504}$*
$P(C)=\frac{4}{9}$	$P(B \mid C)=\frac{3}{8}$	$P(B \mid C\cap B)=\frac{2}{7}$	
		$P(C \mid C\cap B)=\frac{3}{7}$	
		$P(A \mid C\cap C)=\frac{2}{7}$	
	$P(C \mid C)=\frac{3}{8}$	$P(B \mid C\cap C)=\frac{3}{7}$	
		$P(C \mid C\cap C)=\frac{2}{7}$	$P(C\cap C\cap C)=\frac{4}{9}\times\frac{3}{8}\times\frac{2}{7}=\frac{24}{504}$‡

*p (All different suppliers) $=6\times\frac{24}{504}=\frac{144}{504}=\frac{2}{7}$ (0.2857)

‡P (All same supplier) $=0+\frac{6}{504}+\frac{24}{504}=\frac{30}{504}=\frac{5}{84}$ (0.0595)

Fig. 4.3 Probability tree for intersection of conditional events.

different suppliers? By following the possibilities at the first, second and third selections through the tree, the problem is considerably simplified, as in Fig. 4.3.

4.5 Probability distributions

We again distinguish between the two basic types of numbers which comprise most statistical data. First, the frequencies of occurrence of events or attributes give rise to **counted** data, for which 'discrete distributions' (related to the scale of numbers 0, 1, 2,...) are appropriate. Secondly, from interval or ratio levels of measurement, we obtain **variables**, notionally on a continuous scale. Thus continuous distributions are used in conjunction with data where some property of each item is represented by a numerical value (its height, weight or tensile strength for example). Figure 4.4 illustrates the difference between the two types.

Often the distinction between the two types is blurred: continuous variables are rounded to discrete steps, or counted data involve large numbers, or are converted to proportions, so that we regard them more conveniently as continuous. It is

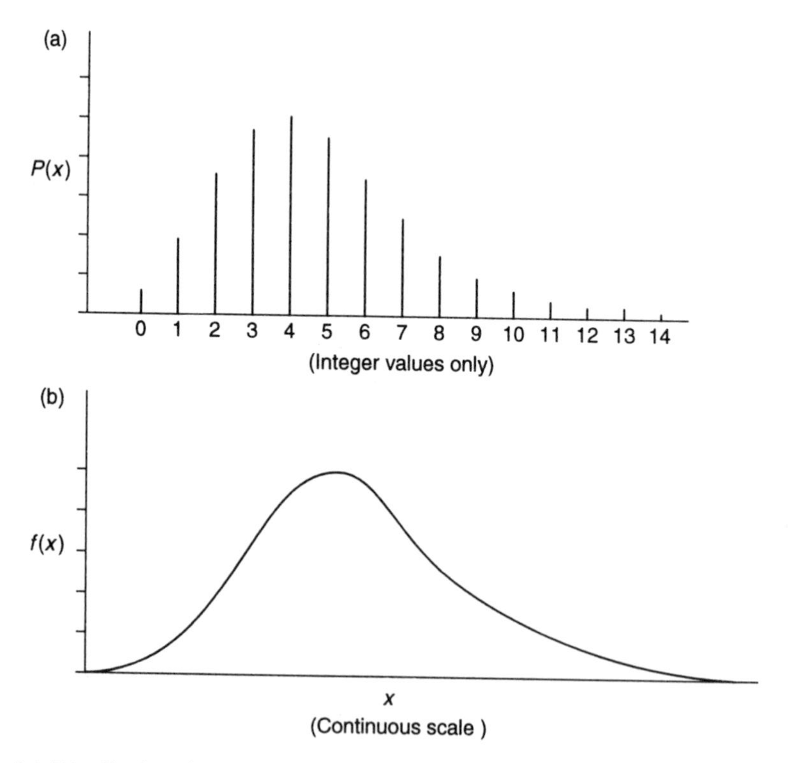

Fig. 4.4 Distributions for counting and measurement. (a) Counting: events and attributes; (b) Measurement.

the **process** by which the numbers arise – counting or measurement – that is important.

4.6 Measurements

Several distribution models for measured data are noted (though not in detail) in Chapter 14 in connection with capability assessment. However, the most widely used model for measurements made on a (conceptually) continuous interval or ratio scale is the **Normal** or Gaussian distribution. It is characterized by its mean, μ, and standard deviation, σ. (In fact the variance, σ^2, is the theoretical parameter, but σ is more useful in actual applications).

The 'normal distribution of errors' is modelled on the assumption that overall variation around a true or central value results from aggregating the contributions of numerous independent sources of error, with no one source dominating the system. The concept matches many natural and industrial processes reasonably well.

The shape of the distribution is given by plotting its density function $f(x)$; probabilities of occurrence of values between given limits (e.g. $-\infty$ to x_1, or x_1 to x_2, or x_2 to ∞) are obtained from the probability integral or distribution function, $F(x)$. The expressions for $f(x)$ and $F(x)$ are shown in Fig. 4.5, but it is unnecessary to use them in most applications as adequate tables are available. In fact, even some calculators provide Normal probabilities as standard functions.

Any practical Normal distribution is reduced to standard form by calculating

$$u(\text{or } z) = \frac{x - \mu}{\sigma} \tag{4.8}$$

the standard Normal variable. The symbol u is in common use in Europe and the UK, while American and Japanese texts use z.

However, as μ and σ are generally unknown, it is necessary to use the best available estimates, $\bar{x}$ and s, giving

$$u(\text{or } z) = \frac{x - \bar{x}}{s} \tag{4.9}$$

For the Normal distribution:

(i) about two-thirds of the values lie in the range $\mu \pm \sigma$ (or roughly $\bar{x} \pm s$);
(ii) 95–96% of the values lie in the range $\mu \pm 2\sigma$ (or roughly $\bar{x} \pm 2s$);
(iii) the vast majority (all but 0.25%) lie in the range $\mu \pm 3\sigma$ (or $\bar{x} \pm 3s$);
(iv) only a few items in 100 000, or about 30 p.p.m. at each end of the distribution, lie beyond $\mu \pm 4\sigma$.

As well as providing a useful tool for describing many sets of data in real life, the normal distribution is important in sampling theory, covered later in this book.

The example in section 2.2.1 listed a set of 52 weights of valve liners. For these data, we had:

$$\bar{x} = 239.51, \; s = 0.934$$

We would **estimate** that:

(i) about two-thirds of the weights lie between 238.58 and 240.44 (in fact 38 of the 52, or 73%, are listed between 238.6 and 240.4, inclusive);
(ii) 95–96% are likely to lie between 237.64 and 241.38 (in fact 48 of the 52, or 92%, are listed between 237.6 and 241.3 inclusive);
(iii) all values in this sample lie within $\bar{x} \pm 3s$, i.e. 236.71 to 242.31;
(iv) in a moderate sized sample, one would not expect to find any values outside $\bar{s} \pm 4s$ unless a 'rogue' has appeared. There do not appear to be any such wild values in this set.

Using the standard Normal variable and Table A.1 we may also predict the proportions at or beyond, say, 237.5 and 242.5 in the 'population' from which our sample was drawn. Thus:

(a) for $x = 237.5$, u(or z) $= \dfrac{(237.5 - 239.51)}{0.934} = -2.15$, giving an estimated 1.6% below 237.5 g;

(b) for $x = 242.5$, u (or z) $= \dfrac{(242.5 - 239.51)}{0.934} = 3.20$, giving an estimated 0.07% above 242.5 g.

4.7 Attributes and events

4.7.1 *Binomial distribution*

Where samples of size n are drawn at random from a population with a proportion p having some characteristic of interest (e.g. they are non-conforming), the numbers of items having that characteristic will vary from sample to sample. These numbers, often denoted by np, will (for a stable population) have the **binomial** distribution.

The probability of occurrence of exactly r 'defectives' is given by

$$P(r) = p^r(1 - p)^{n-r} \frac{n!}{r!(n - r)!} \tag{4.10}$$

However, for ease of calculation, it is best to use a recursive ('bootstrap') method: First,

$$\text{calculate } P(0) = (1 - p)^n \tag{4.11}$$

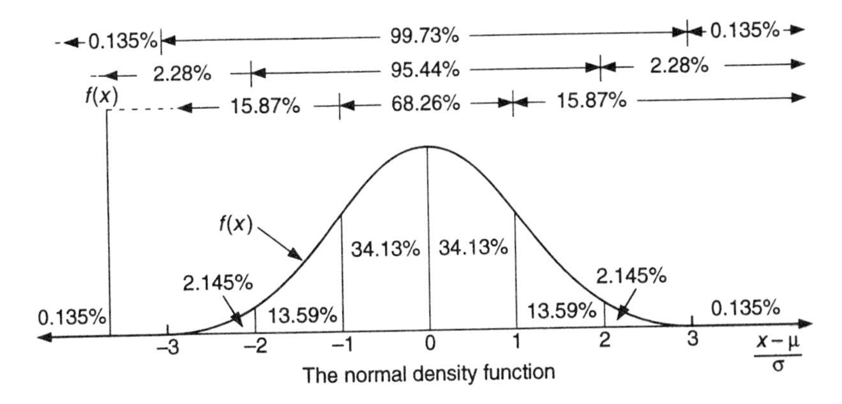

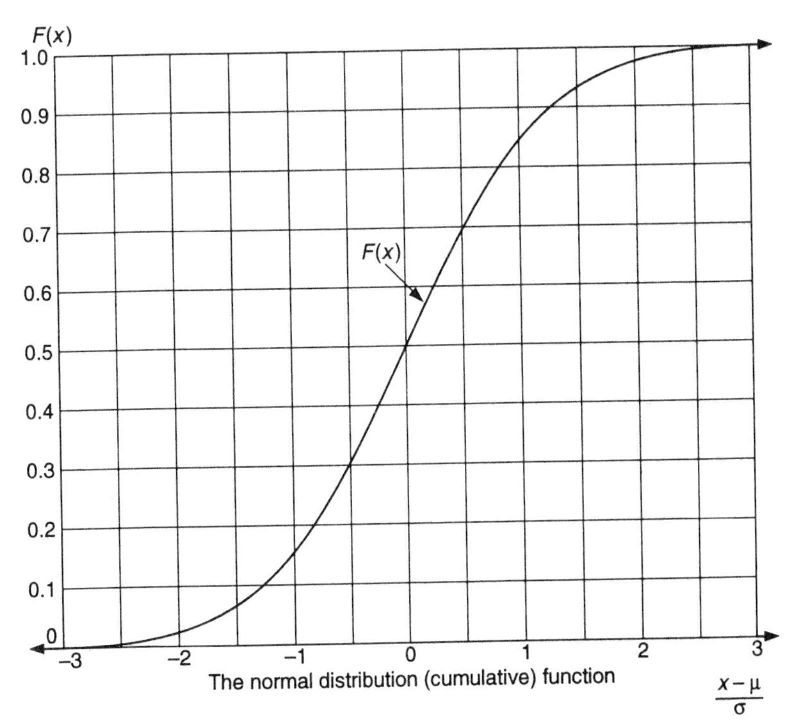

Fig. 4.5 The normal distribution. (a) $f(x) = \dfrac{1}{\sigma\sqrt{(2\pi)}}\exp\left[-\dfrac{1}{2}\left(\dfrac{x-\mu}{\sigma}\right)^2\right]$

(b) $F(x) = \displaystyle\int_{-\infty}^{x} \frac{1}{\sigma\sqrt{(2\pi)}}\exp\left[-\frac{1}{2}\left(\frac{x-\mu}{\sigma}\right)^2\right]dx.$

where μ = mean, and σ = standard deviation.

(all other terms in (4.10) cancel when $r = 0$), then

$$P(r) = P(r-1) \times \frac{p}{(1-p)} \times \frac{n-r+1}{r} \tag{4.12}$$

This method is simple to use (especially when working with a calculator with one or more memories), and is readily programmed.

Consider the case with $n = 50$, $p = 0.08$, $1 - p = 0.92$. Then:

$$P(0) = 0.92^{50} = 0.015\,466;$$
$$P(1) = P(0) \times \frac{0.08}{0.92} \times \frac{50-1+1}{1} = 0.067\,245\,5;$$
$$P(2) = P(1) \times \frac{0.08}{0.92} \times \frac{50-2+1}{2} = 0.143\,262;$$

and so on.

Also the mean number of defectives (per sample of n) is np, and the standard deviation is

$$\sigma_{np} = \sqrt{np(1-p)} \tag{4.13}$$

So, in this examples:

$$\mu_{np} = \text{Mean} = 50 \times 0.08 = 4.0$$
$$\sigma_{np} = \sqrt{50 \times 0.08 \times 0.92} = 1.918.$$

As for the Normal distribution, the parameters (in this case p) usually have to be estimated from sample data, so that $\hat{p} = np/n$ or $\bar{p}$ (the average of p from several samples) is substituted for p in (4.10) to (4.13).

4.7.2 Poisson distribution

Often events occur in a 'continuum', like accidents over time, nicks in the insulation along a length of cable, micro-organisms in the area viewed under a microscope, etc. Then if the events occur independently at random and at a steady **average** rate governed by the system or process, the numbers of events in various elements (of the same size) sampled from the continuum have the **Poisson** distribution.

The probability of exactly r occurrences in a sample element is:

$$P(r) = \frac{\lambda^r}{r!} \exp(-\lambda) \tag{4.14}$$

where λ is the average rate of occurrence per element. The symbols m or $\bar{c}$ are also in widespread use for this average rate. It happens that the variance for the Poisson distribution is also λ, so there is only one parameter, and the standard deviation is $\sqrt{\lambda}$.

As for the binomial, the simplest method is to build $P(r)$ up from $P(0)$ by recursion. We then have:

$$P(0) = \exp(-\lambda), \quad \text{since } \lambda^0 \text{ and } 0! \text{ are both } 1.0 \tag{4.15}$$

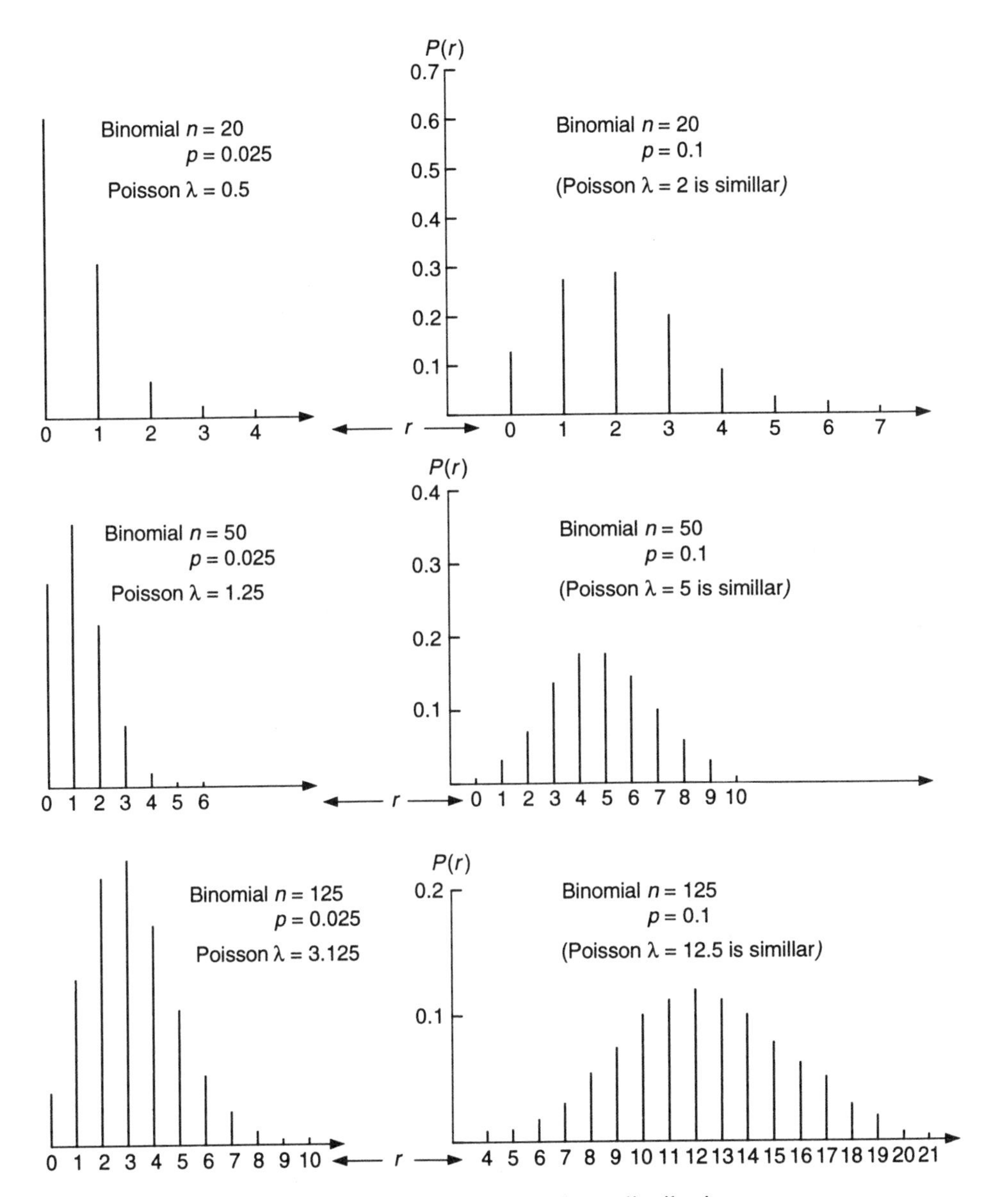

Fig. 4.6 Some binomial and Poisson distributions.

and

$$P(r) = P(r-1) \times \frac{\lambda}{r} \tag{4.16}$$

The estimated parameter m or $\bar{c}$ is substituted for λ as appropriate.

4.7.3 The Poisson distribution is often used as a good (and simpler) approximation for the binomial when $\bar{p} < 0.1$, substituting $n\bar{p}$ for λ in the above expressions. If we re-work the previous example using the Poisson approximation, we have:

$\lambda = n\bar{p} = 4.0$;
$\sigma = \sqrt{n\bar{p}} = 2.0$ (c.f. the correct value 1.918).

Also

$P(0) = \exp(-4.0) = 0.0183$;
$P(1) = P(0) \times \frac{4}{1} = 0.07326$;
$P(2) = P(1) \times \frac{4}{2} = 0.146525$.

Comparing these values with the true binomial probabilities (calculated in section 4.7.1) the match is quite good. This is a case close to the suggested upper limit of $\bar{p}$ for valid use of the approximation.

Apart from its use as an approximation, the Poisson distribution provides a model for the occurrence of **non-conformities**, whereas the binomial models non-conforming **items**.

Figure 4.6 gives examples of some binomial and Poisson distributions. Many collections of tables give cumulative or term-by-term probabilities for a selection of binomial and Poisson distributions, and nomograms or charts are also available.

4.8 Normal approximations

4.8.1 The Normal distribution can provide a useful approximation for the binomial or Poisson in many cases. A reasonable condition is that $n\bar{p}$ (for the binomial) or $\bar{c}$ (the estimate of λ for the Poisson distribution) should be at least ten, and preferably 15 or more.

Then μ for the normal distribution is equated to $n\bar{p}$ or $\bar{c}$, and σ is replaced by $\sqrt{[n\bar{p}(1-\bar{p})]}$ or $\sqrt{\bar{c}}$. Percentage points (e.g. for use with control charts) or probabilities of occurrence of tail values are then obtained via the standard normal variable, u or z.

As an example where the approximation is *not* appropriate, consider again the case of $n\bar{p} = 4$, $\sqrt{[n\bar{p}(1-\bar{p})]} = 1.918$, as in section 4.7.1. We may require an upper control limit at the 0.1% probability level with $u = 3.09$. This is obtained as:

$$n\bar{p} + 3.09\sqrt{[n\bar{p}(1-\bar{p})]} = 4 + 3.09 \times 1.918 = 9.9266$$

or approximately 10. In fact, $r \geqslant 10$ occurs with probability 0.56% for a binomial distribution with $n = 50$, $p = 0.08$, and even $r \geqslant 11$ has probability 0.17%.

A lower limit at 0.1% probability does not exist, since $P(0) = 0.015\,466$ exceeds 0.001. The normal approximation gives the absurd value of $4.0 - 3.09 \times 1.918 = -1.9266$.

The approximation is much better when $n\bar{p}$ is 10, for example with $n = 200$, $\bar{p} = 0.05$. Then:

$$n\bar{p} = 10, \ \sqrt{[n\bar{p}(1-\bar{p})]} = 3.0822$$

For 0.1% limits, we have:

$$n\bar{p} \pm 3.09\sqrt{[n\bar{p}(1-\bar{p})]} = 10 \pm 9.524 = 0.476, \ 19.524$$

These would be interpreted as implying that occurrences of 0 or 20 are 'out of control'. The true binomial probabilities for $r = 0$, and $r \geqslant 20$ are 0.000 035 and 0.0027, respectively.

4.8.2 Conversely, if we require **approximate** values for the probabilities of $r = 0$ and $r \geqslant 20$, we may use

$$u_1 = \frac{0 - 10}{3.0822} = -3.244$$

$$u_2 = \frac{20 - 10}{3.0822} = +3.244$$

giving $P(u) = 0.000\,59$, reasonably close to the true probabilities above.

4.8.3 In various applications, the normal distribution provides a useful model for sample statistics such as $\bar{x}$, especially in large samples when even R and s have approximately normal distributions. Some of these ideas are introduced in Chapter 5, and the use of the normal distribution via data transformations is briefly noted in Chapter 14.

4.9 Conclusion

The subjects covered in these notes – levels of measurement, probability, distribution models – along with the basic ideas of tabulation, tallies, diagrams and measures of location and dispersion covered in previous chapters, provide the foundations of SPC methods for control charting and capability evaluation. They also underpin the concepts of estimation and hypothesis testing to be developed shortly, and the separation of components of variation covered in Chapter 16.

5

Sampling, estimation and confidence

5.1 Introduction

The object of sampling is usually to estimate some feature(s) of the population, process or system from which the sample is taken. Sampling forms an essential part of many data collection activities, for example in surveys and sampling inspection, as well as process control and experimentation.

The purpose of sampling can often be narrowed to obtaining estimates of population or process features, such as averages, the extent of variation, proportions of items having some characteristic, or the rate of occurrence of events. If repeated samples were drawn from the same 'population' (a population, in various circumstances, may be a production run, a time period, a batch of material, the output of an experiment, etc.), differing estimates of the characteristic would result. Thus sample means, ranges, proportions, would vary, forming a **sampling distribution**. A sampling distribution is thus the pattern yielded by the values of some sample statistic ($\bar{x}$, R, p, etc.) when sampling repeatedly under the same conditions from a stable system.

5.2 Sampling distributions

5.2.1 To introduce the idea of sampling distributions, consider a large population containing equal proportions of 1's, 2's and 3's. Such a population might be a box of numbered discs (1,2,3), a pack of cards consisting only of 1's, 2's and 3's, or an ordinary six-sided die with the faces 4,5,6 replaced by an additional set of 1,2,3.

Now over the course of many samples, the proportions of 1's, 2's and 3's will each be close to one-third, so the mean score (per sample drawing) is:

$$\tfrac{1}{3}(1+2+3)=2$$

and the variance is

$$\tfrac{1}{3}[(1-2)^2+(2-2)^2+(3-2)^2]=\tfrac{2}{3}$$

Table 5.1 Sampling distribution for mean of two scores

Scores	Mean $\bar{x}$	Relative frequency f	$f\bar{x}$
(1, 1)	1.0	1	1.0
(1, 2), (2, 1)	1.5	2	3.0
(1, 3), (3, 1), (2, 2)	2.0	3	6.0
(2, 3), (3, 2)	2.5	2	5.0
(3, 3)	3.0	1	3.0
		$\sum f = 9$	$\sum fx = 18.0$

Overall mean $\bar{\bar{x}} = \sum fx / \sum f = 18/9 = 2.0$
$\equiv \mu$ for individual scores

Thus $\mu = 2.0$, $\sigma^2 = 0.\dot{6}$ and $\sigma = 0.8165$ approx. These are the true or theoretical population parameters.

5.2.2 Now suppose that a sample of size two is drawn with replacement (either there are many cards, or one card is replaced before drawing another; in the case of dice, of course, the scores are simply noted for each set of two throws). If this operation is repeated many times, there are nine possible sample configurations, each equally likely to occur. They are: (1, 1), (1, 2), (1, 3), (2, 1), (2, 2), (3, 1), (3, 2), (3, 3).

As far as sample means are concerned, pairs like (1, 2), (2, 1) are indistinguishable. If we are interested in sample means, the sampling distribution is therefore as shown in Table 5.1 in the form of frequency table. As indicated in Fig. 5.1(b), this is a triangular distribution (compared with the flat-topped or rectangular shape of the distribution of individual scores.

The additional calculations in Table 5.1 indicate that the mean of all the sample means is identical to ($\equiv$) the mean of the individual scores. Thus, in the long run, the arithmetic mean of many sample means will be close to the true mean of a process, and $\bar{x}$ is therefore said to be an unbiassed estimator of the population mean, μ.

5.2.3 If we rearrange the samples of Table 5.1 to group together those that give particular variances, we have a frequency table as in Table 5.2. In this case, we have listed both the σ_n^2 and s^2 values for the variance, following the formulae (3.14) for σ_n^2 and (3.15) for s^2.

From the foot of Table 5.2, it appears that the mean of the σ_n^2 values from samples of two items does not equal the true population variance. However, the mean of the s^2 does equal the true variance, so in this case s^2 yields the unbiassed estimate, accounting for its preferred use in general statistical applica-

Table 5.2 Sampling distributions for variances, σ^2 and s^2 (Samples of $n = 2$ from uniform 1,2,3 distribution)

Scores	Relative frequency f	σ_n^2	s^2	$f\sigma_n^2$	fs^2
(1, 1), (2, 2), (3, 3)	3	0	0	0	0
(1, 2), (2, 1), (2, 3), (3, 2)	4	0.25	0.5	1.0	2.0
(1, 3), (3, 1)	2	1.0	2.0	2.0	4.0
	9 ($\sum f$)			3.0 ($\sum f\sigma_n^2$)	6.0 ($\sum fs^2$)

Mean of $\sigma_n^2 = 3.0/9 = \frac{1}{3}$ (biassed)
Mean of $s^2 = 6.0/9 = \frac{2}{3}$ ($\equiv \sigma^2$ for individual scores)

tions – the use of σ_n^2 via formula (3.14) tends to underestimate the true process variation.

5.2.4 It can be shown, although the reader is referred to theoretical texts for mathematical proof, that the effect of the n-1 divisor holds for any sample size. To facilitate further development, we consider in Table 5.3 the sampling distributions for means and variances for sample of $n = 3$. The distribution of $\bar{x}$ is again illustrated in Figure 5.1, along with that for $n = 5$ and 10.

Table 5.3 Sampling distributions of means and variances with $n = 3$

Scores	Frequency f	Mean $\bar{x}$	Variances σ_n^2	s^2
3 @ 1	1	1.0	0	0
3 @ 2	1	2.0	0	0
3 @ 3	1	3.0	0	0
2 @ 1, 1 @ 2	3	$1.\dot{3}$	$0.\dot{2}$	$0.\dot{3}$
1 @ 1, 2 @ 2	3	$1.\dot{6}$	$0.\dot{2}$	$0.\dot{3}$
2 @ 1, 1 @ 3	3	$1.\dot{6}$	$0.\dot{8}$	$1.\dot{3}$
1 @ 1, 2 @ 3	3	$2.\dot{3}$	$0.\dot{8}$	$1.\dot{3}$
2 @ 2, 1 @ 3	3	$2.\dot{3}$	$0.\dot{2}$	$0.\dot{3}$
1 @ 2, 2 @ 3	3	$2.\dot{6}$	$0.\dot{2}$	$0.\dot{3}$
1 @ 1, 1 @ 2, 1 @ 3	6	2.0	$0.\dot{6}$	1.0

$\sum f\bar{x} = 54$; $\sum f\sigma_n^2 = 12$; $\sum fs^2 = 18$
Mean $\bar{x} = 2.0$; mean $\sigma_n^2 = 0.\dot{4}$ (biassed); mean $s^2 = 0.\dot{6}$ ($\equiv \sigma^2$ individuals)

We note two features from these sampling distributions. First, for variances, the mean of the s^2 values is again exactly two-thirds, demonstrating its absence of bias as an estimator. The use of σ_n^2 still leads to bias, although slightly reduced for $n = 3$ as compared with $n = 2$. This bias would continue to diminish as the sample size increases.

Secondly, with particular reference to Fig. 5.1, the sampling distribution for $\bar{x}$ with $n = 3$ begins to take on the shape of a Normal curve (though it is not a strictly continuous curve—$\bar{x}$ can only take values 1, $1\frac{1}{3}$, $1\frac{2}{3}$ etc., but not intermediate values). For $n = 5$ and 10 the resemblance to the Normal curve is even stronger. In fact many sampling distributions approach Normality for statistics based on large samples, but the approximate Normality is reached with quite small samples in the case of $\bar{x}$, unless the parent distribution is appreciably non-normal. In many practical SPC applications, $n = 5$ is a sufficient sample size to normalize the distribution of $\bar{x}$. This is one reason for its popularity in control chart use.

This important property of sampling distributions is known as the 'central limit theorem', and we now move on to deal with another of its implications.

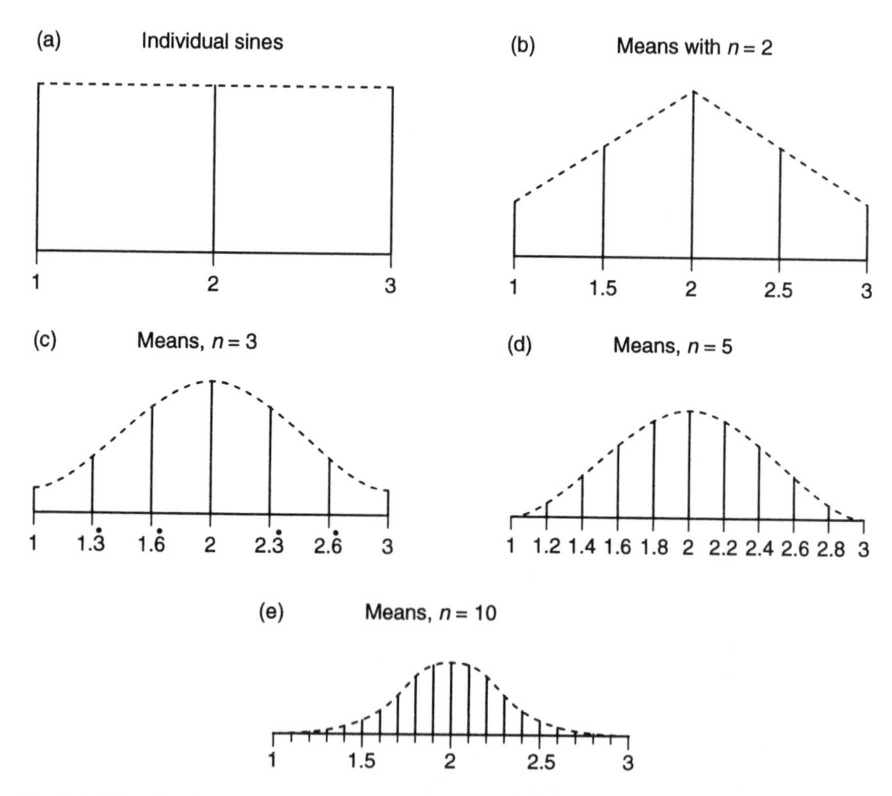

Fig. 5.1 Distribution of mean scores (from 1, 2, 3). (a) Individual scores; (b) means with $n = 2$; (c) means with $n = 3$; (d) means with $n = 5$; (e) means with $n = 10$.

5.3 Standard errors of sampling distributions

5.3.1 Reference to Fig. 5.1 will indicate that although the sampling distributions for sample means all have the same overall range (from 1 to 3), those with larger sample sizes tend to cluster progressively more closely to the centre as the sample size increases. If we calculated the variances of these sampling distributions, we would find them diminishing for the larger sample sizes.

Now for the two distributions of $\bar{x}$ fully listed in Table 5.3, we have:

Variance of $\bar{x}$ for $n = 2$: $\sigma_{\bar{x}}^2 = 0.\dot{3} = \frac{1}{3}$
Variance of $\bar{x}$ for $n = 3$: $\sigma_{\bar{x}}^2 = 0.\dot{2} = \frac{2}{9}$

In fact, for sample size n, the variance of $\bar{x}$ is σ^2/n, where σ^2 is the variance of the individual values in the population. For most practical purposes, we require the square root of this quantity which (to distinguish it from the standard deviation of individual observations) is termed the standard error. Thus we have the very important result that:

$$\text{Standard error of } \bar{x} = \sigma_{\bar{x}} = \sigma/\sqrt{n} \tag{5.1}$$

The standard error describes the variation, from sample to sample, of a **sample statistic** in a manner similar to that of the standard deviation. Combining this measure of variation with other properties of the sampling distribution of $\bar{x}$ noted in section 5.2.4, we note that for samples of 'reasonable' size from populations of individual values whose distribution is not grossly non-normal, the means ($\bar{x}$) in samples of size n have a distribution close to the normal with mean μ (as for the individual values) and standard error $\sigma/\sqrt{n}$. This is often written as:

$$\bar{x}_n \sim N\left(\mu, \frac{\sigma}{\sqrt{n}}\right)$$

(or $N(\mu, \sigma^2/n)$ in more rigorous notation using variance rather than standard error).

5.3.2 Each statistic, e.g. median, range, standard deviation, and also those obtained from counted data, such as p, u (and even np, c) has a standard error. Many are related to the parameters of the distribution in fairly simple ways. Table 5.4 gives some of the simpler cases, and the expressions will be somewhat familiar to many control chart users, in that the formulae for control chart limits use them in conjunction with the multiplier $3\times$; the logic for this will be covered in Chapter 7.

Other statistics have more complicated relationships to the parent population parameters: these include the median, range and standard deviation. In some cases, especially for the larger sample sizes, there are good numerical approximations, otherwise it is necessary to consult tables.

Table 5.4 Standard errors of 'control chart statistics'

Parameter in population	Sample statistic (estimator)	Standard error
Mean, u	$\bar{x}$	$\frac{\sigma}{\sqrt{n}}$
Number non-conforming	np	$\sqrt{\{n\pi(1-\pi)\}}$
Proportion non-conforming	p	$\frac{\pi(1-\pi)}{n}$
Number of non-conformities	c	$\sqrt{C}$
Rate of non-conformities	u	$\sqrt{\left(\frac{U}{n}\right)}$

μ, π, C, U generally estimated via $\bar{x}$, $\bar{p}$, $\bar{c}$, $\bar{u}$, respectively.
For estimators of σ, see Chapter 13.4.

5.3.3 We now summarize these points for the case of the sampling distribution of $\bar{x}$:

(i) For increasing sample sizes, the distribution of $\bar{x}$ progressively narrows. For large samples, few values of $\bar{x}$ are very far from the true mean, μ, of the population.
(ii) The spread of the sampling distribution of $\bar{x}$ (and indeed many other sample statistics) can be measured by the standard error, σ_e, in a sense similar to that in which the standard deviation, σ, measures the spread of the individual values.
(iii) The standard error of the sampling distribution of $\bar{x}$ is $\sigma/\sqrt{n}$, where n is the size of sample from which μ is estimated (via $\bar{x}$).
(iv) The mean of the sampling distribution of $\bar{x}$ is said to be an unbiassed estimate of μ.
(v) Unless the distribution of individuals is appreciably non-normal, the sampling distribution of $\bar{x}$ is close to the normal.

For a normal distribution of individual values, these points are illustrated in Fig. 5.2.

5.4 Reliability of estimates: confidence limits

5.4.1 As noted in 5.3.1, the standard error provides a means of predicting the extent of sampling variation of a statistic or parameter estimate in repeated sampling from a stable population. For example, if many samples, each of size 5,

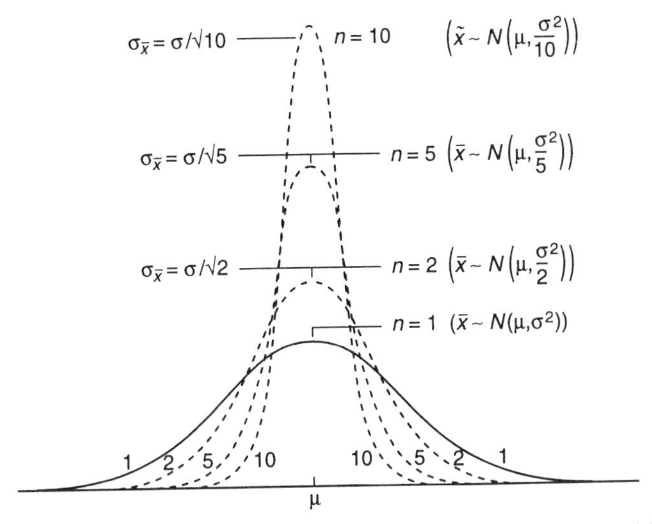

Fig. 5.2 Sampling distributions of $\bar{x}$ for normal distribution of individuals.

are drawn from an in-control process (a practical example of a stable population) with say, a mean of 240 and a standard deviation of 1.0, then only 4–5% of sample $\bar{x}$ values will lie outside

$$\mu \pm 2\frac{\sigma}{\sqrt{n}} = 240 \pm 2 \times 1.0/\sqrt{5} = 239.11,\ 240.89.$$

Even fewer, only about 1 in 400, will lie outside

$$\mu \pm 3\frac{3}{\sqrt{n}} = 240 \pm 3 \times 1.0/\sqrt{5} = 238.66,\ 241.34$$

In control chart theory, these statements have important implications for the positioning of control limits, when values in the unlikely region outside these limits are taken as indicators of disturbance or 'special causes'.

5.4.2 Now even where only one sample of n items is drawn, the theoretical standard error provides a guide to the reliability or precision of the estimate, assuming it is based on a random sample from the population concerned. This reliability is usually expressed as a **confidence interval** or a pair of **confidence limits**, within which the true (but unknown) value of the parameter is believed to lie.

Let us consider the sample mean, $\bar{x}$ and its relation to the population mean, μ. Suppose the true mean of the population is 240.0 (but this is not known by the person drawing the sample), and further that the standard deviation of individual values can be assumed to be 1.0.

One sample of size $n = 10$ is drawn, and $\bar{x}$ is calculated.

5.4.3 The argument in 5.4.1 implies that although individual values of x may well range from around 237 to 243 ... in the example given, $\bar{x}$ values will generally lie much closer to the true mean of 240.0.

Now suppose that for various product batches, perhaps made on different days or from different deliveries of raw material, the mean value μ varies from day to day or batch to batch. We will assume for the moment that the standard deviation is not affected by these fluctuations, and remains stable at $\sigma = 1.0$. From one particular batch, we draw a sample of, say, ten items, and find $\bar{x} = 239.5$. What can we say about the (unknown) batch mean, μ?

Using the preceding arguments, we can say that very few values of $\bar{x}$, only 2.5% in fact, will be less than $\mu - 1.96\sigma_{\bar{x}}$, using $z = 1.96$, the 2.5% point of the normal distribution. Similarly, only 2.5% of $\bar{x}$ values would exceed $\mu + 1.96\sigma_{\bar{x}}$. Now whilst we do not know the value of μ, it appears unlikely that our sample comes from a population with μ less than $\bar{x} - 1.96\sigma_{\bar{x}}$, otherwise we would have been unlikely to observe such a large value of $\bar{x}$. Conversely, a value of μ greater than $\bar{x} + 1.96\sigma_{\bar{x}}$ would be unlikely to yield such a low value of $\bar{x}$ as the one we have observed. Setting limits for μ consistent with our sample information, we have

$$L_1 + 1.96\sigma_{\bar{x}} = \bar{x}$$
$$L_2 - 1.96\sigma_{\bar{x}} = \bar{x}$$

Only L_1 and L_2 are unknown, as in the present example we have $\bar{x} = 239.5$ and $\sigma_{\bar{x}} = \sigma/\sqrt{n} = 1.0/\sqrt{10}$, i.e. 0.316. Rearranging the above equations to obtain L_1 and L_2 in terms of $\bar{x}$, $\sigma_{\bar{x}}$ gives us

$$L_1, L_2 = \bar{x} \pm 1.96\sigma_{\bar{x}} \tag{5.2}$$

in this case $= 239.5 \pm 1.96 \times 0.316$, i.e. 238.88 to 240.12.

Thus we assert, with 95% confidence, that the population mean from which our sample was drawn lies between these limits. In **repeatedly** making such

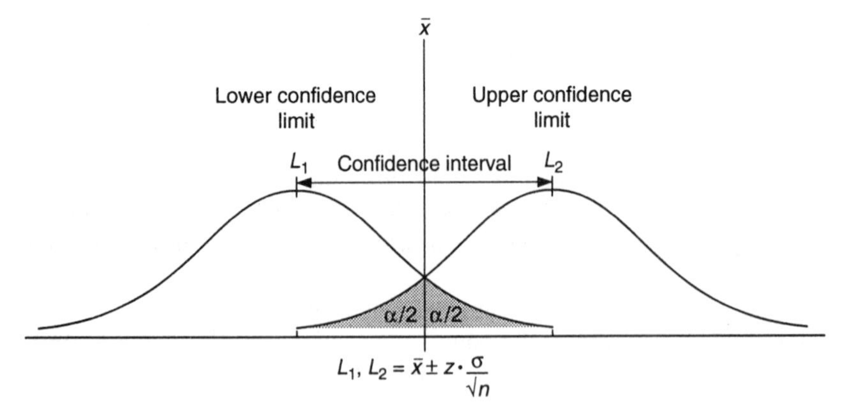

Fig. 5.3. Confidence interval and confidence limits (for μ, based on $\bar{x}$). $L_1, L_2 = \bar{x} \pm 2\sigma/\sqrt{n}$.

Table 5.5 Factors for $100(1-\alpha)\%$ confidence limits for μ using $\bar{x}$ (known σ)*

Confidence level $100(1-\alpha)\%$	$\frac{1}{2}\alpha$	Factor u or z
50	0.25	0.675 ($\simeq 2/3$)
80	0.1	1.282 ($\simeq 1.3$)
90	0.05	1.645 ($\simeq 5/3$)
95	0.025	1.96 ($\simeq 2$)
99	0.005	2.576 ($\simeq 2.6$)
99.9	0.0005	3.290 ($\simeq 3.3$)

*Confidence limits: sample estimate $\pm u$ (or z) $\times$ standard error of estimate.

assertions based on sample information, we shall be correct in 95% of cases in the long run. In the other 5% of cases, the true mean will actually lie outside the confidence interval.

The confidence interval concept is illustrated in Fig. 5.3. The areas labelled $\alpha/2$ measure the risk of error in each direction, so that with a total error risk of α we have a $100\ (1-\alpha)\%$ confidence interval.

5.4.4 For some purposes, we may require either more or less confidence than 95% (i.e. less or more than a 5% risk of error). This is simply achieved by narrowing or widening the interval, choosing $\alpha/2$ so as to yield $100\ (1-\alpha)$ of 90, 99, 99.9% or whatever is required. We may even require a 50%, 'as-likely-as-not' interval. While appropriate values of z can readily be obtained via a table of the normal distribution, the factors in Table 5.5 will cover most applications.

Using Table 5.5, in conjunction with the example in 5.3.3, some of these limits and intervals are:

50%	$239.5 \pm 0.21 = 239.29$ to 239.71;
95%	$239.5 \pm 0.62 = 238.88$ to 240.12;
99%	$239.5 \pm 0.81 = 238.69$ to 240.31;
99.9%	$239.5 \pm 1.04 = 238.46$ to 240.54.

(Evidently a 100% interval of $-\infty$ to $+\infty$, or even 0 to infinity for a positive-valued characteristic, is uninformative; some risk of error is necessary if a useful measure of reliability is to be obtained.

5.5 Confidence limits for μ, where σ is estimated by s

5.5.1 Unfortunately the true value of σ, and hence of $\sigma_{\bar{x}}$, is not usually known, and has to be estimated – usually from the same sample as is used to estimate the mean. The usual estimate, s, (via 3.15) obviously cannot be regarded as an exact parameter: different samples will yield somewhat differing values of s. In fact we shall consider confidence limits for σ (based on s) in section 5.6.

To allow for this uncertainty about the value of σ, the confidence interval needs to be widened, and the correct procedure is to use the 'Student's t' statistic in place of z. The degree of widening of the interval depends on the sample size, or more exactly on the number of degrees of freedom available for estimating s. A full account of degrees of freedom is not required here, but the concept can be envisaged in terms of the number of 'units of information' available for estimating variation. A sample of $n = 1$ provides a crude indication of location, but no measure of variation. For $n = 2$, the difference between the two values contributes one unit of information on variation, and each further addition of one item to the sample gives a further degree of freedom. For one sample of n items there are thus $\nu = n - 1$ degrees of freedom; $n - 1$ also, of course, relates to the divisor used in the expression (3.20) for s.

Tables of 'Student's t' have the complication that for various uses, either one- or two-tail probabilities are required. To avoid both this complication, and also the fact that t-values for degrees of freedom 49, 99 (as would occur with samples of 50 or 100) are rarely given, some useful t-values are listed in Table 5.6, indexed by sample size and level of confidence. The table is specifically designed for

Table 5.6 Values of t for estimation of confidence limits for μ (based on $\bar{x}$ and s)

Sample size	Confidence level (%)					
n	50	80	90	95	99	99.9
3	0.82	1.89	2.92	4.30	9.92	31.6
4	0.77	1.64	2.35	3.18	5.84	12.9
5	0.74	1.53	2.13	2.78	4.60	8.61
6	0.73	1.48	2.02	2.57	4.03	6.87
8	0.71	1.42	1.90	2.36	3.50	5.41
10	0.70	1.38	1.83	2.26	3.25	4.78
12	0.70	1.36	1.80	2.20	3.11	4.44
15	0.69	1.34	1.76	2.14	2.98	4.14
20	0.69	1.33	1.73	2.09	2.86	3.88
25	0.685	1.32	1.71	2.06	2.80	3.75
30	0.68	1.31	1.70	2.04	2.76	3.66
40	0.68	1.30	1.69	2.02	2.71	3.56
50	0.68	1.30	1.68	2.01	2.68	3.50
60	0.68	1.30	1.67	2.00	2.66	3.46
80	0.68	1.29	1.66	1.99	2.64	3.42
100	0.68	1.29	1.66	1.98	2.63	3.39
(∞)*	0.675	1.28	1.645	1.96	2.58	3.29

*Values for $n = \infty$ are identical to u (or z).

estimation of confidence intervals, and is not intended for the many other applications of Student's t. A more conventional presentation appears in Table A.4.

The procedure for obtaining confidence intervals and limits, where s is used to estimate σ, is simply to substitute s for σ as an estimate of $\sigma_{\bar{x}}$, viz:

$$\hat{\sigma}_{\bar{x}} = s/\sqrt{n}, \tag{5.3}$$

and to replace z by t in the expressions for confidence limits,

$$L_1, L_2 = \bar{x} \pm t\left(\frac{s}{\sqrt{n}}\right) \tag{5.4}$$

Values of t are selected from Table 5.6 according to the sample size and the level of confidence required.

5.5.2 We are now in a position to consider further the data for valve liners in 2.2.1. Here, for a sample of $n = 52$ items, we had $\bar{x} = 239.5$, and $s = 0.934$. Within what region may we assume the true batch or process mean to lie, based on these estimates?

Following (5.3) and (5.4), we have

$$\hat{\sigma}_{\bar{x}} = 0.934/\sqrt{52} = 0.1295$$

and for 95% confidence,

$$L_1, L_2 = 239.51 \pm 2.01 \times 0.1295 = 239.25, 239.77$$

Among other things, we may note that it seems unlikely that the true mean is at or above the mid-specification value of 240.0. For other levels of confidence, we have:

As likely as not, μ lies between 239.42 and 239.60, (50% confidence);
Almost certainly, μ lies between 239.06 and 239.96., (99.9% confidence).

5.6 Confidence limits for σ, based on s

5.6.1 If we require a confidence interval for the true **mean** of a population, based on a sample estimate $\bar{x}$, it is evident that an interval for the true standard deviation, based on the estimate s, will also be useful.

Although s, like $\bar{x}$, has a standard error (it is actually $\sigma/\sqrt{\{2(n-1)\}}$ for moderate sized samples), the sampling distribution of s is fairly skewed, especially in small samples. The method of obtaining interval estimates for σ uses the χ^2 (chi-squared) distribution, but is somewhat indirect. Table 5.7 therefore gives, for various values of n and levels of confidence, pairs of multipliers derived from χ^2. To obtain the upper and lower confidence limits for σ, the sample estimate s is multiplied by each of the factors listed in the appropriate row and column.

Table 5.7 Multipliers for confidence limits for σ*

Sample size n	Confidence level, 100(1 − α)% 50	90	95	99
3	0.85, 1.87	0.58, 4.41	0.52, 6.29	0.43, 14.14
4	0.85, 1.57	0.62, 2.92	0.57, 3.73	0.48, 6.47
5	0.86, 1.44	0.65, 2.37	0.60, 2.87	0.52, 4.40
6	0.87, 1.37	0.67, 2.09	0.62, 2.45	0.55, 3.48
8	0.88, 1.28	0.71, 1.80	0.66, 2.04	0.59, 2.66
10	0.89, 1.24	0.73, 1.65	0.69, 1.83	0.62, 2.28
12	0.90, 1.20	0.75, 1.55	0.71, 1.70	0.64, 2.06
15	0.90, 1.17	0.77, 1.46	0.73, 1.58	0.67, 1.85
20	0.915, 1.14	0.79, 1.37	0.76, 1.46	0.70, 1.67
25	0.922, 1.12	0.81, 1.32	0.78, 1.39	0.73, 1.58
30	0.927, 1.11	0.83, 1.28	0.80, 1.34	0.74, 1.49
40	0.936, 1.09	0.85, 1.23	0.82, 1.28	0.77, 1.40
50	0.941, 1.08	0.86, 1.20	0.84, 1.25	0.79, 1.34
60	0.946, 1.072	0.87, 1.18	0.85, 1.22	0.81, 1.30
80	0.952, 1.061	0.89, 1.15	0.87, 1.18	0.83, 1.25
100	0.957, 1.054	0.90, 1.13	0.88, 1.16	0.84, 1.22

*Multiply s by each factor for appropriate sample size and confidence level to obtain lower and upper limits for σ.

5.6.2 The factors listed in Table 5.7 are based on two main assumptions.

(i) The estimate s is based on a single random sample of n observations or items from a homogeneous system;
(ii) the individual data values have (at least approximately) a normal distribution.

In the case of violations of the second of these assumptions, it is much more difficult to obtain valid confidence intervals for σ, but a transformation to normality may be possible. Some possible transformations are suggested in section 14.6.1.

The first assumption is often invalidated by the fact that an estimate $\hat{\sigma}$ is used, based on values of s or R in a number of subgroups; say k subgroups each of size n. Approximate confidence intervals may then be obtained by entering Table 5.6 with a pseudo sample-size of $k \times (n-1)$. Strictly, the dummy sample size should be obtained as

$$\{k(n-1) \times \text{discount factor}\} + 1$$

for cases where $\bar{s}$ or $\bar{R}$ is used to obtain $\hat{\sigma}$. The discount factors are listed in Table 16.1.

5.6.3 Finally, we apply the procedure of this section to the valve liner data, where we had $n = 52$, $s = 0.934$. Interpolating in Table 5.6 between $n = 50$ and

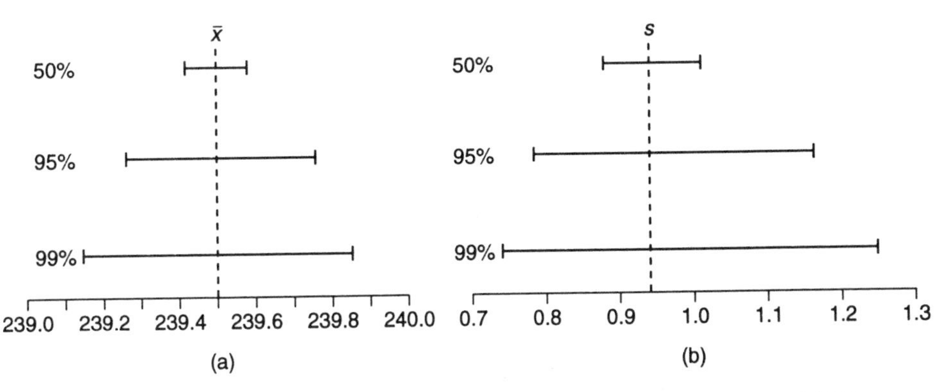

Fig. 5.4 Confidence intervals for (a) μ, and (b) σ. ($n = 52$; $x = 239.51$; $s = 0.934$.)

$n = 60$, we have pairs of factors for 50, 95 and 99% confidence approximately as follows:

50% 0.942, 1.078
95% 0.84, 1.25
99% 0.79, 1.33.

These yield confidence intervals for σ:

50% 0.880 to 1.007
95% 0.785 to 1.168
99% 0.738 to 1.242.

It thus appears quite plausible, for example, that our data may have come from a process with $\sigma = 1$ or $\sigma = 0.8$. It seems unlikely, however, that the true standard deviation is as low as 0.7 or as high as 1.25.

The confidence intervals for both μ and σ, obtained in this section (for σ) and in section 5.4.4 (for μ) are illustrated in Fig. 5.4, showing clearly that the greater the % confidence required, the wider the intervals need to be.

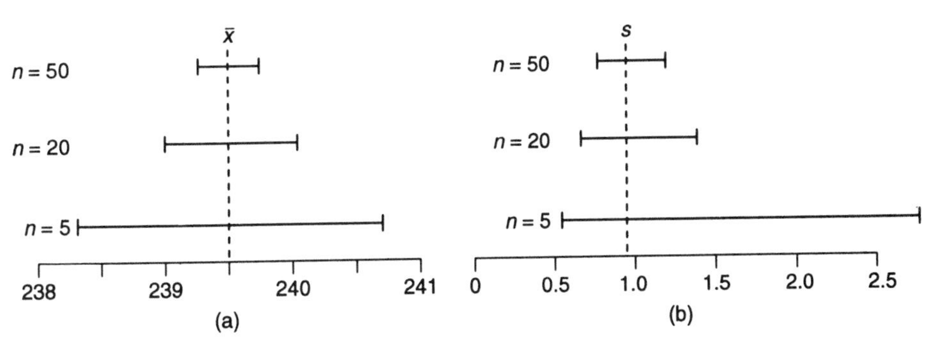

Fig. 5.5 95% confidence, interval for various sample sizes.

Also illustrated, in Fig. 5.5, is the effect of sample size. It shows the confidence intervals that would apply for the same values of $\bar{x}$ and s, but with sample sizes of 5, 20 and 50, each with the conventional confidence level of 95%.

5.7 Confidence limits for proportions and rates of occurrence

5.7.1 When dealing with limits for many measured characteristics, the use of methods based on the Normal distribution is often justified by the effect of the 'central limit theorem'. As noted in section 5.6, however, limits for the standard deviation need to take into account the non-Normality of the distributions of s or s^2.

Similar problems arise with attributes and events. The binomial and Poisson distributions often assumed as models for counted data are positively skewed, and this in turn means that asymmetrical confidence limits are appropriate if exact confidence levels are required. Figure 5.6 illustrates the situation for the case where a sample of, say, 100 metres of plastic extrusion is inspected for surface blemishes, and four such non-conformities are observed.

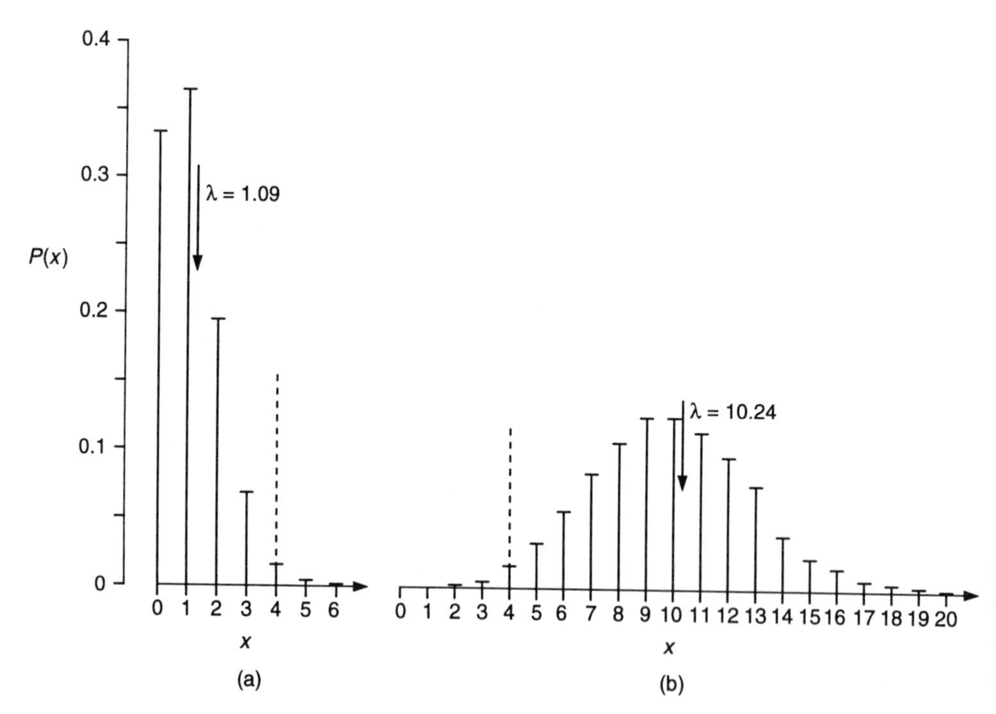

Fig. 5.6 Exact 95% confidence interval for rate of occurrence. $C = 4$ occurrences observed.
(a) For $\lambda = 1.09$, $P(x = 4 \text{ or more}) = 0.025$; (b) for $\lambda = 10.24$, $P(x = 4 \text{ or less}) = 0.025$.

Although the most plausible situation is that the sample came from a system operating at a level of about four defects per 100 metres, it could have been a 'lucky' sample from a system with a rather higher defect rate or a pessimistic one from a system with a lower defect rate. If, following the logic of section 5.3, we find the values of λ such that there are probabilities of 0.025 in the tails cut off by the observed number of defects, then we have

$$\tfrac{1}{2}\alpha = 0.025, \quad \alpha = 0.05, \quad 1 - \alpha = 0.95,$$

so that the distance between the upper and lower values of λ comprises a 95% confidence interval. This 'exact' interval turns out to be 1.09 to 10.24.

Unfortunately the determination of λ_L and λ_U require statistical methods beyond the scope of this book, so that approximations are useful. Confidence intervals based on the Normal distribution give useful results under certain conditions, and the use of z-values along with the standard errors noted in Table 5.4 will now be described.

5.7.2 It is apparent, in Fig. 5.6, that whereas the distribution of $\lambda = 1.09$ is strongly skewed, that for $\lambda = 10.24$ is much less so. For even larger values of λ, the shape of the Poisson distribution approaches that of the Normal distribution. Combined with the fact that the variance of the distribution is also λ (and hence the standard deviation is $\sqrt{\lambda}$), **approximate** confidence intervals may be based on the normal model.

Similar considerations apply to the binomial distribution for binary attributes. For reasonable validity, there should be a **total** of 15 or more non-conforming items or non-conformities in the sample (or combination of samples) used for estimation.

The most useful form for confidence intervals is in terms of the proportion or rate of occurrence. Where the estimate of the rate or proportion is based on a **total** of n items (for attributes) or n units of observation (for events), we then have

(i) for attributes:
Total number of non-conforming items $= np$

Estimate of proportion $\hat{p} = \dfrac{np}{n}$

$$\text{Confidence limits for } p \doteqdot \hat{p} \pm z.\sqrt{\left(\frac{\hat{p}(1-\hat{p})}{n}\right)} \tag{5.5}$$

(ii) for events:
Total number of occurrences $= C$
Estimate of occurrence rate $\hat{u} = C/n$

$$\text{Confidence limits for } U \doteqdot \hat{u} \pm z.\sqrt{\left(\frac{\hat{u}}{n}\right)} \tag{5.6}$$

As earlier in the book, z is the standard Normal score associated with the required tail probability. For 95% confidence and a $2\frac{1}{2}\%$ error risk in each tail, the exact value of z is 1.96, but in practice (and partly because many normal-based methods are in any case approximate) this is often rounded to 2.0. Other approximate values for 50, 80 and 99.9% confidence are included in Table 5.5.

5.7.3 As an example for attributes, consider a random sample of 450 items drawn from a large batch, of which 31 are found to have surface imperfections. Then,

$$\hat{p} = 31/450 = 0.068$$

and we may use (5.5) to obtain approximate confidence limits. For 50% confidence, $z = 0.675$, so that there is a probability of 0.5 that the interval

$$0.06\dot{8} \pm 0.675 \times \sqrt{\left(\frac{0.06\dot{8} \times 0.93\dot{1}}{450}\right)}$$

contains the true batch proportion of faulty items. The 50% confidence interval is thus 0.06083 to 0.07695, or 0.061 to 0.077 approximately.

Similarly for 95% confidence, taking the rounded value of $z = 2.0$, the interval becomes 0.045 to 0.093 – it appears very likely that somewhere between 4.5% and 9.3% of items have missing washers; as likely as not there are between 6.1 and 7.7% and our point estimate is 6.9%.

5.7.4 To illustrate the method for events, we suppose that 71 defects were noted in a total of 10 000 metres of plastic extrusion. The rate **per kilometre** is

$$\hat{u} = 71/10 = 7.1 \text{ defects per km}$$

since there are ten one-kilometre units represented in the sample. Then for 50% confidence we have limits

$$7.1 \pm 0.675\sqrt{\left(\frac{7.1}{10}\right)}, \quad \text{i.e. } 6.53 \text{ to } 7.67 \text{ per km}$$

For 95% confidence the limits become

$$7.1 \pm 2.0\sqrt{\left(\frac{7.1}{10}\right)}, \quad \text{or} \quad 5.41 \text{ to } 8.79 \text{ per km}$$

Alternatively, we might prefer to express the rate of occurrence *per 100 metres*, in which case the total sample can be regarded as 100 units of 100 metres.

Then $\hat{u} = 71/100 = 0.71$ defects per 100 m, and, for example, the 95% confidence interval becomes

$$0.71 \pm 2\sqrt{\left(\frac{0.71}{100}\right)} = 0.541 \text{ to } 0.879 \text{ per } 100 \text{ m}$$

Note that the relative width of the interval does not depend on the way in which the total sample information is divided into number of units and size of unit.

Space, and in some, cases complexity, preclude inclusion of methods for dealing with other estimates such as ranges, medians, quartiles and other percentage points, as well as the handling of grossly non-Normal data. Chapter 13.6 will touch on the subject of confidence intervals for capability indices.

6

Simple tests of hypotheses: 'signifiance tests'

6.1 Significance tests: general principles

6.1.1 Chapter 5 covered one major aspect of statistical inference – the **estimation** of parameters using sample statistics, and indicating via confidence intervals the reliability of these estimates.

Another important aspect is that of **testing** statements about a process, system or population. Familiar examples include:

Is the process stable ('in control')?
Does the batch satisfy a specification requirement?
Has an attempted process improvement really worked?
Is there a real difference between alternative supplies of a material?
Did the safety campaign genuinely reduce the accident level?

These are, of course, questions about the **system**. For testing purposes, they need to be cast as statements or hypotheses. Such statements, for statistical testing, need to be in a 'null' form, and the above examples would require the proviso 'unless there is good evidence to the contrary' before rewriting as:

The process *is* stable.
The batch *does* satisfy the specification.
The process is *unaffected* by the attempted improvement.
There is *no* real difference between supplies.
The safety campaign had *no* effect.

In this form, the data can be subjected to procedures which yield a **test statistic** whose distribution, in the null case, can be formulated. If the observed test statistic appears unlikely to have arisen from the null distribution, i.e. the probability of its occurrence (under the initial hypothesis) is low, then we may regard the initial hypothesis as discredited, and adopt a suitable alternative. Of course, often we are hoping that the data will 'prove' the effectiveness of an improvement or safety campaign, or a clear difference between supplies, but rigorous scientific method (like the legal system) requires that we assume no change or difference exists until clear evidence emerges to the contrary.

Some of the questions or hypotheses may concern averages, others concern variability, proportions, rates of occurrence or the presence of rogue values. The tools for dealing with these questions are almost identical to those for confidence interval estimation – sampling distributions and (in many cases) standard errors. To illustrate the ideas, we develop a simple example.

6.1.2 Suppose that the process producing the valve liners of section 2.2.1 is intended to operate at a mean level of 240 g. We will assume that the standard deviation is stable at 1.0 (tests involving **estimated** standard deviation will follow shortly).

Now this particular batch appears to have a mean of 239.51 g, based on a sample of 52 items. If we repeatedly sampled 52 items from a large population, we would expect (via Table 5.4 and equation (5.1)) that values of $\bar{x}$ would be normally distributed with mean – according to our initial hypothesis – of 240.0 and standard error $1.0/\sqrt{52} = 0.1387$. Therefore most samples (i.e. 95%) should have means within $1.96\sigma_{\bar{x}}$ of the hypothetical value, and almost all (99.9%) within $3.29\sigma_{\bar{x}}$.

When we actually calculate the discrepancy between the observed and hypothetical means, and express this as a multiple of the standard error, we find: $(239.51 - 240.0)/0.1387 = -3.533$.

The observed mean appears to lie much further from 240 than can be attributed to variation arising from random sampling, and we are bound to conclude that the process was not, in fact, operating at 240.0 but at some lower mean level that is more consistent with our observations.

6.1.3 Now let us frame this problem in the terminology of hypothesis testing. The initial or null hypothesis is:

$$H_0 : \mu_0 = 240$$

We should also consider the alternatives if this hypothesis should be untrue. If there is a prior reason (not based on observation of the sample data, but on knowledge of circumstances) to presume that the mean may have moved upwards, we might have the alternative

$$H_1 : \mu > 240$$

Should the prior circumstances suggest a downward slippage, we would have

$$H_1 : \mu < 240$$

In the most general case, where until we saw the data we were not aware of any reason for a shift in mean, we formulate the alternative as

$$H_1 : \mu \neq 240$$

The difference between the first pair of these alternatives and the third lies in the allocation of the probabilities for judging 'significance'.

To continue with the example, we now summarize the data and state any assumptions, in this case:

$n = 52$, $\bar{x} = 239.51$;
σ_x assumed to be 1.0;
whence $\sigma_{\bar{x}} = 1.0/\sqrt{52} = 0.1387$.

We consider the level of significance required to justify rejection of H_0 in favour of H_1. The conventional levels are:

(i) 5% ($\alpha = 0.05$) – justifying the serious questioning of H_0, perhaps the collection of further data to confirm, or the acceptance of an improvement action as having been successful.
(ii) 1% ($\alpha = 0.01$) – often termed 'highly significant'. There is little reason to doubt the failure of H_0 to represent the situation, and actions of a more serious nature might be justified – changing from one supply source to another, carrying out major overhaul on a process or launching a problem-solving activity.
(iii) 0.1% ($\alpha = 0.001$). The probability of the occurrence of a test statistic (such as the one we obtain from our data) under H_0 is so low – less than a 1 in 1000 chance – that the evidence against H_0 and in favour of H_1 appears overwhelming.

These significance levels are often coded in computer output as:

5% significant or *;
1% highly significant or **;
0.1% very highly significant or ***.

Occasionally values approaching significance, such as 10–5%, are labelled as *? On some systems, the tail probability for occurrence of the test statistic under H_0 is given as a 'P-value', e.g. $z = -2.77$, $P = 0.00280$; this would fall within the 'highly significant' or ** region between $\alpha = 0.01$ and $\alpha = 0.001$.

From the significance level, we might deduce a **critical value** for the test statistic, using tables to find the value of z, t or some other statistic corresponding to the significance level. For example, with $\alpha = 0.05$ and a normal test statistic, we would choose either $z = 1.96$ in the general case of a two-sided alternative (so that $\frac{1}{2}\alpha = 0.025$) or $z = 1.645$ for a one-sided alternative.

Finally, we calculate the appropriate test statistic, and either compare it with the critical value, or more often in a computer-aided system, inspect its P-value, drawing the appropriate conclusion about H_0 versus H_1. We should note that a non-significant result does not **prove** the validity of H_0, but indicates that there is insufficient evidence to discredit it. The legal parallel is that acquittal does not **prove** innocence, only that evidence against innocence was not strong enough to prove guilt 'beyond reasonable doubt'.

For the valve liner data, we continue thus:

Significance level: say 5% ($\alpha = 0.05$);
Test statistic: $(\bar{x} - \mu_0)/\sigma_{\bar{x}}$;
Sampling distribution: standard normal;
Critical value: for H_1: $\mu \neq 240$, i.e. a 2-tail test;
$\frac{1}{2}\alpha = 0.025$, $z = 1.96$;
Value of test statistic: $(239.51 - 240.0)/0.1387 = -3.533$.

This lies beyond the critical value, so that H_0 is rejected. This would be stated, in practice, as 'the mean differs significantly (beyond the 5% level) from 240.0'.

In fact, even if we had chosen the more cautious 0.1% level, with $z = 3.29$, we would still have established the discrepancy as significant. Many practitioners consider that the P-value provides more information, permitting a judgment as to **how** significant (or non-significant!) the result actually is. In this case, with $z = -3.533$, the P-value, via the Normal distribution, is 0.0002 205 4, or about 0.02%. This, of course, is simply an alternative way of indicating that the mean value in the sample differs very significantly from the hypothetical 240.0.

6.2 Testing $\bar{x}$ against μ_0

6.2.1 We now summarize procedures for several specific types of test. Each follows the general lines of section 6.1.3, viz:

(i) formulate H_0 and H_1
(ii) identify a sampling distribution
(iii) calculate a test statistic
(v) either compare the test statistic with a critical value for the chosen significance level, or obtain its P-value and consider the weight of evidence.

The first case we describe is that of comparing the mean of a sample, $\bar{x}$, with an assumed standard value – perhaps a nominal, or a historically-based average. The example in section 6.1.3 was of this type, where the standard deviation of the parent distribution was assumed to be known; this in turn implied the standard error of $\bar{x}$ as $\sigma_x/\sqrt{n}$. We must also deal with the more common situation where the only information on σ is contained in s, the estimate (with $n-1$ degrees of freedom) from the same sample of n items used to estimate $\bar{x}$. As we have seen in section 5.6, this leaves us uncertain of the true value of σ, and this uncertainty must be allowed for by using the t-distribution in place of the normal distribution.

Details of the test procedure are as follows.

H_0: $\mu = \mu_0$
H_1: 2-tail test: $\mu \neq \mu_0$
1-tail test: *either* $\mu > \mu_0$
or $\mu < \mu_0$
based on 'external' knowledge, i.e. *not* by inspection of the sample values.

We must now consider whether σ is known or is estimated by s.
For known σ, we have:

$$\left.\begin{aligned} \text{Test statistic} \quad z &= \frac{\bar{x} - \mu_0}{\sigma/\sqrt{n}} \\ \text{or} \quad & \frac{(x - \mu_0)\sqrt{n}}{\sigma} \end{aligned}\right\} \qquad (6.1)$$

with sampling distribution Normal.
For σ estimated by s, these become:

$$\left.\begin{aligned} \text{Test statistic} \quad t &= \frac{\bar{x} - \mu_0}{s/\sqrt{n}} \\ \text{or} \quad & \frac{(\bar{x} - \mu_0)\sqrt{n}}{s} \end{aligned}\right\} \qquad (6.2)$$

in a more convenient form.

Sampling distribution t, with $\nu = n - 1$ degrees of freedom.

6.2.2 An obvious example of the test with σ estimated by s is to re-work the data for valve liners using the more realistic t-test. Using $\bar{x} = 239.51$, $s = 0.934$, and noting that with $n = 52$ there are $\nu = 51$ degrees of freedom, we have:

H_0: $\mu = 240$;
H_1: $\mu \neq 240$;
(2-tailed test)

$$\text{Test statistic; } t = \frac{(239.51 - 240.0)\sqrt{52}}{0.934} = -3.783.$$

If we require to set up a critical value corresponding to a 5% 2-tail risk, then $\alpha = 0.05$, $\frac{1}{2}\alpha = 0.025$, and from Table A.4 we see that the critical t-value is 2.01. The (absolute) value of t calculated from the data considerably exceeds the critical value, so H_0 is rejected in favour of H_1. Further action would, of course, depend on circumstances – rejection of a batch, adjustment of process parameters, etc. One useful step would be to obtain a confidence interval for μ ($\neq 240$) based on the data – as calculated in section 5.4.3, the 95% confidence limits are 239.25 and 239.77. Note that the value 240 lies outside this interval, so the same conclusion would be reached as via the estimation approach of section 5.4.

6.3 Comparison of two sample means

6.3.1 Another question that arises frequently is whether the mean values in two samples can be regarded as compatible in the statistical sense, or whether they differ to an extent that cannot be attributed to chance alone. In other words, do they differ significantly, so that we must conclude that they represent different conditions?

Typical applications include those of assessing whether two operators, or machines, are producing output to the same standard; of assessing whether two sources of supply, nominally meeting the same specification, are virtually identical or whether they really differ in their properties or in effects on the process that uses them. 'Before and after' comparisons in improvement effects may also fall into this category.

We again identify the procedure in terms of hypotheses and test statistics, dealing first with the situation where standard deviations are already known. First, we note that if we repeatedly draw pairs of samples, with sizes n_1 and n_2, from the same distribution, and form the difference $\bar{x}_1 - \bar{x}_2$, then the standard error of this difference is given by

$$\sigma(\bar{x}_1 - \bar{x}_2) = \sqrt{\left(\frac{\sigma_1^2}{n_1} + \frac{\sigma_2^2}{n_2}\right)} \tag{6.3}$$

Of course, where σ_1 and σ_2 are not known, they need to be estimated by s_1 and s_2. This results in a slight complication dealt with in section 6.3.3.

6.3.2 Taking first the simpler case of known σ's, we have:

Initial hypothesis H_0: $\mu_1 = \mu_2$
Alternatives:
one-tail test H_1: $\mu_1 > \mu_2$
or H_1: $\mu_1 < \mu_2$

(The direction of the difference must be specified, without reference to the data, by the circumstances).

two-tail test H_1: $\mu_1 \neq \mu_2$

(this is the preferred and more general case).

Test statistic: $$z = \frac{\bar{x}_1 - \bar{x}_2}{\sqrt{\left(\frac{\sigma_1^2}{n_1} + \frac{\sigma_2^2}{n_2}\right)}} \tag{6.4}$$

reducing to $(\bar{x}_1 - \bar{x}_2)/\{\sigma\sqrt{1/n_1 + 1/n_2}\}$ if $\sigma_1 = \sigma_2$.

Sampling distribution: standard Normal.

For the one-tail H_1, an observed difference in the 'wrong' direction obviously lends no support to H_1, but may indicate some problem or anomaly to be investigated.

Tests of this nature often provide an approximate means of handling data from attributes or events, as will be noted in section 6.6, and we therefore defer an example to that section.

6.3.3 The more common case is that where our samples of sizes n_1 and n_2 yield mean values $\bar{x}_1, \bar{x}_2$ and **estimated** standard deviations s_1 and s_2. As for the parallel one-sample case in section 6.2.1, the t-test is appropriate, but requires a stronger initial hypothesis that not only does $\mu_1 = \mu_2$ but the distributions are identical in other respects also. The important implication is that $\sigma_1 = \sigma_2$ and the two estimates s_1 and s_2 should not differ significantly. The test of section 6.5 is often appropriate to establish that s_1 and s_2 are reasonably compatible. If they are, a pooled estimate (subscript 'p') is obtained as

$$s_p = \sqrt{\left(\frac{(n_1 - 1)s_1^2 + (n_2 - 1)s_2^2}{(n_1 - 1) + (n_2 - 1)}\right)} \qquad (6.5)$$

In terms of variances, the pooled s^2 is a weighted average of s_1^2 and s_2^2, the weights being the numbers of degrees of freedom (i.e. $\nu_1 = n_1 - 1$ and $\nu_2 = n_2 - 1$, respectively).

Having obtained s_p, the procedure is similar to that for known and identical σ's, but using t in place of z as the test statistic, viz:

$$t = \frac{\bar{x}_1 - \bar{x}_2}{s_p \sqrt{\left(\frac{1}{n_1} + \frac{1}{n_2}\right)}} \qquad (6.6)$$

Table A.4 is then used either to define a critical value, or to assess the 'p-value' for the observed t. The t-statistic uses the combined degrees of freedom in both samples, so that

$$\nu = \nu_1 + \nu_2 = n_1 + n_2 - 2$$

6.3.4 As an example of the two-sample t-test, let us suppose that for another batch of valve liners we obtain, via a sample of 36 items, $\bar{x} = 240.08$, $s = 1.006$. Is there evidence of a change in mean? We will assume that no deliberate changes have been made, so that the two-tail alternative H_1: $\mu_1 \neq \mu_2$ applies (rather than $\mu_1 < \mu_2$ after inspection of the data). We have:

	Sample 1	Sample 2
Sample size	$n_1 = 52$	$n_2 = 36$
Mean	$\bar{x}_1 = 239.51$	$\bar{x}_2 = 240.08$
Standard deviation	$s_1 = 0.934$	$s_2 = 1.066$

We should first assess the compatibility of s_1 and s_2, and in section 6.5.2 this is implemented and indicates that it is reasonable to pool the two estimates. Then,

$$s_p = \sqrt{\left(\frac{51 \times 0.934^2 + 35 \times 1.066^2}{51 + 35}\right)} = 0.989\,85$$

and

$$t = \frac{239.51 - 240.08}{0.98985\sqrt{\left(\frac{1}{52} + \frac{1}{36}\right)}} = -2.656$$

Inspecting Table A.4, we see that 2.656 lies near the 0.5% percentage point for t with either 80 or 100 degrees of freedom, and is thus judged highly significant. We conclude that there is a real difference between the means of the two samples that cannot be attributed to random sampling effects.

6.4 Testing s against σ_0

6.4.1 A case that arises less frequently is that of assessing whether the observed variation in a set of data is compatible with an assumed value of σ. A test that is useful even for quite large departures from normality is based on the χ^2 distribution.

H_0: $\sigma^2 = \sigma_0^2$

H_1: $\sigma^2 \neq \sigma_0^2$ (two-sided)

$\sigma^2 > \sigma_0^2$ *or* $< \sigma_0^2$ (one-sided).

Test statistic $\dfrac{(n-1)s^2}{\sigma_0^2}$

Sampling distribution χ^2 with $\nu = n - 1$ degrees of freedom.

Unfortunately most tables of χ^2 do not cater for $\nu = 49, 99$ etc. which would be convenient for samples of sizes 50, 100 etc. Additionally, the test statistic relates to variances rather than to standard deviations. Table A.2 therefore provides percentage points for the ratio

$$\sqrt{\left(\frac{\chi^2}{\nu}\right)} \qquad (6.7)$$

permitting direct evaluation of significance of s/σ_0. The Table includes points for those values of ν that correspond to convenient sample sizes. Linear interpolation is sufficiently accurate for other values of ν.

(The interested reader may note the relationship between Tables A.2 and 5.7. For any value of ν, the factors for C% confidence in Table 5.7 are the reciprocals of the entries for $\nu = n - 1$ in Table A.2 with tail percentages of $\frac{1}{2}(100 - C)$. Thus for $n = 50$ and 95% confidence, Table 5.7 gives 0.84, 1.25; for $2\frac{1}{2}$% points with $\nu = 49$, Table A.2 gives 0.803 and 1.197. Taking reciprocals yields 1.2453 and 0.8354, respectively).

6.4.2 We may now ask whether the standard deviation of 0.934 for the valve liner data is consistent with a long-established standard of 1.0. Some process improvement activity has been directed at reducing variation, and a one-tail test is therefore considered appropriate.

H_0: $\sigma = \sigma_0$
H_1: $\sigma < \sigma_0$ (a one tail test because of the process improvement activity, *not* the fact that s is less than 1.0)
Test statistic: s/σ_0
Data: $s = 0.934$, $n = 52$.

Then $s/\sigma_0 = 0.934$, and interpolating in Table A.2 we find this lies near the lower 25% value for this sample size. Although we evidently cannot say there has been *no* improvement, there is not enough evidence (yet) to reject H_0 in favour of H_1. There is a legal analogy to the 'not proven' or jury disagreement situations. A similar ratio of s to σ based on several hundred items might have been conclusive. The example illustrates that large sample sizes are often required to yield precise estimates or tests involving measures of variation.

6.4.3 A further and important application of (6.7) is in tests of variation or stability for attributes. Consider the following numbers of non-conformities obtained with a view to setting up a c-chart:

6, 11, 8, 2, 13, 12, 4, 8, 13, 2, 6, 11, 3, 13, 10, 7, 8, 10, 4, 9.

Now $\bar{c} = 8$, and control limits would be obtained as $\bar{c} \pm 3\sqrt{\bar{c}}$ (see Chapter 10.3), giving $UCL = 16.5$. Thus none of the counts violates UCL, and other conventional rules for interpreting a c-chart give no positive indications, although nearly half of the points lie in the outer thirds of the space between the control limits.

We note that the underlying model for the c-chart is the Poisson distribution, which has variance equal to its mean, and hence $\sigma = \sqrt{\mu}$. We may therefore check whether the observed standard deviation of the counts, s_c, is approximately equal to $\sqrt{\bar{c}}$. In fact $s_c = 3.67065$, compared to $\sqrt{\bar{c}} = 2.8284$. The $\sqrt{(\chi^2/\nu)}$ test may be applied here, and with 20 counts of non-conformities gives $\nu = 19$ with $s_c/\sqrt{\bar{c}} = 1.298$. This lies between the 5% and $2\frac{1}{2}$% points in Table A.2, and suggests the counts are too 'noisy', i.e. that the process is unstable or not strictly in control. Instabilities of this kind are common, and a rigid c-chart (or np, p or u-chart) may sometimes be inappropriate.

The dispersion tests for p and u values with varying sample sizes are rather more complicated than for np and c with regular samples. The expressions for np and c are detailed below.

6.4.4 Dispersion tests for np and c values (non-conforming items and non-conformities) are as follows.

(i) np values. Each count of non-conforming items is based on a sample size n. There are k such counts, and we require $n\bar{p}$, $\bar{p}$ and s_{np} for the whole set.

Then via the binomial distribution model,

$$\sqrt{\left(\frac{\chi^2}{\nu}\right)} = \frac{s_{np}}{\sqrt{\{n\bar{p}(1-\bar{p})\}}}, \qquad \nu = k - 1 \tag{6.8}$$

(ii) c values. Each count of non-conformities is based on the same sample size, quantity of material or area of opportunity. For k counts, we require $\bar{c}$ and s_c. Then via the Poisson distribution model,

$$\sqrt{\left(\frac{\chi^2}{\nu}\right)} = \frac{s_c}{\sqrt{\bar{c}}}, \qquad \nu = k - 1 \tag{6.9}$$

Further examples of these tests are given in Chapter 16.8.

6.5 Testing s_1 against s_2

6.5.1 One of the most important tests is that for comparing two sample estimates of standard deviation. This may be to justify the t-test of section 6.3.3, but in many cases differences in variation may be even more important than in mean level. With the emphasis on reduction in variation as a means of quality improvement, there are many cases where before-and-after studies involve comparisons of standard deviations.

The standard statistical procedure for samples drawn from normally distributed populations is the F-ratio test for variances. In line with earlier tests, we adapt this to $\sqrt{F}$ to permit direct evaluation of standard deviations. The formal steps are then:

H_0: $\sigma_1 = \sigma_2$
H_1: $\sigma_1 \neq \sigma_2$ (two-sided)
$\left.\begin{array}{l}\sigma_1 < \sigma_2 \\ \sigma_1 > \sigma_2\end{array}\right\}$ (one-sided)

the second is appropriate for before/after process improvement studies. Test statistic: s_1/s_2 or s_2/s_1 to yield $\sqrt{F} > 1$ (i.e. larger s divided by smaller s).

Sampling distribution: $\sqrt{F}$, as shown in Table A.3, using column headings for the sample size corresponding to the larger s, row labels for n corresponding to the smaller s.

As for previous tests, the percentage values should be doubled for two-tail tests.

6.5.2 In section 6.3.4 we considered two samples of valve liners and tested the difference between their means. It was noted that the values of s_1 and s_2 should be checked for compatibility before pooling, and we are now able to implement this. We had

$$n_1 = 52, \quad s_1 = 0.934$$
$$n_2 = 36, \quad s_2 = 1.066.$$

We shall treat this as a two-tail test for illustration. The ratio $s_2/s_1 = 1.141$, and by interpolation in Table A.3 this just exceeds the 25% value for $\sqrt{F}$. Doubling this for the two-tailed test indicates that some 50% of such pairs of samples might yield $\sqrt{F}$ as large as or larger than this, so the discrepancy is not statistically significant. In fact the exact p-value (for both tails) is 0.4126.

6.5.3 Suppose now that the following data were obtained before and after a factory move during which machines had been refurbished, and working conditions and other 'housekeeping' aspects improved.

	Before	After
n	40	25
$\bar{x}$	24.31 N	25.45 N
s	1.676	0.973

The data refer to tensile strength of specimens of a plastic material, for which the specification requires a minimum of 20 newtons (N).

Then $\sqrt{F} = 1.676/0.973 = 1.7225$, which lies between the 0.5 and 0.1% values in Table A.3. The reduction in variation is highly significant.

We may also wish to see if the improvements have also resulted in a significant increase in mean level. Here there is a problem, because it is now not valid to pool the two standard deviations for a t-test. However, it *is* possible to test the hypothesis $\mu_1 = \mu_2$ even if s_1 and s_2 differ significantly, and a convenient form is the t' test:

$$t' = \frac{\bar{x}_1 - \bar{x}_2}{\sqrt{\left(\frac{s_1^2}{n_1} + \frac{s_2^2}{n_2}\right)}} \tag{6.10}$$

Unfortunately the disparity between s_1 and s_2 'wastes' some of the degrees of freedom, and the t-test must therefore be interpreted with a degree of caution. The approximate effective degrees of freedom are

$$\nu' = \frac{\left(\frac{s_1^2}{n_1} + \frac{s_2^2}{n_2}\right)^2}{\frac{1}{\nu_1}\left(\frac{s_1^2}{n_1}\right)^2 + \frac{1}{\nu_2}\left(\frac{s_2^2}{n_2}\right)^2} \tag{6.11}$$

It is worth noting, in order to avoid this calculation in many cases, that ν' cannot be greater than $\nu_1 + \nu_2$ nor less than either ν_1 or ν_2, whichever has the larger value of s^2/n.

Applying this test to the data above,

$$t' = \frac{24.31 - 25.45}{\sqrt{\left(\frac{1.676^2}{40} + \frac{0.973^2}{25}\right)}} = -3.47$$

Now v' cannot be less than 39 nor more than 63, but even for the smaller degrees of freedom t' exceeds the 0.1% one-tail value. It is thus unnecessary to calculate v' exactly, but for completeness, it is (using (6.11)), 62.75; in this case the wastage of degrees of freedom is small – it only becomes serious for large discrepancies between s_1 and s_2.

6.6 Tests for attributes and events

6.6.1 Rigorous tests for attributes and events are rather complicated and necessitate special tables. Some useful approximate tests are available, and work well when the numbers of non-conforming items or non-conformities are not less than 15 in any of the groups being compared. For non-conforming items, we rely on the properties of the binomial distribution, which has only one parameter (p) requiring estimation. Once p is known or estimated, both the mean (np) and variance ($np(1-p)$) of the numbers of non-conforming items are implied. The tests are therefore based on the Normal, rather than the t, distribution.

Similarly for non-conformities, the Poisson model has one parameter λ which implies both mean and variance. When data from attribute or event applications are expressed as proportions or rates of occurrence, these also have theoretical variances which can be used in expressions giving an approximate z-value.

It is therefore only necessary to list the formulae for the test statistics, all of which are referred to tables of the Normal distribution.

6.6.2

(i) Test of observed proportion, $\hat{p}$, against an assumed value, π.

$$\text{Test statistic, } z = \frac{\hat{p} - \pi}{\sqrt{\left(\frac{\pi(1-\pi)}{n}\right)}} \qquad (6.12)$$

(ii) Test of the difference between two proportions, $\hat{p}_1$ and $\hat{p}_2$.

$$\text{Form } \hat{p}_{12} = (n_1\hat{p}_1 + n_2\hat{p}_2)/(n_1 + n_2) \qquad (6.13)$$

$$\text{then } z = \frac{\hat{p}_1 - \hat{p}_2}{\sqrt{\left[\hat{p}_{12}(1-\hat{p}_{12})\left(\frac{1}{n_1} + \frac{1}{n_2}\right)\right]}} \qquad (6.14)$$

Another, more approximate, method eliminates the need to pool $\hat{p}_1$ and $\hat{p}_2$ via

$$z = \frac{\hat{p}_1 - \hat{p}_2}{\sqrt{\left[\dfrac{\hat{p}_1(1-\hat{p}_1)}{n_1} + \dfrac{\hat{p}_2(1-\hat{p}_2)}{n_2}\right]}} \tag{6.15}$$

(iii) Test of event rate $\hat{u}$ against an assumed value, U.

$$\text{Test statistic, } z = \frac{\hat{u} - U}{\sqrt{(U/n)}} \tag{6.16}$$

(v) Test of the difference between two event rates $\hat{u}_1$ and $\hat{u}_2$.

$$\text{Form } \hat{u}_{12} = (n_1\hat{u}_1 + n_2\hat{u}_2)/(n_1 + n_2) \tag{6.17}$$

$$\text{then } z = \frac{\hat{u}_1 - \hat{u}_2}{\sqrt{\left[\hat{u}_{12}\left(\dfrac{1}{n_1} + \dfrac{1}{n_2}\right)\right]}} \tag{6.18}$$

Again a more approximate form, avoiding the need to combine $\hat{u}_1$ and $\hat{u}_2$, is

$$z = \frac{\hat{u}_1 - \hat{u}_2}{\sqrt{\left(\dfrac{\hat{u}_1}{n_1} + \dfrac{\hat{u}_2}{n_2}\right)}} \tag{6.19}$$

In each of (6.16) to (6.19), the n's refer to the number of sampling units available for estimating $\hat{u}$. These units may be items, but may also be 100's of square metres, 1000's of kilogrammes, etc.

6.6.3 As an example of tests for proportions, we take some data for two machines, nominally identical, producing flanged discs by stamping metal blanks. The discs are inspected for burrs and other surface imperfections.

Machine A 450 inspected, 31 defective;
Machine B 750 inspected, 36 defective.

Then $\hat{p}_A = 0.06\dot{8}$, $\hat{p}_B = 0.048$,

and $\hat{p}_{AB} = \dfrac{31 + 36}{450 + 750} = 0.0558\dot{3}$.

Using the more rigorous form (6.14),

$$z = \frac{0.06\dot{8} - 0.04\dot{8}}{\sqrt{\left[0.0558\dot{3} \times 0.9441\dot{6}\left(\dfrac{1}{450} + \dfrac{1}{750}\right)\right]}} = 1.526$$

$p_z = 0.0635$, and for a two-tail test this would not reach any reasonable significance level.

The approximate form (6.15) gives

$$z = \frac{0.06\dot{8} - 0.048}{\sqrt{\left[\frac{0.06\dot{8} \times 0.93\dot{1}}{450} + \frac{0.048 \times 0.952}{750}\right]}} = 1.464$$

in this case slightly more conservative than the previous value.

6.6.4 For events, we consider again the case of Chapter 5.5.4, where 71 defects were noted in a total length of 10 000 metres of extrusion. Is this compatible with a standard of 5 defects per kilometre?

Here $\hat{u} = 7.1$ and $n = 10$ (the units are kilometres), with the hypothetical value $U = 5$. Then,

$$z = \frac{7.1 - 5}{\sqrt{\left(\frac{5}{10}\right)}} = 2.97$$

using (6.16) as the test statistic. With $p_z = 0.001\,49$, this indicates a highly significant violation of the standard.

It is worth noting that a change of unit size would not affect the value of z. Thus if the standard specified counting defects per 100 metres, the 10 000 metre sample now comprises $n = 100$ sampling units. The mean rate per 100 metres is therefore $71/100 = 0.71$.

For the same standard, the specified rate would become 0.5 defects per 100 metres, and thus, as before, we have

$$z = \frac{0.71 - 0.5}{\sqrt{\left(\frac{0.5}{100}\right)}} = 2.97$$

6.7 Tests for outliers

6.7.1 A problem of perennial interest is whether one or more of the extreme values in a data set is a misfit or ‘rogue’. There are many tests for objectively assessing outliers, but the great majority are based on assuming a Normal distribution for individual values. Skewed distributions often give rise to apparent outliers, so the general shape of a tally or histogram should be taken into account before plunging into routine tests for outliers.

A simple rule-of-thumb procedure is sometimes linked to the blob diagram or box-and-whisker plot. The semi-interquartile range, $\frac{1}{2}(Q_3 - Q_1)$ in the notation of section 2.3.3, is calculated, and inner and outer ‘gates’ are measured outwards from the quartiles.

Inner gates	$Q_1 - 3 \times SIQR$,	$Q_3 + 3 \times SIQR$;
Outer gates	$Q_1 - 5 \times SIQR$,	$Q_3 + 5 \times SIQR$.

Values beyond the inner gates (which correspond to about 2.7 standard deviations for the normal distribution) are mildly suspect, values beyond the outer gates (4.05 standard deviations) are strongly suspect.

These simple criteria may need to be modified to take account of the facts that *SIQR*, like s, is a sample estimate, and that genuine extreme values from the parent population are more likely to occur in larger samples than in small ones. The test of the next section achieves these aims.

6.7.2 Grubbs (1969) proposed a test for a single suspected outlier based on the criterion

$$T = \frac{x_n - \bar{x}}{s} \quad \text{or} \quad \frac{\bar{x} - x_1}{s}$$

for an upper-end or lower-end outlier, respectively. Tables exist for some percentage points of T, but the following expressions are sufficiently accurate for practical purposes over the ranges of sample sizes indicated.

5% points	$T = 1.427 + 0.3956 \ln(n - 3)$, n from 5 to 100.
$2\frac{1}{2}$% points	$T = 1.510 + 0.4174 \ln(n - 3)$, n from 5 to 100.
1% points	$T = 1.369 + 0.5339 \ln(n - 3)$, n from 5 to 25.

6.7.3 For the valve liner data, the lower extreme value was 236.9 with $\bar{x} = 239.51$ and $s = 0.934$. Then $T = 2.794$. For $n = 52$, the approximate 5% point for T is 2.97 (Grubbs gives 2.96 for $n = 50$), so the lowest value does not appear to be an outlier. Had this lowest value been 235.9, with all 51 other values unchanged, we would have $\bar{x} = 239.49$, $s = 0.9966$. Then T would become 3.60, and would indicate a significant lower outlier.

6.7.4 A test when outliers are suspected at *both* ends of a sample (supposedly from a Normal distribution) uses the ratio of range to standard deviation, R/s. This test, due to Quesenberry and David, is also tabulated by Grubbs. Approximations for the 5% and 1% points are:

5%	$2.10 + 0.838 \ln(n - 3)$,	$n = 5$ to 100
1%	$2.075 + 0.95 \ln(n - 3)$,	$n = 5$ to 100.

For the same data, the upper extreme was 241.6, so that $R = 241.6 - 236.9 = 4.7$. The ratio R/s is thus $4.7/0.934 = 5.032$. The 5% criterion from the above formula is 5.36 (Grubbs gives 5.35 for $n = 50$), and again outliers are *not* indicated by the test.

6.8 Other aspects of significance tests

6.8.1 Tests exist for many other situations. Some are suitable for use when the interval or ratio property is not achieved, and comprise non-parametric or robust tests. They are often based on ranking and order-statistics such as medians and quartiles.

Extensions of the t and F tests to deal with several groups (rather than two) are useful, but would require more space than can be allotted here. The analysis of variance (ANOVA) for several means, and tests of the 'Bartlett-box' type for several variances, provide for these multi-sample situations. A simplified form of analysis of variance is presented in Chapter 16 for the detailed analysis of control chart data. Gauge capability analysis (Chapter 15) is another technique that uses ANOVA or approximations to it.

6.8.2 One final point deserves attention. Suppose that it is possible to obtain very large samples, and in connection with a process improvement activity we have the following data:

	Before	After
n	4850	3575
$\bar{x}$	100.23	100.25
s	0.248	0.240.

For the $\sqrt{F}$-test, the conventional tables do not extend to these enormous degrees of freedom, but via other techniques the two standard deviations would appear significantly different with P approximately 0.02. Then t' (since one should not pool the two standard deviations) would become 3.73, and is significant beyond the 0.1% level. Yet such small differences in both mean and standard deviation are unlikely to be of practical importance! There is obvious 'overkill' here.

Whilst the general subject of **power** of tests lies beyond the scope of this book, some guidelines are possible in straightforward cases. One needs to establish the magnitude of difference between means, or ratio of standard deviations, that will be of **practical** interest, and also the level of significance at which such a discrepancy should be signalled. The sample sizes required to establish significant differences can then be approximately assessed, especially if (for differences between means) a standard deviation can be postulated (or guessed at from previous experience). We denote differences between means by D, and ratios between standard deviations, proportions of attributes or rates of events, by B.

(i) Difference between a mean and a standard value.
Difference of interest D

Critical value	z
(where z is chosen with tail probability corresponding to significance level α)	
Postulated standard deviation	σ
n should exceed	$\left(\frac{z\sigma}{D}\right)^2$

(ii) Difference between two means.

Difference of interest	D
Critical value	z
Postulate equal standard deviations	σ
For equal samples, n_1 *and* n_2 should exceed	$2\left(\frac{z\sigma}{D}\right)^2$
For known n_1, n_2 should exceed	$\frac{1}{\left(\frac{D}{z\sigma}\right)^2 - \frac{1}{n_1}}$

(iii) Ratio of a standard deviation to a standard value.

Ratio of interest	B
n, z as defined above	
n should exceed	$\frac{z}{[\ln(B)]^2}$

(iv) Ratio of two standard deviations, equal sample sizes.

n_1 and n_2 should both exceed	$\frac{2z}{[\ln(B)]^2}$

By inserting the expressions for binomial or Poisson standard deviations, these formulae can also be adapted to estimate sample sizes for attributes and events.

6.8.3 As examples of the use of these formulae, we will suppose that for the situation in section 6.8.2 a difference between means of 0.2 is of interest, and a ratio of standard deviations of 1.3:1 is also interesting.

Using the formula in (iv) above, and an intended significance level of 5% (2-tail):

$$n_1 \text{ and } n_2 \text{ should both exceed } \frac{2 \times 1.96}{[\ln(1.3)]^2} = 57.$$

Now if sample sizes of around 60 *are* used, a ratio of 1.3 between s_1 and s_2 will exceed 1.257, the $\sqrt{F}$ value for 59, 59 degrees of freedom at $\alpha = 0.025$.

If the standard deviations in the two samples are of the order of 0.25, then for a difference between means of 0.2 to be significant at 5% (two-tail),

$$n_1 \text{ and } n_2 \text{ should both exceed } 2\left(\frac{1.96 \times 0.25}{0.2}\right)^2 = 12$$

so that the samples of 60 would be more than adequate for this purpose.

Finally, taking the data of section 5.5.4 with a defect rate of 7.1 per kilometre based on a sample of 10 000 metres, we might ask what sample size would be necessary to confirm an improvement to 5 per kilometre as significant at the 5% one-tail-level?

Here $z = 1.645$ for $\alpha = 0.05$, and if $\hat{u}_2 = 5$ based on n_2 sample units,

$$z = 1.645 = \frac{7.1 - 5}{\sqrt{\left(\dfrac{7.1}{10} + \dfrac{5}{n_2}\right)}}$$

giving $(7.1/10 + 5/n_2) = (2.1/1.645)^2$. Then

$$\frac{5}{n_2} = \left(\frac{2.1}{1.645}\right)^2 - \frac{7.1}{10} = 0.9098$$

and the minimum value for n_2 is 5/0.9098, i.e. about 5500 metres.

Other formulae for significance tests, in conjunction with those for standard errors of sample statistics, may be similarly used to estimate sample sizes so as to avoid inadequate or excessive samples for specific purposes.

7

Control charts for process management and improvement

7.1 Introduction

7.1.1 It is important to recognize that process control is not solely concerned with manufacturing processes and specifications. It is applicable wherever any process (manufacturing, clerical, administrative, medical, etc.) is to be operated so as to achieve its maximum potential performance. Such performance may be measured in many ways, of which the most common in industry are yield, strength, dimensional accuracy, stock levels, profit levels, cash flow, safety performance, orders booked, etc.

These examples clearly illustrate that the word control is not confined to the detection and correction of **manufacturing** conditions but has company-wide significance.

Control chart methods provide continuous, up-to-date process information for the management of many aspects of the process, for identifying opportunities for improvement, and for monitoring the effectiveness of corrective or improvement actions.

7.1.2 Because SPC is a management tool, it is an absolute requirement that **management must be committed both to the philosophy and practice of SPC**. Many 'quality' programmes fail because QC departments are given the responsibility of assessing quality and of notifying any lapses, but lack the authority to instigate positive action for fundamental improvements. SPC must become an integral part of process management if control charting is to be effective. Broadly, there are four aspects to the use of control charts.

(i) To **measure the performance** of the process, indicate the extent to which it is stable or 'in control', and signal the presence of assignable (special) causes of variation.
(ii) To **signal departures from target conditions** permitting prompt correction or investigation of off-standard performance.

(iii) To **provide pointers** to means of improving the performance of the process (whether in terms of process yield, capability to a specification, reduced levels of interruption, or other aspects).
(iv) To **continue the flow of information** relevant to maintaining control and making process management decisions.

7.2 Pre-requisites

7.2.1 Once management commitment is established, several points need careful consideration before instituting the use of control charts.

(a) Establish priorities
Control charts need to be introduced on a structured basis. Priorities can be established through problem-solving techniques (Pareto analysis, cause/effect diagrams) or various other means.

(b) Decide what to measure or observe
A few pointers to selection include:

(i) Early operations usually have the greatest impact (thus control of input materials will be more beneficial than recording finished product properties).
(ii) Control of **process** variables gives earlier feedback than measurement or inspection of output.
(iii) Choose parameters for relevance to end-use, not ease of measurement.
(iv) Some key variables may usefully predict other aspects of performance.
(v) Measurements generally provide more information than counts of occurrences.

(c) Set up a system for data collection
Consider:

(i) *How to measure or observe*: Take into account equipment needs, training, setting of standards for sensory judgement.
(ii) *Where to measure or observe*: Action near the process is often preferable to activity in offices or laboratories, but quality of measurement considerations may outweigh this factor for tests involving apparatus or instruments.
(iii) *When to measure*: There may be limitations of access, e.g. customer complaints (only when they occur), end of run (roll, coil, batch) in continuous processes, etc.
(iv) *Setting up procedures for measuring (observing) and recording*: Routine record sheets should, wherever possible, include provision for a **process log**. This will facilitate identification of special causes when signals occur on the control charts.

(v) *Whether 100% observation or sampling is necessary or appropriate*: Critical factors may require 100% checks, but in measurement of process variables or product characteristics, sampling is usually necessary (most obviously where testing or measurement is destructive).

(d) Obtain initial data on the process

This should preferably be in the format that will be used for routine SPC records; existing data from miscellaneous sources is often usable, but may require special analysis and cautious interpretation.

7.2.2 Sources of variation

Before considering the details of control chart operation, it is necessary to distinguish the two sources of variation with which they are concerned, and also two others which are not within the scope of control charting.

First, the control chart is not concerned with variations in performance which arise from external influences. These include variation due to differences between users, differences in the environment in which items or services may have to operate – factors which cannot be controlled within the manufacturing or operational part of a system. Secondly, it is not concerned with deterioration over time due to wear, corrosion, damage, fatigue, infection (even computer viruses!). Such potential sources of what are often called 'outer noise' and 'inner noise' need attention at the design end of the system – establishing customer needs, design for purpose, identifying potential failure modes, specifying suitable materials, establishing correct manufacturing, service or operational conditions.

7.2.3 Control charts *are* concerned with two sources of manufacturing or operational variation. Their purpose is to identify any **assignable** causes of variation in the system, so that they can be dealt with by correction or elimination. Some degree of **inherent** variation in any system is unavoidable at a particular 'state of the art', and the extent of this inherent variation must be measured and allowed for. Sources of assignable variation are then identified as 'signals' above the background 'noise' of the inherent variation.

Other terms applied to these sources are **common cause** variation and **special causes**. The former refers to the inherent variation, produced by minor perturbations in the system to which all of its output is exposed; the latter affects the output only when the special, or assignable, cause is present.

7.2.4 The two kinds of variation have different features which may be summarized as follows:

Inherent variation (Common cause)	**Assignable variation** (Special cause)
Many small contributions	A few major contributions
Often not individually identifiable	Usually readily identified
Stable overall level of variation	Irregular in occurrence
Predictable overall effects	Often unpredictable in effect
Requires fundamental action to achieve reduction	Corrected by local action
Used as basis to signal assignable causes	Indicated by control chart signals

Although the actual nature of these contributions will vary enormously from one application to another, some examples from various fields may be useful. One must remember, though, that in a programme of process improvement, some sources of inherent variation may be identified for attention, so that they may become a subset of assignable causes during this phase of development.

Examples of contributions to inherent variation	**Examples of sources of assignable variation**
Machinery vibration	Faulty incoming materials
Thermostat cycles	Tool wear
Operator influences	Process stoppage
Poor method design	Human error
Unclear instructions/method	Equipment malfunction or breakdown
Ambient temperature/humidity	Power failure/surge
Non-homogeneous materials	Errors in formulating ingredients
Mains voltage fluctuations	Incorrect test or inspection
Noisy environment	Over or underheating
Uneven liquid flows	Job functions not performed
Variable particle size	Wrong method/tool/equipment used
Poor maintenance	Exceptional weather conditions
Variations in mixed or blended materials	Change in method, procedure or operator
Insufficient time allowance for operations	Editing or tampering with data

7.3 Basis of control charts

7.3.1 Control charts are based on a few simple assumptions.

(a) Observations on **rational subgroups** from the process provide a means of establishing the extent of inherent (common cause) variation. These subgroups should provide opportunity only for inherent system variation to occur, and must exclude 'special' causes. Thus for a stream of discrete items, a rational subgroup might contain five consecutive items, provided that these do not extend over a change of raw materials, shift change-over, etc. In other cases, a rational sub-group might comprise a day's records of process input/output, or a weekly or monthly accident or absence record.
(b) From the pattern of common-cause variation, a **target distribution** is set up. For measurements this will often be a normal distribution representing the variation in sample averages to be expected under conditions of process stability. For measures of variation in measured data (range R, standard deviation s), for counted data or for situations where conditions necessitate subgroups comprising only one item, other target distributions may be necessary.
(c) Based on the target distribution, reasonable limits are set to the extent of common cause variability.

7.3.2 The use of such limits brings the risk of two types of error:

I Due to a chance combination of circumstances, the limits may be violated even though *no* special cause is present. This will lead to a false alarm, and any corrective or investigative action will later prove to have been unnecessary.
II A special cause may be present, but may not be signalled against the inherent background variation. This will mean that the need for corrective action will not (immediately) be recognized, and substandard process performance will continue until a signal *does* occur.

7.3.3 It is important to recognize the existence of these risks, rather than to assume that the control system gives no trouble when the process is in control, but responds immediately to any slippage in standards. Neither part of this statement is true, though it is correct to state that the greater the departure from target conditions, the sooner a signal is likely to appear.

The level of type I risk is usually set at one false alarm per 200–1000 subgroups for standard SPC procedures, though it may be varied by deliberate re-design of control parameters where operating conditions require a different approach. The level of type II risk is not, in most cases, directly controlled, but it may be noted that:

(i) the larger than sample size, the smaller the level of type II risk associated with any particular extent of slippage;

(ii) the more frequently that subgroups are taken the more rapidly any out-of-control situation will be signalled (though the rate of occurrence of false alarms will also be increased);

(iii) large slippages will be detected more rapidly than small ones; where necessary, procedures can be designed for sensitivity to particular levels of slippage or out-of-control conditions.

7.3.4 The type I risk is directly related to the significance level of the hypothesis tests considered in Chapter 6. In fact, when using a control chart, the hypothesis 'H_0: process is stable' is tested each time a fresh subgroup is entered on the chart. However, a significance level of, say, 5%, would lead to 1 in 20 subgroups producing false alarms, i.e. tending to produce overcontrol. Thus the significance level for any one sample or subgroup statistic is more stringent, but a **set** of rules for interpreting patterns provides greater power of detecting instability, albeit somewhat increasing the false alarm risk.

Partly because of the use of such sets of rules, but also because the control chart is continuously active rather than a 'one-shot' assessment like a hypothesis test, it is more useful to measure performance via average run length (*ARL*) criteria than by significance probabilities. A false alarm risk of 1 in 200 can be expressed as an *ARL* of 200 samples when the process is in-control. This is often written as $L_0 = 200$. In general, L_0 values of 100 to 1000 are chosen for practical control schemes. The speed of reaction to a change in conditions is likewise measured by an *ARL*, but it is necessary to specify the condition to which the *ARL* refers. Many control charts have a typical reaction time of about 10–20 samples if the process mean shifts by about one standard error. This might be represented by the statement $L_1 =$ (say) 10 for $\Delta = 1.6\,\sigma_e$, where Δ is used to indicate the shift in mean level.

Obviously a good control system has a fairly long L_0 (to avoid overcontrol when the process is stable) and a fairly short *ARL* at some undesirable process condition. Some comparisons between various control systems are given in Table 12.2.

7.4 Overcontrol and undercontrol

7.4.1 The twin aims of control charts are to avoid both overcontrol and undercontrol. Overcontrol results from premature and erroneous response to what is, in fact, inherent or common cause variation. It is a remarkable fact that an over-controlled process, where any small random departure from the target is countered by an unnecessary corrective action, actually deteriorates in performance. Consider the following 20 values observed on a stable process, with a mean around 5 and a standard deviation of about 0.01:

4.986, 4.996, 5.013, 4.986, 4.985, 5.005, 5.012, 4.988, 4.990, 5.009,
5.004, 4.976, 5.025, 5.001, 4.993, 4.994, 5.007, 5.007, 4.992, 4.995

The mean, $\bar{x}$, is 4.998 2 and the standard deviation, s, is 0.011 97.

Now suppose the operator has been instructed to keep the values as close as he can to 5.000. Mistakenly, he attempts to do so by readjusting the process after each item is measured. Thus after item 1, he increases the setting by 0.014 to 'correct' the system; the next item would then have become 5.010 instead of 4.996, and he would consider another adjustment, this time downwards, would be required. A net adjustment of $+0.014-0.010=+0.004$ means that the third item would become 5.017, and so on. The 'corrected' values would then be:

4.986, 5.010, 5.017, 4.973, 4.999, 5.020, 5.007, 4.976, 5.002, 5.019, 4.995, 4.972, 5.049, 4.976, 4.992, 5.001, 5.013, 5.000, 4.985, 5.003

Although $\bar{x}$ is now 4.999 75 (closer to 5), the standard deviation has increased to 0.01914, up by some 60%. In fact 100% **overcontrol** of this kind will, in the long run, produce an increase by a factor of $\sqrt{2}$, about 41%, in variability. Similar effects are produced by control instruments with insufficient damping, producing what is termed 'hunting' – a form of automatic overcontrol.

7.4.2 Evidently overcontrol should be avoided. The use of averages rather than individual values, and logical rules for deciding when adjustments or corrections are required, provide safeguards against false alarms.

At the other extreme, uncontrolled variation from assignable causes must be avoided, and not merely attributed to inherent process variation. Objective rules are required to avoid differences of interpretation among individuals. The control chart is fundamentally a run chart with control limits applied to provide guidance on necessary actions. There are several types of control limit, of which the two most important are action limits and warning limits.

7.5 Control limits and rules

7.5.1 Action limits

These limits are almost universally used in conjunction with control charts. In American control chart literature they are referred to simply as control limits. As noted previously, they usually correspond to values of the observed characteristic which may be encountered about once per 200–1000 subgroups when the process is in a state of statistical control. They thus avoid overcontrol, i.e. rarely produce false alarms, but violation of the limit for any subgroup is taken as a **signal** for action or investigation.

This action may, depending on circumstances, range from taking a re-check sample to the stopping of a production line; or from an instant corrective action to a painstaking investigation of reasons for lack of control.

In most applications, upper and lower action limits (UAL and LAL) are used; they are drawn on the control chart, along with the target value, and each subgroup result is plotted on the chart and assessed against the limits.

For some characteristics, violation of a limit in one direction may indicate an unexpectedly favourable situation. Some examples are:

(i) Controlling to satisfy a one-sided specification. Violation of an **opposite** limit will then indicate an apparent improvement which may be exploited, or may imply that the process is being operated at an unnecessary level for the particular application, perhaps wasting time, money or materials.
(ii) Control for events or attributes. Violation of a **lower** limit suggests an apparent reduction in the level of undesirable occurrences. It is wise to check first that calculations or assessments have been correctly carried out; if they have, then the identification of the special cause may give useful guidance towards a permanent process improvement.
(iii) When monitoring ranges or standard deviations violation of the lower limit for an R or s chart may indicate an unexpected reduction in common cause variation. If true, this again deserves exploitation.

Control limits on the favourable end of the scale are sometimes ignored on the grounds that it does not matter if product or process are better than expected. However, *any* signals from a chart should be investigated, partly to check the validity of the data and partly because serendipity (the knack of making useful discoveries by chance) is a talent to be encouraged.

7.5.2 Warning limits

A common practice, especially in the UK and Europe, is to include a pair of warning lines (UWL, LWL) on the charts. These are often set at a level such that about one in 20–50 subgroups will plot at or beyond the warning line (though usually within the action line) when the process in control.

Because of the risk of overcontrol, violation of a warning line is *not* taken as a signal for immediate action. The usual practice is either to take a re-check sample, or if this is not possible, to wait until the next subgroup result is available. Then if the two successive results both yield warning values, the necessary corrective or investigative action is taken. In American and Japanese practice, warning lines are less often used. When they are, it is often with the rule that two warning values in any three successive samples constitute a signal. Rules of the kind noted in 7.5.3(c) have a similar function.

7.5.3 Other rules

Whilst a profusion of rules is to be avoided, both on grounds of complexity and risk of overcontrol, some broad features or patterns on control charts may give useful indications. These include:

(a) Runs of more than seven or eight consecutive points above or below the target value suggest a shift in average level, even though no single subgroup yields an action signal.

(b) Four or more warning values within the space of ten subgroups, or more than four in twenty consecutive subgroups, should be investigated. If the offending values are a mixture of upper and lower warning values, the indication is of excessive variation. If they are all on one side of the target value, a shift in average level is indicated.

(c) Without formally marking the zones on the chart, check that approximately two-thirds of the plotted points fall within the middle third of the space between the action limits. The rationale for this rule is that since these limits represent roughly a three standard error spread, the middle third corresponds to about $\pm$ one standard error around the target value, and for a normal distribution about 68% of values should fall within this range. Too many points in the outer thirds signal out-of-control conditions such as a trend, a shift in mean, overcontrol, a 'noisy' process with between-sample common causes, etc. This rule is sometimes quantified in terms of four in any five values beyond one standard error from the target constitutes a signal.

Too many points in the middle third ('centre-line hugging') may indicate (among other things) that the measure used for common cause variation is too large, that measurement is too coarse, or that some assignable causes of variation are being included **within** samples. Chapter 11 takes up this point.

(d) Other patterns, such as trends running from below target to above target, or vice versa, should also be checked, especially if seven or more values progressively move in one direction. Similarly, **regular** cycles may be detected, though experience is necessary to distinguish regular cycles from random oscillations, unless a cyclic feature in the underlying process conditions, e.g. a day/night ambient temperature pattern, or a temperature cycle due to thermostatic 'hunting', can be related to apparent cycles on the control chart.

(e) Some recent research has investigated *ARL* properties of the combination: Any single value beyond 3σ, two in any three beyond 2σ, four in any five beyond 1σ and any run of 8 or more on one side of the centre-line. The two-sided L_0 is then about 92 samples, demonstrating the risk of false alarms when several rules are used together.

7.6 Choosing the target value

7.6.1 An often neglected aspect of control chart practice is that of choosing a rational target value as the centre-line for the chart. A sensible choice requires a proper consideration of the objectives of the control chart and an understanding of the nature of the process. Some questions to be addressed are:

(i) Is there a logical and obvious 'aim' value for the application? Candidates for consideration include:
—nominal value for a two-sided specification;

—a mean level comfortably (e.g. several standard deviations) clear of a specified minimum or maximum limit;

—a technologically sound aim value for a process variable that is known to give good operational performance (e.g. an optimum temperature or proportion of ingredient in a material formulation).

Note that arbitrary targets such as a hoped-for level of non-conforming items or of customer returns do *not* fall under this category.

(ii) In conjunction with (i), when the control chart indicates departure from target, is adjustment a straightforward matter without disturbing other features of the system? 'Chosen' target values are appropriate where correction to a dimension, ingredient or process setting can be promptly and precisely implemented without other undesirable effects, such as increasing variation or destabilizing other important variables.

(iii) Is one of the primary objects of the control chart to register shifts or drifts away from a **specific** value, in addition to the general objective of identifying, reducing and eventually eliminating the influence of special causes? Thus in a dimensional application, a centreline based on a recent (and satisfactory) average, say $\bar{\bar{x}}$, will serve to signal any change from $\bar{\bar{x}}$, but may fail to signal that the process is not centred on the nominal value. Eventually, of course, the calculation of a new $\bar{\bar{x}}$ or capability assessment will reveal any long-term drift from nominal, but the chart can be used to give pro-active control if the nominal value is used as the centre-line. Similar considerations apply to many non-manufacturing applications, such as in budgetary control when a target value of zero (i.e. no long-term deviation from budget) is applicable. Use of a logical target value here can give early warning of any overspending or underspending tendency.

7.6.2 For applications other than those covered by section 7.6.1, a target based on recent performance is usually applicable. This is especially the case for parameters such as range, standard deviations, proportions of non-conforming items and rates of occurrence of events such as non-conformities or accidents. In such applications one cannot re-adjust to an arbitrary target by simple actions on the process; any special causes need investigation and identification of reasons for the change before any steps to correct the situation can be implemented. Thus $\bar{R}$, $\bar{s}$, $\bar{p}$ or $\bar{u}$ will almost always provide the target values for their respective control charts, but for individual values or $\bar{x}$ applications, the points in section 7.6.1 should be considered before statistics like $\bar{\bar{x}}$ are adopted as routine target values.

7.7 In-control/Out-of-control

7.7.1 The terms in-control, out-of-control refer simply to the process **stability**. In-control (sometimes, confusingly, called 'under control'!) means the process is performing as well as can be expected in the light of its inherent variation. Out-of-control means that some effect has disturbed this stability, but occasionally the effect may be a beneficial one rather than a deterioration. In particular, note that no reference has so far been made to process capability. In fact, even out-of-control processes may not necessarily be violating specification limits, and conversely the fact that a process is in-control is not an absolute statement of good quality – a statistically stable process may be too variable to provide adequate capability to a specification.

Control charts may be used even with non-capable processes, where they will at least help to make the best of a poor system. They are also relevant where there is *no* specification, as in many **process** (rather than **product**) and non-production applications where the object is to detect any change (adverse or beneficial) in the performance of the system.

7.7.2 Examples of in-control and out-of-control

Details of the calculations involved in setting up and operating control charts will be presented in the following chapters. In Figs. 7.1–7.9, examples are given of the appearance of control charts in several widely occurring situations.

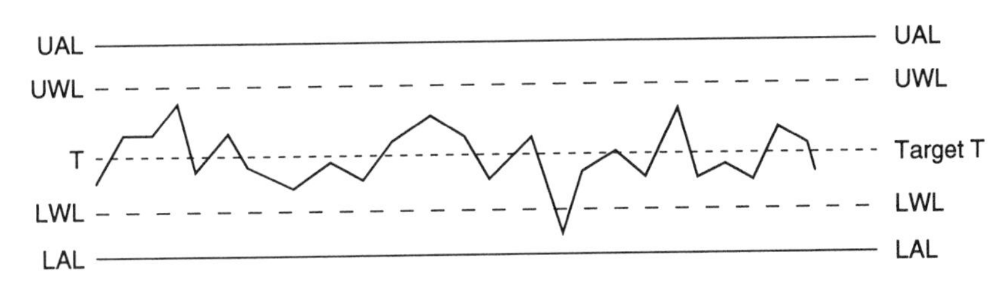

Fig. 7.1 Process in-control. *All* points are with UAL and LAL. Most are within UWL and LWL, but *occasional* isolated values between warning and action lines are to be expected. Indeed, if 'warning' values never occur, there is probably some error in calculation of the limits. Also, there are no distinctive patterns (runs, trend, cycles) merely evidence of random or common cause variation.

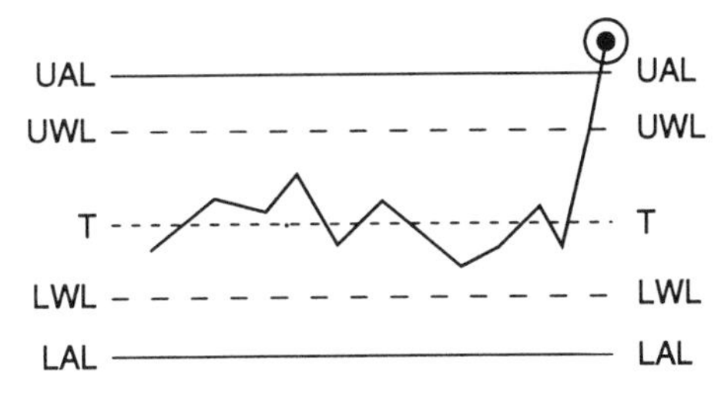

Fig. 7.2 Out of control: violation of control limit. As soon as a value beyond an action line is plotted, the cause must be investigated and, where possible, corrected.

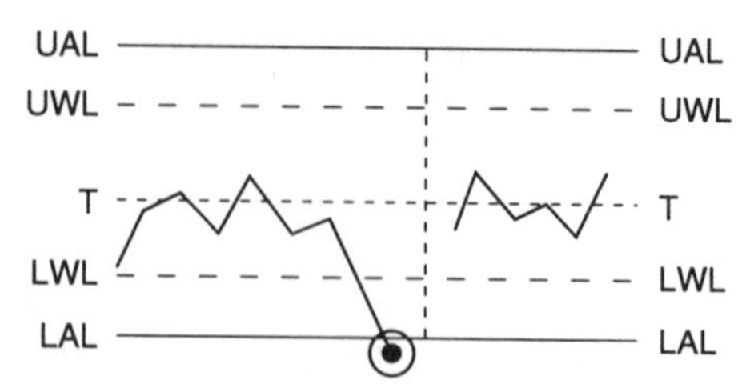

Fig. 7.3 Out of control: subsequently corrected. A point below LAL signals an out-of-control condition. If investigation reveals a removable cause, the subsequent values should indicate restoration of control.

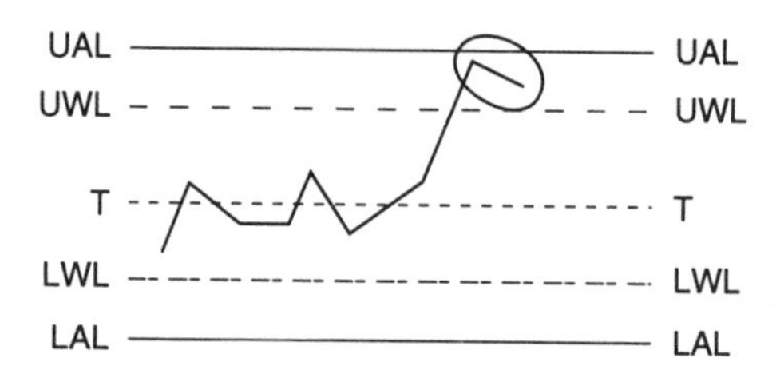

Fig. 7.4 Out of control: two warning values. Here the 'signal' arises from two successive values beyond a warning limit. Investigation and corrective action are required. (Some US practitioners adopt a two-in-any-successive-three rule for warning values).

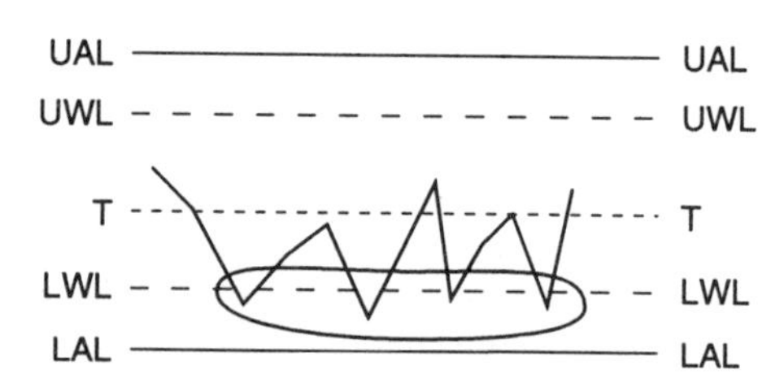

Fig. 7.5 Out of control: frequent values beyond warning line. Frequent values beyond a warning line suggest a shift in process average. A correction to a temperature setting, or material formulation etc. may be necessary.

The patterns illustrated may occur on any kind of control chart, e.g. $\bar{x}$, R charts; np, u, etc., and they demonstrate the role of control charts in **diagnosis** of process problems. The charts themselves do not produce solutions to the problems; this requires a blend of technical expertise, commonsense, perseverance, and above all management commitment to the pursuit of continuous improvement in all aspects of performance.

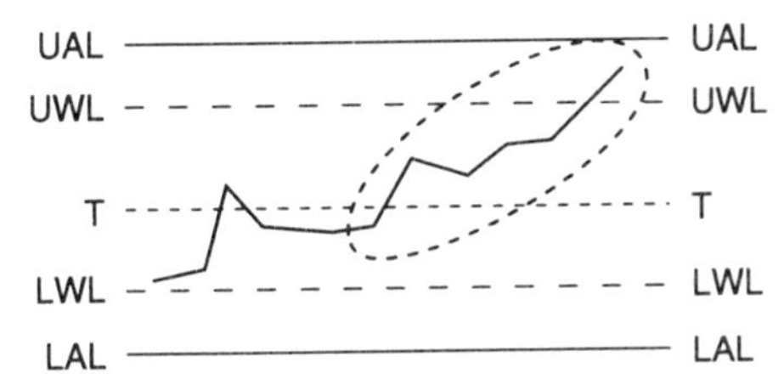

Fig. 7.6 Out of control: persistent trend. A run of values in a gradually rising or falling trend may signal a deterioration (or less likely, an unexpected gradual improvement) in performance.

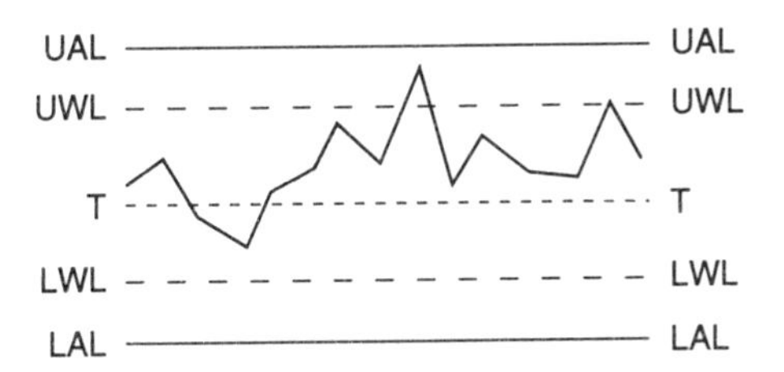

Fig. 7.7 Out of control: run of seven or more on one side of target. A run of seven or more values in succession on the same side of the target may signal a shift in process average. See also example of Fig. 7.5.

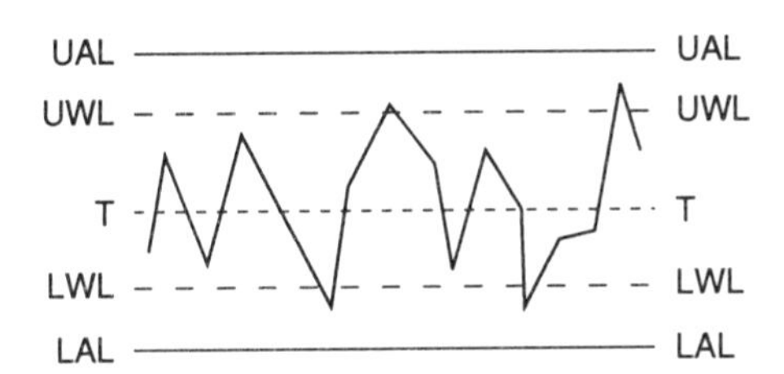

Fig. 7.8 Out of control: apparent overcontrol. The excessive zig-zagging, with few points near the target line and many near the control limits indicates 'over-control'. This may arise from unnecessary intervention by process operators, or simply that the real level of common cause variation has not been adequately evaluated, i.e. the process is 'noisier' than indicated by the 'within-sample' variation.

7.8 Setting up the control chart – general

7.8.1 We have covered the theory and concepts of control charts. We now turn to the routine aspects of setting up and operating the various kinds of control charts.

Each type of chart should have a worksheet specifically designed for its effective use. As well as suitable spaces for recording data and the results of certain

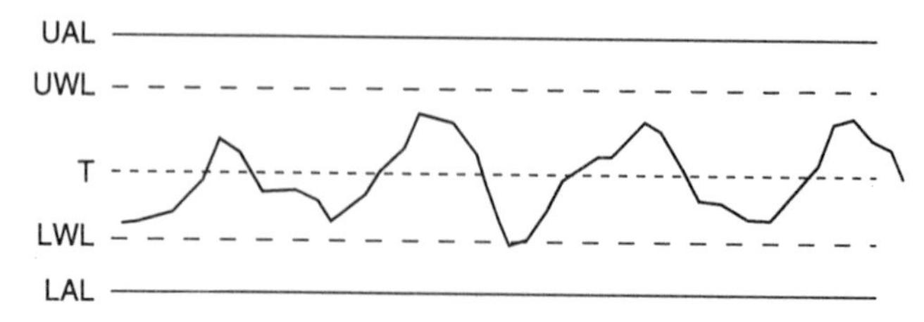

Fig. 7.9 Out of control: cyclic pattern. Although not violating control limits, producing runs of points above and below the target etc., there is a distinctive pattern of variation over time, which is likely to be related to some feature of the process. Possibilities may be day-night temperature effects, shift working patterns, start-up/run-down problems, and so on. Other distinctive patterns might show up as once a week high (or low) values, switches from runs above target to runs below and back again, etc.

calculations, provision should be made for other records which are equally important:

(i) administrative details concerning the operation, department, characteristic to be observed, method of measurement, units, frequency of checking, target value (aim or specification), etc.;
(ii) the dates and times for each process check or sample;
(iii) a process log for recording information on process changes and actions on assignable causes.

These items of information are vital to the effective use of the control charts in problem solving and process improvement. Failure to record all relevant background data will reduce the effectiveness of SPC in controlling and improving the operation.

7.8.2 Step-by-step approach to completing the control chart

The sequence of operations is the same for all the types of control charts described in the following chapters. The differences lie in the details of the calculations, which will be covered in subsequent chapters. The basic operations are:

Step 1: Obtain process evaluation data in the same format as will be used for process control, i.e. same sample size and frequency. We shall denote sample size by n and the number of samples or subgroups in the evaluation data by k, so that the total number of data elements is $k \times n$.

Step 2: It is usually necessary to calculate some 'statistics' from the data in each sample (or in each subgroup where data is based on 100% observations). Such statistics may include the sample mean $\bar{x}$, range R, standard deviation s, or (for attribute and event applications) the proportion of non-conforming items or rate of occurrence of non-conformities.

Step 3: When sufficient samples or subgroups have been obtained, some estimates of parameters are required, e.g. process average, or overall proportion of non-conforming items, etc.

Step 4: **Trial** control limits are next calculated, using the parameter estimates from step 3 along with the appropriate factors from the tables accompanying the relevant sections in the following Chapters.

Step 5: Once the trial limits are known, it is possible to scale the control chart in preparation for plotting. In some cases, where there is previous experience to draw upon, scaling may be effected earlier in the sequence (even prior to step 1), so that plotting may take place at the time of data collection.

Step 6: Draw the trial limits and any appropriate parameter estimates on the chart.

Step 7: Plot the evaluation data.

Step 8: Study the plotted data in order to assess the state of the process and decide if it is in control or whether any special causes are present. (Rules for the evaluation of charts and examples of the types of variation encountered in practice were covered earlier in this Chapter.)

Step 9: If special causes are present, they must be investigated and corrected. New evaluation data may be necessary following any process improvement action, and steps 1 to 8 should then be repeated.

Step 10: Where conformity to a specification is required, determine the process capability. The practical details of capability assessment will be covered in Chapter 13. It should be noted that non-capability does not preclude the use of control charts. For non-capable processes they may highlight areas for improvement, though their use will not render the process immediately capable. Where the application does not involve a specification, this step is not formally relevant, though summaries of performance may still have a useful function.

Step 11: Consider whether any adjustment of process **setting** is necessary to improve performance, and **define a target value for on-going control** (see section 7.6).

Step 12: Set up on-going control limits based on the appropriate target value, and enter them on the chart ready for evaluation of future data as collected.

Step 13: Run the process, collecting data at the prescribed times and entering the necessary statistics on the chart. For every sample or subgroup plotted, re-evaluate the state of control and take appropriate action. Maintain the process log.

Step 14: Periodically update estimates and control limits, and review capability. At each update, summarize suitable aspects of performance, and look for means of reducing variation.

It will often be found that several steps can be merged, or that their positions in the sequence may vary depending on details of application, previous experience, state of the process, etc. Remember that the control chart should be updated each time a new sample or subgroup is taken. It ceases to have much control or improvement value if data is collected and entered in 'batches' once a week or some other interval. Whenever practicable charts should also be displayed near the process and the people who run it, rather than filed in cabinets out of sight, though there may be exceptions in some cases due to environment or other circumstances.

7.9 Calculators and computers in SPC

The rapid developments in electronic computers, calculators, automatic data capture and storage, video display equipment, etc., have had a considerable influence on SPC applications, and on the development of new methods of control.

Small calculators are particularly useful in reducing the tedium of calculations for standard deviation, proportions, etc. It is wise to include checks that results are reasonable in the light of the input data – for instance that the calculated mean lies between the smallest and largest values in the sample.

More sophisticated equipment also has its place, especially where a process embodies appreciable automatic control. However, where operators have an important level of input, contact between the operator, the data and the control system is most desirable, even essential. This may be achieved by conventional pencil-and-chart methods, or by a judicious choice of computer and screen display. Removal of decision-making to persons or departments not deeply involved with running the process should be avoided.

8

Control charts for average and variation

8.1 Introduction

8.1.1 Under this heading, we include control charts, based on small subgroups of observations, for **measured** characteristics. The differences among individuals within the subgroups (or samples) provide a measure of the inherent variation in the process and a means of monitoring for changes, adverse or advantageous, in that variation. The subgroup ranges, R, or standard deviations, s, are the appropriate statistics for this purpose.

Based on the estimated inherent variation, control limits may be established for an R or s chart, and also for the other part of the chart used to monitor the average level of the process, the mean ($\bar{x}$) or median (M) chart. The former is the most widely used, and generally uses the available data more effectively than the median chart. The latter may have some advantages where calculation is a problem (rarely so since the advent of the electronic calculator) or where $\bar{x}$ may be influenced by 'wild' values – but here there is a need to investigate the reasons for occurrence of such outliers, and to eliminate their causes.

8.1.2 Thus the principal variants of control charts in this Chapter are $\bar{x}$, R and $\bar{x}$, s. Some mention is also made of M, R and the plotting of all the individual values in the subgroup – the 'multivari' chart. The combination M, s is rather illogical and is not included. Attention is also confined to control based on observation of **performance**; methods based on specification limits (tolerance based charts and precontrol) are not covered, but the possibility of **relaxed control** is noted in Chapter 11.

8.1.3 Finally, it must be stressed that **rational** subgrouping is essential. Failure to achieve this may result in serious over or under-estimation of inherent variation, and hence in establishing control limits which are either too wide to detect assignable variation or so tight that they either produce over-control or are so evidently unusable that control charting is abandoned or becomes a paperwork formality. Methods for detecting this situation and for dealing effectively with it are noted in Chapter 11.3.

8.2 UK and US practice

8.2.1 Unfortunately, we have to recognize that early in the history of the control chart, there was a divergence in practice between the USA on the one hand and the UK on the other. This difference has little significant effect on control charts for $\bar{x}$, M or for individual values from a Normal distribution. The effect is more serious for R and s charts (as well as for attributes and events where small numbers of non-conformities occur), and can be critical when extending the concepts of control charts to appreciably non-Normal distributions.

Although the actual calculation of the limits can be reduced to simple formulae combined with the use of factors from standard tables, it is worth considering the origin of this divergence. Largely on an empirical basis, Shewhart (1935) used, for the $\bar{x}$ chart, control limits set at three standard errors ($3\sigma_e$) on either side of the target value. This, for the Normal distribution, results in a 'false alarm' rate of 1 in 740 samples (in either direction: 1 in 370 where two-sided control is applicable, as it generally is). In the UK, Pearson (1935, BS 600) modified this to 1 in 1000, or $3.09\sigma_e$, and also adopted the 1 in 1000 criterion for other sample statistics like R and s, which have non-Normal sampling distributions. He (and colleagues) added 'warning lines' at the 1 in 40 level, corresponding to $1.96\sigma_e$ for a Normally distributed statistic.

8.2.2 American practice, which has also been adopted where SPC has been influenced by US-based companies or practitioners (as in Japan), extended the $3\sigma_e$ approach to R, s and other charts, This often has the effect of yielding negative (and therefore unusable) limits for essentially positive statistics, and also of increasing the false alarm rate for an apparent increase in variation (or non-conformities in the case of attributes and events). Whether this has also contributed to the non-use of warning lines in US practice is not clear, but they have not been widely adopted although other rules (runs, proportions of values in the middle third of the space between control limits) have a similar intention.

8.2.3 This author's view is that while the effect on the $\bar{x}$ chart, and others based on a Normal sampling distribution, is trivial, the use of 'probability limits' in other cases has appreciable advantages, especially where lower limits are concerned.

This is not to say that these limits have any strict probability interpretation. The value of the probability approach is simply to obtain logically asymmetrical limits when dealing with values from an obviously skewed distribution. For sample averages, evidently $3\sigma_e$ for Action lines and $2\sigma_e$ for warning lines provide a simple and reasonable compromise. However, for completeness, factors based on both UK and US practice are included in the following account.

8.3 Derivation of the control chart factors

8.3.1 As noted earlier, the actual calculation of control limits requires only the substitution of estimates like $\bar{\bar{x}}$, $\bar{R}$ or $\bar{s}$, and of appropriate factors, into standard formulae. Some explanation of the derivation of these factors is appropriate here.

Strictly, the control limits for sample means, assuming rational subgrouping, will be at $\pm z$ standard errors (of $\bar{x}$) on either side of the mean value at which the process is intended to run, say a target value T. Using British practice, we have therefore $z = 3.09$ or 1.96, giving:

$T \pm 3.09\sigma_{\bar{x}}$ for action lines;
$T \pm 1.96\sigma_{\bar{x}}$ for warning lines.

Now for random sampling from a stable population (here interpreted as rational subgroups from an 'in-control' process)

$$\sigma_{\bar{x}} = \sigma_x/\sqrt{n}$$

where σ_x is the standard deviation of the individual values and n is the subgroup size. Hence the control limits correspond to:

$$T \pm 3.09\sigma_x/\sqrt{n}, \quad T \pm 1.96\sigma_x/\sqrt{n}$$

8.3.2 Now in practical applications, σ_x will not be known exactly, but must be estimated from a suitable evaluation study – often the first 'trial' control chart or a review of recent data. As discussed in Chapter 6, the fundamental parameter is in fact the variance, σ^2, rather than the standard deviation, and can be most efficiently estimated from the subgroup data as the average of the subgroup s^2 values. The **estimate**, σ_x, is then obtained, for k subgroups, as

$$\hat{\sigma}_x = \sqrt{\text{Average } (s^2)} = \sqrt{\left(\frac{1}{k}\sum_{j=1}^{k} s_j^2\right)}$$

However, to avoid the squaring and square-rooting operations (albeit they are no longer a problem with the $\bar{x}$, s functions of a scientific calculator: see Chapter 16), simpler estimates of σ_x were developed. For example, the mean range of samples of size n from a Normal distribution is found to be $d_n \cdot \sigma_x$, where d_n is a factor depending on sample size. Values of d_n (denoted by d_2 in US practice, irrespective of sample size) are listed in Table 13.1. Thus for $n = 5$, d_n is 2.326 and since, over a large number of samples of five items,

$$\bar{R} \rightarrow 2.326\sigma_x$$

we obtain the estimate of σ_x by reversing this relationship,

$$\hat{\sigma}_x = \bar{R} \div 2.326 \quad \text{or} \quad \bar{R}/d_n$$

in general.

Note that the d_n factors apply to **average** range, and do *not* imply that $R = 2.326s$ in individual subgroups. This becomes an important point in Chapter 15 when dealing with evaluation of measurement precision. A minimum of 20 subgroups ($k \geqslant 20$) is reasonable for using this method of estimating σ_x.

8.3.3 Alternatively, subgroup standard deviations, s, may be averaged. Now $\bar{s}$, the mean of the s-values, will always be smaller than the $\sqrt{\text{Average}\,(s^2)}$ estimate considered above. It is found that, again for samples of size n from a Normal distribution, $\bar{s} = c_n.\sigma_x$, so that we obtain the estimate

$$\hat{\sigma}_x = \bar{s}/c_n$$

Table 13.1 contains the values of c_n (denoted by c_2 in American practice). For $n = 5$, c_n is very close to 0.9400, so that $\hat{\sigma}_x$ then equals $\bar{s}/0.94$.

These estimates are quite robust to non-Normality (i.e. for various non-Normal distributions the factors would not differ much from those for the Normal distribution). They thus form a useful basis for simplified estimation procedures.

8.3.4 Re-writing the expressions for control limits in terms of estimates, we thus have

$$T \pm z\sigma_{\bar{x}} \rightarrow T \pm z\frac{\sigma_x}{\sqrt{n}} \rightarrow T \pm z\frac{\bar{R}}{d_n\sqrt{n}} \quad \text{or} \quad T \pm z\frac{\bar{s}}{c_n\sqrt{n}}$$

Now if we take out and evaluate $z/(d_n\sqrt{n})$ or $z(c_n\sqrt{n})$ for any chosen z and a particular sample size n, we have the standard control chart factors for $\bar{x}$. Thus, for example, in UK practice, with $n = 5$ and $z = 3.09$ (action), 1.96 (warning), using $\bar{R}$ we have

$A = 3.09/(2.326 \times \sqrt{5}) = 0.594$
$W = 1.96/(2.326 \times \sqrt{5}) = 0.377.$

Using $\bar{s}$, the factors become

$A = 3.09/(0.94 \times \sqrt{5}) = 1.470$
$W = 1.96/(0.94 \times \sqrt{5}) = 0.933.$

Similarly for American practice with $z = 3$, the factors based on $\bar{R}$ and $\bar{s}$ become 0.577 and 1.427, respectively.

8.3.5 For control limits for R and s, British practice uses the 0.1% and 2.5% points in each tail of the sampling distribution of R or s. Because of the skewness of these sampling distributions, these limits are not symmetrical about $\bar{R}$ or $\bar{s}$. American practice does use symmetrical limits, ignoring the lower limit when the approximation leads to a negative factor. Thus the disparities between the two systems are more obvious for R or s charts than for $\bar{x}$, as is seen for the case with $n = 5$, as follows.

Control chart factors

	for range, R		for standard deviation, s	
	UK	US	UK	US
Upper action	2.34	2.114	2.286	2.089
Upper warning	1.81	Not used	1.776	Not used
Lower warning	0.37	Not used	0.370	Not used
Lower action	0.16	Not used	0.160	Not used

The factors (for both R and s) are simply multiplied by $\bar{R}$ or $\bar{s}$, respectively, to obtain the control limits. Table 8.1 lists the standard factors for $\bar{x}$, $\bar{R}$ and $\bar{x}$, s charts, for both UK and US practice.

8.4 'Trial' control chart and 'on-going' control

The first step in implementing a control chart is to see whether or not the process is stable, or in control. These items of jargon simply mean that only inherent variation is present, and that no assignable or special causes are present. They do not necessarily imply that the process is good when 'in control', nor necessarily bad when 'out-of-control'. As noted in Chapter 7, an 'out-of-control' signal may actually indicate that the process has changed for the better! The terminology may appear inappropriate, but it is standard practice.

To evaluate initial stability, control limits are first set up using the observed process average as a provisional target. Thus the expressions for trial $\bar{x}$ charts usually involve

$$\bar{\bar{x}} \pm \text{factor} \times \bar{R} \quad (\text{or } \bar{s})$$

It may then be found that the process is stable (in control), but that $\bar{\bar{x}}$ is not the intended average level. For on-going control, it may be more appropriate to choose a **target** value, say T, and to set up the limits based on

$$T \pm \text{factor} \times \bar{R} \quad (\text{or } \bar{s})$$

The choice of T is discussed in section 7.6.

8.5 Control charts for median values

An alternative to the use of $\bar{x}$ for monitoring the average level of a process or system is provided by the sample median, M. To exploit the advantage of simplicity that this may offer, it is advisable that:

Table 8.1 Factors for $\bar{x}, R$ and $\bar{x}, s$ charts

Part 1: $\bar{x}, R$ charts. UK Practice

	$\bar{x}$ chart		*R* chart			
Sample size	Action $A'_{0.001}$	Warning $A'_{0.025}$	LAL $D'_{0.999}$	LWL $D'_{0.975}$	UWL $D'_{0.025}$	UAL $D'_{0.001}$
2	1.937	1.229	–	0.04	2.81	4.12
3	1.054	0.668	0.04	0.18	2.17	2.98
4	0.750	0.476	0.10	0.29	1.93	2.57
5	0.594	0.377	0.16	0.37	1.81	2.34
6	0.498	0.316	0.21	0.42	1.72	2.21
7	0.432	0.274	0.26	0.46	1.66	2.11
8	0.384	0.244	0.29	0.50	1.62	2.04
9	0.347	0.220	0.32	0.52	1.58	1.99
10	0.317	0.202	0.35	0.54	1.56	1.93
11	0.294	0.186	0.38	0.56	1.53	1.91
12	0.274	0.174	0.40	0.58	1.51	1.87

Part 2: $\bar{x}$, *R* charts. US practice

	$\bar{x}$ chart	*R* chart	
Sample size	A_2 UCL, LCL	D_3 LCL	D_4 UCL
2	1.880	–	3.268
3	1.023	–	2.574
4	0.729	–	2.282
5	0.577	–	2.114
6	0.483	–	2.004
7	0.419	0.076	1.924
8	0.373	0.136	1.864
9	0.337	0.184	1.816
10	0.308	0.223	1.777
11	0.285	0.256	1.744
12	0.266	0.284	1.717

(continued)

Table 8.1 (*Contd.*)

Part 3: $\bar{x}$, s charts. UK practice

Sample size	$\bar{x}$-chart		s-chart			
	Action $A_{0.001}$	Warning $A_{0.0025}$	LAL $D_{0.999}$	LWL $D_{0.975}$	UWL $D_{0.025}$	UAL $D_{0.001}$
2	2.739	1.737	–	0.039	2.809	4.124
3	2.013	1.277	0.036	0.180	2.167	2.966
4	1.677	1.064	0.098	0.291	1.916	2.527
5	1.470	0.933	0.160	0.370	1.776	2.286
6	1.326	0.841	0.215	0.428	1.684	2.129
8	1.132	0.718	0.303	0.509	1.567	1.932
10	1.005	0.637	0.368	0.563	1.495	1.809
12	0.912	0.579	0.418	0.602	1.444	1.725
15	0.812	0.515	0.474	0.645	1.390	1.635
20	0.700	0.444	0.541	0.694	1.332	1.539

Part 4: $\bar{x}$, s charts. US practice

Sample size	$\bar{x}$ chart	s chart	
	UCL, LCL(A_3)	LCL(B_3)	UCL(B_4)
2	2.659	–	3.267
3	1.954	–	2.568
4	1.628	–	2.286
5	1.427	–	2.089
6	1.287	0.030	1.970
8	1.099	0.185	1.815
10	0.975	0.284	1.716
12	0.886	0.354	1.646
15	0.789	0.428	1.572
20	0.680	0.510	1.490

For other sample sizes, see relevant British, American, International or Industry standards.

Part 5: M, R (median and range) charts (Section 8.5.2)

Sample size	UK practice		US practice
	Action limits* ($M_{0.001}$)	Warning limits* ($M_{0.025}$)	Control limits* ($3\sigma_M$)
3	1.222	0.775	1.187
5	0.712	0.452	0.691
7	0.524	0.333	0.509
9	0.424	0.269	0.412

*Limits obtained as Target value $\pm$ Factor $\times$ $\bar{R}$.

(i) the sample size should be an odd number (otherwise the median becomes the average of the **two** middle values, requiring a calculation);
(ii) the sample size should be small (3, 5 or 7), otherwise too much effort (and risk of error) is involved in locating the median.

8.5.2 It has been noted that the median is less sensitive to extreme values than the mean: this can be either an advantage where 'rogue' values have less effect, or a disadvantage where the extremes represent some feature that ought to be signalled. There is a more important aspect to the use of medians: their sampling distribution is some 15–25% wider than for $\bar{x}$, resulting in the standard error, say σ_M, being larger than $\sigma_{\bar{x}}$ (see Appendix A.7.2 (iii)).

This implies that, in order to achieve the same risk of false alarms as the $\bar{x}$ chart, the control limits need to be set wider – with the disadvantage of slower response to real changes in average performance. Conversely, some users prefer to use the same limits for M as for $\bar{x}$ in order to improve the identification of assignable causes, but with appreciably greater risk of false alarms than for the $\bar{x}$ chart.

Control limit factors for M-charts, where the false alarm risk is held at the same level as for $\bar{x}$ charts, are given in Table 8.1, Part 5.

8.5.3 The M-chart can be operated in conjunction with a chart for sample range, the M, R combination, but for simplicity many users prefer to plot all the sample values, strung along a vertical line, with the median (central) value emphasized as a star, or with a box or circle surrounding it. No formal limits for range are used, but the general level of variability and the presence of rogues are visible on this chart. However, it is not easy to identify changes in the amount of inherent variation, and the use of a range chart is strongly recommended to enhance the effectiveness of this procedure, often termed the multivari chart. For a fuller account see Owen (1989).

8.6 Examples of the use of $\bar{x}$, R; $\bar{x}$, s; and M with individual values

8.6.1 Suppose that the data shown in Table 8.2 have been obtained from an initial evaluation. The data comprise 25 subgroups of five items each, so that we have (following the steps of section 7.8.2):

Step 1: ($k = 25$, $n = 5$, 125 values in all)

This step applies to all the charts considered in this chapter. From Step 2 onwards, the procedures will differ, and we illustrate those for the $\bar{x}$, R chart, the $\bar{x}$, s and the multivari with medians identified.

Step 2: The sample statistics required are listed in the final columns of Table 8.2; $\bar{x}$, R, s and M. Generally only two of these would be required, but for illustration of the alternative methods all are shown for this example.

Table 8.2 Data for trial control chart

Sample	Weights of valve liners Observations ($n = 5$)					$\bar{x}$	M	R	s
1	238.9	239.1	239.3	239.7	240.6	239.52	239.3	1.7	0.672
2	238.1	241.5	242.4	240.5	239.9	240.48	240.5	4.3	1.638
3	239.6	239.2	239.0	237.9	238.4	238.82	239.0	1.7	0.672
4	238.1	239.4	240.6	239.4	239.8	239.46	239.4	2.5	0.904
5	238.9	237.8	237.6	238.8	240.6	238.78	238.8	2.8	1.145
6	238.9	239.7	239.7	239.2	239.8	239.46	239.7	0.9	0.391
7	240.8	241.4	239.8	239.5	240.6	240.42	240.6	1.9	0.769
8	239.7	240.2	238.7	239.8	239.4	239.56	239.7	1.5	0.559
9	239.5	239.1	238.5	240.7	240.4	239.64	239.5	2.2	0.910
10	240.5	238.8	238.4	238.4	239.4	239.10	238.8	2.1	0.883
11	240.7	242.1	240.0	241.0	238.0	240.36	240.7	4.1	1.521
12	239.0	240.8	240.0	238.9	238.0	239.34	239.0	2.8	1.081
13	239.4	237.9	237.8	238.8	238.7	238.52	238.7	1.6	0.669
14	242.0	242.1	240.2	242.1	240.3	241.34	242.0	1.9	0.996
15	240.9	238.4	240.1	240.0	237.7	239.42	240.0	3.2	1.322
16	238.7	242.4	236.9	239.9	240.9	239.76	239.9	5.5	2.097
17	238.0	240.5	239.9	239.9	238.3	239.32	239.9	2.5	1.101
18	239.6	239.6	239.2	239.4	239.0	239.36	239.4	0.6	0.261
19	240.1	238.8	238.4	239.4	240.2	239.38	239.4	1.8	0.789
20	237.7	239.6	240.1	237.7	238.4	238.70	238.4	2.4	1.102
21	237.9	238.9	239.2	239.1	241.2	239.26	239.1	3.3	1.201
22	240.7	239.5	241.1	241.3	238.7	240.26	240.7	2.6	1.117
23	237.5	240.0	239.7	238.1	229.6	236.98	238.1	10.4	4.258
24	238.6	238.0	239.9	239.1	240.4	239.20	239.1	2.4	0.967
25	240.0	239.6	239.8	239.0	238.2	239.32	239.6	1.8	0.729

Table 8.3 Estimates of parameters from data of Table 8.2

Estimate	Including sample 23	Excluding sample 23
$\bar{\bar{x}}$	239.43	239.53
$\bar{M}$	239.57	239.63
$\bar{R}$	2.740	2.421
*$\hat{\sigma}$ via $\bar{R}$	1.178	1.041
$\bar{s}$	1.1102	0.979
*$\hat{\sigma}$ via $\bar{s}$	1.181	1.041

*Via d_n and c_n factors of Table 13.1.

Step 3: Even a cursory inspection of the data in Table 8.2 reveals a glaring anomaly. A very low $\bar{x}$ and large values of both R and s in sample 23 can be traced to an x-value of 229.6. Is this a genuine weight, or a misrecording for 239.6? Its inclusion in the calculations for $\bar{\bar{x}}$, $\bar{R}$ and $\bar{s}$ would distort the resulting control limits, and sample 23 is therefore omitted from these estimates as used for the trial chart, though the statistics including this sample are listed in Table 8.3. Note that the median for this subgroup, although the lowest in the data set, is less affected than $\bar{x}$ and does not violate control limits (UCL or UAL). For consistency with the $\bar{x}, R$ and s analysis, it is omitted from further calculations.

This illustrates both a virtue and a disadvantage of the median: it is less affected by outlying values (statistically more robust) than $\bar{x}$, but can therefore be less effective at signalling their presence unless combined with an R or multivari chart.

The parameter estimates are listed in Table 8.3.

Step 4: Trial control limits can now be calculated. They are **trial** limits because (even after deleting sample 23) there is no guarantee that the process is in control at this stage. Using the appropriate factors from Table 8.1, we obtain the limits in Table 8.4. Note that for the M chart, one would use either the limits shown in Table 8.4 which allow for the wider sample-to-sample variations in the median than in $\bar{x}$, or the same limits as for $\bar{x}$. The latter procedure gains sensitivity but increases the risk of false alarms.

Table 8.4 Trial control limits from data of Table 8.2

Chart	UK practice		US practice	
$\bar{x}$ chart	UAL	240.97	UCL	240.93
(via $\bar{R}$ or $\bar{s}$)	UWL	240.44	LCL	238.13
	LWL	238.62		
	LAL	238.09		
M-chart	UAL	241.35	UCL	241.30
	UWL	240.63	LCL	237.95
	LWL	238.45		
	LAL	237.91		
R-chart	UAL	5.665	UCL	5.118
	UWL	4.382	LCL	Not available
	LWL	0.896		
	LAL	0.387		
s-chart	UAL	2.238	UCL	2.045
	UWL	1.739	LCL	Not available
	LWL	0.362		
	LAL	0.157		

Steps 5–7: Once the control limits are obtained, the trial chart can be drawn up, scaled and the data plotted. The resulting $\bar{x}$, M, R and s charts are shown in Fig. 8.1. Figure 8.2 illustrates the multivari chart.

Step 8: Apart from sample 23, the R and s charts give no cause for concern. Sample 16 gives both R and s between their respective UWL and UAL, and sample 18 shows both R and s between LWL and LAL, but no other warning (or action) indications appear, and there are no sustained runs above or below $\bar{R}$ or $\bar{s}$.

Both median and $\bar{x}$ charts appear to be 'noisy', with several points beyond or very close to warning lines, and sample 14 beyond UAL. Note that the

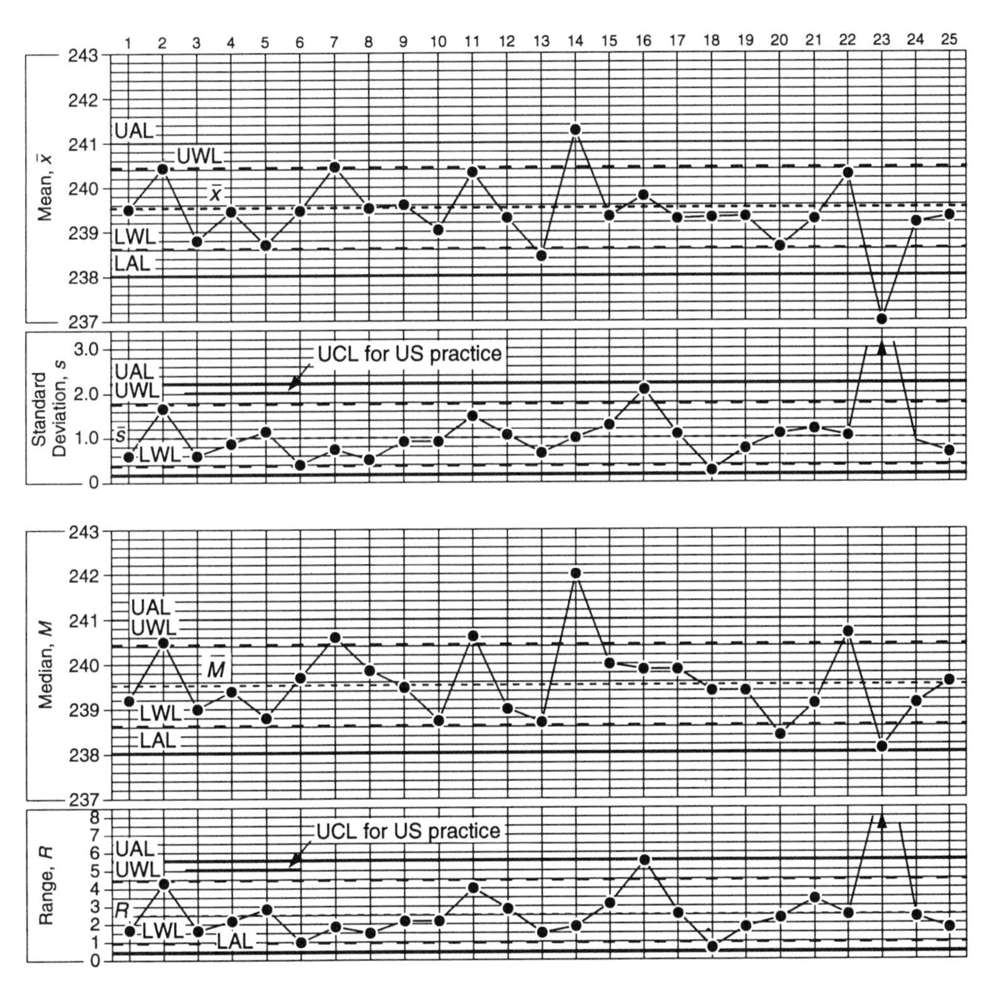

Fig. 8.1 Trial control charts for data of Table 8.2 with $\bar{x}$, s; M, R as ordinates. (UK limits are indicated; US limits are shown where appreciably different). Control limits for M are identical to those for $\bar{x}$ (see section 8.5.2).

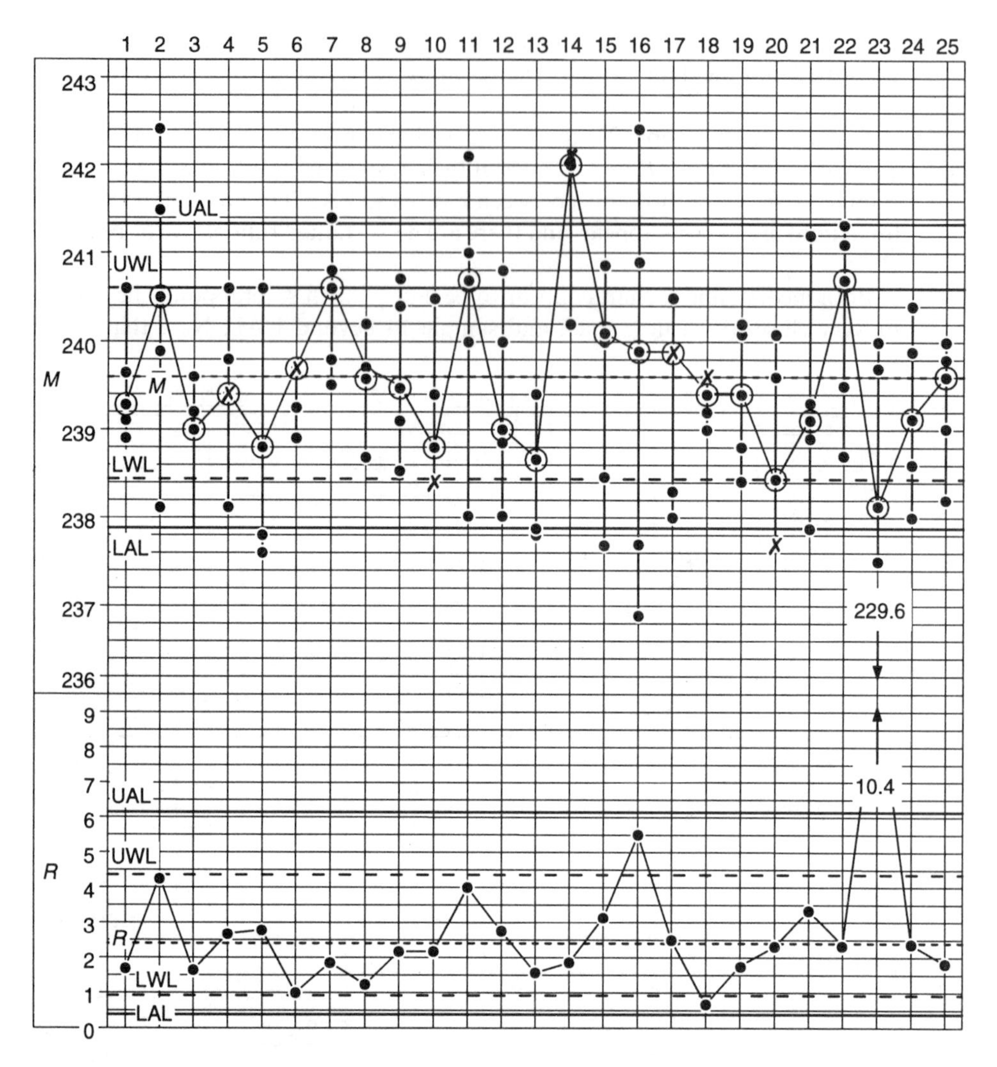

Fig. 8.2 Median and multivari chart for data of Table 8.2. Control limits for M via Table 8.5.

use of $\bar{x}$ limits on the M chart would further question the stability of the sample medians. There is no overall pattern to these variations, and this appears to be a matter of random sample-to-sample variation as further considered in section 11.3.

Steps 9–14: As these data are further examined in Chapter 11.3, we merely note that if the conventional 'strict' control were applied in this case, the on-going limits might be based on a mid-specification target value of 240.0, so that control limits for $\bar{x}$ would become:

UAL	241.438				
UWL	240.913	UK limits	UCL	241.397	US limits
LWL	239.087		LCL	238.603	
LAL	238.562				

Limits for M would be similarly derived, but the trial limits for R and s are suitable for on-going control without modification.

8.6.2 Periodically, updated values of $\bar{R}$ (or $\bar{s}$) and $\bar{\bar{x}}$ ($\bar{M}$) are calculated, and consideration given to changing the control limits to match new levels of performance. Some important points here are:

(i) It is *not* necessary to calculate new limits from every completed chart if $\bar{R}$ (or $\bar{s}$) have not changed appreciably. There is an advantage in maintaining some stability in these limits.

(ii) If $\bar{R}$ (or $\bar{s}$) has **increased**, the limits should not automatically be widened, but the reason for the increase should be sought. It is, of course, possible that the original R (or s) was somewhat optimistic, and it may, after consideration of capability implications, be reasonable to widen control limits to avoid false alarms or overcontrol.

Where $\bar{R}$ (or $\bar{s}$) shows a **decrease**, one should consider whether this apparent improvement can be maintained before narrowing the limits. It may be that a smaller $\bar{R}$ (or $\bar{s}$) is a confirmation that a deliberate improvement action has been successful, and this should then be exploited to achieve closer control and thus better process consistency.

(iii) Review values of $\bar{\bar{x}}$ (or $\bar{M}$) may reveal subtle shifts in average that have not actually yielded out-of-control signals. Tests of significance (Chapter 6) may be appropriate here to avoid unnecessary action based on apparent shifts that are, in fact, within the range of inherent variation.

(iv) Note the more thorough analytical methods described in Chapter 16 which may be useful from time to time. It is not suggested that this form of analysis is necessary on completion of every chart, and experience will indicate when the additional information from these procedures may be useful.

8.7 Further points on average and variation charts

Although this is not a textbook on control charts *per se*, full accounts of which can be found elsewhere, a few further points can be noted.

8.7.1 First, with the increasing use of calculators and computers, the choice of median charts (on grounds mainly of simplicity and avoidance of arithmetic) has less to recommend it than hitherto. Sometimes even the calculation of $\bar{M}$ and $\bar{R}$ are avoided, and control limits can be obtained via **median** range, with a target based on the median of sample medians. Although no actual arithmetic is involved in finding the median of some 20 or 30 medians or ranges, the operation is rather laborious and also error-prone. Further, having found these medians, arithmetic is involved in calculating the control limits and the use of the more efficient $\bar{x}, R$ or $\bar{x}, s$ procedures therefore seems logical.

8.7.2 Secondly, as has been briefly noted, methods exist for drawing up charts based on specification values rather than on actual process data. The main objection to this is that **improvement** is not encouraged, merely conformity. If the process is not actually able to meet the specification, overcontrol may result, whereas a 'performance based' chart, although not rendering the process capable, is making the most effective use of data in identifying opportunities for improvement whilst avoiding overcontrol.

For highly capable or precise processes, some relaxation from a single target value is possible, and this will be one of the aspects covered in Chapter 11.

Precontrol charts are again concerned with operation to specification, and are not included here.

9

Control charts for 'single-valued' observations

9.1 Introduction

9.1.1 Conventional control charts ($\bar{x}, R$; $\bar{x}, s$ or M, R) are based on rational subgrouping of observations into samples whose size (number of items) is denoted by n. For each sample, the required statistics ($\bar{x}$ or M, R or s) are calculated and plotted on the control chart. Control limits are derived from the average range ($\bar{R}$), or some overall estimate of the standard deviation ($\hat{\sigma}$), using factors which depend on the sample size.

In many situations there may be no logical basis for subgrouping, so that conventional charts can only be used by some artifical or arbitrary grouping. This may mask important features of the data, and also defer drawing conclusions about the process (consequently delaying the implementation of corrective action) until a subgroup has been completed. To obtain the most effective use of the data, it is preferable to update the control chart each time a new observation becomes available. There are several procedures for achieving effective control, but some measure of the inherent variation in the process is needed to replace R or s in the absence of natural subgroups.

9.1.2 Examples of the types of data requiring such control procedures are:

(i) analytical determinations of batch properties (with only one analysis per batch);
(ii) values from test or measurement procedures that take some time to execute;
(iii) instrumental observations of process conditions;
(iv) material usage per unit production quantity;
(v) 'non-production' data such as absence rates, energy use, financial data, etc.

The general principles of control chart preparation remain applicable, i.e. an initial process study, the use of trial control limits for assessing stability, setting up limits for on-going control, periodic review and updating.

The methods available include:

'x' charts for individual values
moving average charts

exponentially weighted moving averages
Cusums
adaptive control systems.

The last two are not convered in this Chapter. Cusum methods are separately described in Chapter 12 and adaptive methods (where a mathematical model representing the process uses recent data to predict the next observation, and to calculate an adjustment if an 'off-target' condition is indicated) involve statistical theory beyond that assumed in this book.

9.2 Measuring inherent variation

9.2.1 As will be detailed in later sections, the method of measuring inherent variation may depend on the type of control chart used to monitor the average value of the process. Broadly, the methods use a short sequence of the individual values, varying from two to perhaps ten observations. For three or more values, moving ranges are generally used to measure variation; less commonly a moving standard deviation may be adopted. 'Moving' in this case (as for moving averages described in section 9.4) means that for the first k values in the series, the value of R (or s) is calculated; when the $(k+1)$th value is available, the first is dropped and the corresponding statistic is calculated from observations 2 to $k+1$, then 3 to $k+2$, etc.

When only two observations are used, either moving ranges or their squares may be used. For this specific case with $k=2$, the resulting statistics are often referred to as successive differences (with symbol d or δ), or as a special case of moving range labelled MR_2.

9.2.2 There is theoretical evidence that moving ranges with k between 5 and 10 make the most effective use of the available data. Unfortunately it is not always convenient to use spans of this size, especially if a moving average is adopted for monitoring the process mean. Obviously it will then be useful to use the same k for the MA as for the MR chart.

Further, for processes that exhibit trend, or are prone to occasional 'wild' values, it may be advisable to use short-span moving ranges – often even $k=2$. The successive difference statistic also has other potential uses described in Chapter 16 which may strengthen its claims in some applications. We consider this case first.

9.2.3 For a sequence of single-valued observations

$$x_1, x_2, x_3 \cdots x_{n-1}, x_n$$

we may form the differences

$$\delta_1 = |x_2 - x_1|,\ \delta_2 = |x_3 - x_2|, \text{ etc. up to } \delta_{n-1} = |x_n - x_{n-1}|$$

(where the || indicate that the absolute values, i.e. without sign, are taken).

Now the mean of such differences in a random series of values from a Normal distribution is $\sigma\sqrt{(4/\pi)}$, or approximately 1.128σ. Inverting this relationship, we may estimate σ from $\bar{\delta}$ using

$$\hat{\sigma} = \bar{\delta}/1.128 \text{ (or approximately } \tfrac{8}{9}\bar{\delta}) \tag{9.1}$$

In an alternative notation $\bar{\delta}$ may be written as $M\bar{R}_2$, the mean moving range with span $k=2$.

Another, and more efficient, means of using the δ's to estimate σ is via their squares. It happens that the expected value of the squared successive difference is twice that of the underlying process variance. Again inverting this relationship, we have

$$\hat{\sigma} = \sqrt{(\tfrac{1}{2}MSSD)} = \sqrt{\left(\tfrac{1}{2}\frac{\Sigma\delta^2}{k-1}\right)} \tag{9.2}$$

using $MSSD$ to represent 'mean square successive difference'. This statistic also provides the basis for the useful test of stability described in Chapter 16.

For use with control charts, $\hat{\sigma}$ may now be used to obtain control limits using the general principles in Chapter 7. Some factors for specific cases are detailed shortly.

9.2.4 Extending the idea of moving differences to moving ranges in general, one may prefer in some cases to use MR_k, the difference between the largest and smallest observations within a span of k items. The average MR_k is then calculated, and σ may be estimated from the usual relationships

$$\hat{\sigma} = M\bar{R}_k/d_n$$

Table 9.1 Factors for *MA*, *MR* charts

Span of *MA*, *MR* (k)	*MA* chart Based on $\hat{\sigma}^{\dagger}$	 Based on $M\bar{R}$ (A_2)	*MR* chart UK $D'_{0.001}$	 US (d_4)	 Estimator for $\hat{\sigma}(d_2)$
2	2.121	1.881	4.12	3.268	1.128
3	1.732	1.023	2.98	2.574	1.693
4	1.500	0.729	2.57	2.282	2.059
5	1.342	0.577	2.34	2.114	2.326
6	1.225	0.483	2.21	2.004	2.434
7	1.134	0.419	2.11	1.924	2.704
8	1.061	0.373	2.04	1.864	2.847
10	0.949	0.308	1.93	1.777	3.078
12	0.866	*	*	*	*
15	0.755	*	*	*	*
20	0.671	*	*	*	*

* *MR* not advised for $k > 10$.

† Factors correspond to $3/\sqrt{k}$. For higher control, use $2.58/\sqrt{k}$.

where d_n is the expected range in samples of k values from a Normal distribution. The d_n factors are given in Table 9.1.

9.2.5 As well as using δ or MR to estimate the inherent variation, the short-term variation is also monitored via a δ or MR chart. This is operated in a similar manner, and using the same factors for obtaining control limits, as a conventional R chart. There are two important differences in implementation, however:

(i) Because these are **moving** ranges, one cannot use warning lines with the MR chart. When a fairly extreme individual value produces one large MR, it is likely to influence the next MR as well; indeed, perhaps for MR_k, $k > 2$, it may yield several successive large moving ranges. This leads to the second point.
(ii) Runs of seven or eight successive MR values above or below the centre-line occur not infrequently, so that runs rules need cautious use. A reasonable guide is that for MR_k, a run of at least $k + 7$ values is needed to give a clear signal. This rule will also apply to runs on an MA chart.

The factors for control limits when MA and MR are operated with the same 'span' are the $D'_{0.001}$ or d_4 listed in Table 8.1, Parts 1 and 2.

9.2.6 Moving standard deviations, Ms, may be used in a similar manner to MR. The standard deviation, s, is calculated from each sequence of k values, and the average of these statistics, $M\bar{s}$, is used either to estimate the underlying process standard deviation or to provide control limits for both Ms and MA charts.

Where $M\bar{s}$ is used to estimate the process standard deviation, it must be noted that a bias correction factor, c_n, must be applied to avoid underestimation of σ. Thus the estimated standard deviation, $\hat{\sigma}$, is given by

$$\hat{\sigma} = M\bar{s} \div c_n$$

The c_n values (which are denoted by c_2 in US references) are tabulated in Chapter 13.1. Factors for control limits correspond to those for $\bar{x}, s$ charts with sample size n equal to the span, k, of the Ms chart. These are listed in Table 8.1, Parts 3 and 4 for UK and US practice, respectively.

It has to be said that despite the use of computers in SPC, Ms is not yet widely used in conjunction with control of individual values or moving averages.

9.3 Process control: 'x'-charts

9.3.1 The simplest control chart for 'single-value' situations is the x-chart. The individual values in the series are plotted and evaluated against control limits in the usual way.

For UK control charts (with action and warning lines), the limits are calculated as:

Target value $\pm 3.09\sigma_e$ (or $\pm 2.74\bar{\delta}$) for action lines;
Target value $\pm 1.96\sigma_e$ (or $\pm 1.74\bar{\delta}$) for warning lines.

In practice, these are often rounded to $3\sigma_e$ and $2\sigma_e$, corresponding to $2.66\bar{\delta}$ (or 2.66 $M\bar{R}_2$) and $1.77\bar{\delta}$ (1.77 $M\bar{R}_2$). This brings action lines into agreement with the US control limit, but as with $\bar{x}$, R charts, the US practice does not usually include warning lines. Any other rules normally included for $\bar{x}, R$ (i.e. runs, inner/outer thirds, etc.) are also valid for x-charts.

9.3.2 Some difficulties occur in using x-charts that are less common in applications involving $\bar{x}$. Because there is no averaging over several values, non-Normality or wild values are not smoothed out, and the validity of '3-sigma' limits may often be suspect. A general guide to dealing with non-Normality is given in section 11.6. Process variables such as temperatures, tensions or pressures, material characteristics like particle sizes and small proportions of ingredients, and electrical measurements in general, are especially prone to non-Normality.

Dependencies between successive values (i.e. where the state of the process at one sampling point is still influenced by some hang-over effect from a previous occasion) can also cause difficulties. This is a more difficult problem that may need formulation of an appropriate process model taking account of the dependencies for its solution, and is beyond the scope of this book.

9.4 Moving average charts

9.4.1 Sometimes when using single observations, the absence of any averaging effect may result in false alarms due to occasional wild values, leading to unnecessary corrections or other actions, and hence to over-control. Where the main object of the procedure is to detect (and correct) changes in the underlying process **average**, some smoothing of the individual values may be desirable.

One means of smoothing is to use the average of the most recent k values. Thus if we choose $k = 5$, the first moving average is obtained as

$MA = (x_1 + x_2 + x_3 + x_4 + x_5) \div 5$
the next as $(x_2 + x_3 + x_4 + x_5 + x_6) \div 5$

and so on; each time a new value becomes available, it is included in the average and an earlier value is dropped.

9.4.2 Some experience and judgement are required to choose a suitable value of k. If too few values are included in the moving average, smoothing may be inadequate; if too many are included, response to changes in the process may be slowed by the retention of past data that has become history! Values of k between 2 and 10 are most commonly used, with 3 or 5 as preferred values in the absence of contrary indications.

In some cases, the choice of k may be assisted by considering any cyclic pattern that may exist in the system. This may especially apply to non-manufacturing applications in sales, absence, finance, etc., where obvious annual cycles may suggest a 12-month moving average. Shorter-term cycles may include weekly patterns in output (Mondays and Fridays lower than midweek) or daily cycles due to ambient temperature effects. Thermostats or other control or topping-up devices may introduce cycles of length from a minute or two up to a few hours.

It is important that when the MA chart is used to smooth a cycle, the variation should be measured from the moving averages themselves, calculating their standard error and using an appropriate multiple for the control limit. The limits must **not** be calculated from a MR chart with the same span as the MA, as the cycle amplitude will then expand the control limits. Although a MR chart may be useful to **monitor** the amplitude, the usual control factors for MR will not apply (because a large part of the variation is systematic rather than random), so that control limits will need to be set by judgement and experience.

9.4.3 For MA charts, warning lines are inappropriate, because of the interdependence between successive MA's which contain $k-1$ contributing values in common. Similarly, runs rules need modification, and runs of at least $k+7$ successive values having a common pattern (e.g. above or below target, steadily rising or falling) are needed to give positive indications. To compensate for loss of sensitivity resulting from these restrictions, some users tighten the control limits so that they correspond to a 1 in 200 probability level, i.e. about $2.58\sigma_e$.

The standard error for MA (other than the special case where it is used for smoothing a cycle) is $\sigma/\sqrt{k}$, and control limits can thus be obtained as

target value $\pm 3\hat{\sigma}/\sqrt{k}$ $\quad (2.66\bar{\delta}/\sqrt{k})$,

or

Target value $\pm 2.58\hat{\sigma}/\sqrt{k}$ $\quad (2.28\bar{\delta}/\sqrt{k})$.

where more stringent limits are preferred.

9.4.4 An important subset of MA applications is that where a MR chart of span k is operated in association with the MA chart. The main advantage is that the usual control chart factors (as listed in Table 8.1 Parts 1 and 2) can be applied to the MA chart, using $\overline{MR}_k$ in place of the conventional $\bar{R}$ from subgroups of observations. As noted in section 9.4.2, this simplification can be counterproductive if the MA is used to smooth cycles, and it may also give rise to problems where a process average is subject to trend, or where occasional wild values occur. These aspects deserve careful thought before adopting the MA_k, MR_k approach.

9.5 Exponentially-weighted moving averages (*EWMA*)

9.5.1 Despite their name, these moving averages (sometimes known, more correctly, as geometric moving averages) are simpler to use than the previous *MA* charts. The moving average is constructed so as to place more emphasis on recent observations, and less on the values some distance back in the series. Such a moving average might be constructed as:

$$EWMA \text{ (at stage } i) = 0.4x_i + 0.24x_{i-1} + 0.144x_{i-2} \text{ etc.}$$

each 'weight' being 0.6 times the previous one. In general, we write

$$EWMA = \alpha x_i + \alpha(1-\alpha)x_{i-1} + \alpha(1-\alpha)^2 x_{i-2}, \text{ etc.}$$

but in fact this reduces to

$$EWMA(i) = \alpha x_i + (1-\alpha).EWMA(i-1)$$

Thus only α, $1-\alpha$, the most recent observation and the previous *EWMA* are involved, and the calculation of *EWMA* becomes very simple on a calculator with a memory facility. In one widely used calculator 'language' the steps would be:

$$x_i \times \alpha + \text{memory recall} \times (1-\alpha) = \text{memory in}$$

The updated *EWMA* is then both displayed and in memory ready for the next calculation.

9.5.2 The *EWMA* is thus 'pulled up by its bootstraps' from one observation to the next. Because of the weighting on recent data, it is more 'up-to-date' than an unweighted *MA*, and it is not necessary to delay plotting *EWMA*'s until k values have been accumulated. The first data value is simply used as the first (dummy) *EWMA*, and plotting can reasonably begin after j values have been incorporated, with j being the nearest integer to $1/\alpha$.

9.5.3 The weighting factor α can also be used to assess the degree of smoothing that will be achieved in comparison to a MA chart. The standard error of the *EWMA* is $\sigma\sqrt{(\alpha/(2-\alpha))}$, and by equating this to the standard error of *MA* ($=\sigma\sqrt{1/k}$) and solving for k, the equivalence between *EWMA* and *MA* in terms of smoothing can be obtained, as in Table 9.2.

9.5.4 One should note that *EWMA* charts are not appropriate when smoothing is required to damp out cyclic patterns, but they can be preferable to *MA* when linear trends occur – the greater weighting on recent data assists in detecting a trend more promptly. Also, there is no obvious link between an *EWMA* and an *MR* chart, so that successive differences ($\delta = MR_2$) are a logical choice for monitoring variation.

Table 9.2 Equivalence in smoothing of *EWMA* and *MA*

Values of k for *MA* chart	Eqivalent α for *EWMA**	Suggested rounded values of α†
2	0.6	0.6–0.7
3	0.5	0.5
4	0.4	0.4
5	0.3	0.3, 0.35
6	0.286	0.25, 0.3
8	0.2	0.2, 0.25
10	0.18	0.15, 0.2
12	0.154	0.15
15	0.125	0.1, 0.125 or 0.15
20	0.0952	0.1

*Solutions to $\alpha/(2-\alpha) = 1/k$, i.e. $\alpha = 2/(k+1)$.
†Values of α from 0.2 to 0.5 are commonly used, with a preference for the 0.3 to 0.4 region.

9.5.5 As for *MA* charts, runs rules need to be modified, and a reasonable rule is to require a run of $7 + 1/\alpha$ as evidence of a shift in mean or a trend. Warning lines are also inapplicable, so that the tightening of control limits from $3\sigma_e$ to about $2.58\sigma_e$ may be preferred to gain sensitivity.

Using the expression for the standard error of *EWMA*, the control limits become

$$\text{Target value} \pm 3\sigma\sqrt{\left(\frac{\alpha}{2-\alpha}\right)}$$

$$\left(\text{or } 2.58\sigma\sqrt{\left(\frac{\alpha}{2-\alpha}\right)} \text{ if greater sensitivity is preferred}\right)$$

If an associated δ chart is used with the *EWMA*, these limits may be represented as

$$\text{Target value} \pm 2.66\bar{\delta}\sqrt{\left(\frac{\alpha}{2-\alpha}\right)}$$

$$\left(\text{or } 2.28\bar{\delta}\sqrt{\left(\frac{\alpha}{2-\alpha}\right)} \text{ for tighter control}\right)$$

9.6 Cusum (cumulative sum) charts

For many applications, cusum charts provide the most effective means of control. They offer a number of advantages, such as location of change-points, more reliable estimation of current averages for calculating corrective adjustments, and the facility for extracting periodic averages for routine reporting (e.g. on a daily, weekly or other regular basis). They are covered separately in Chapter 12.

9.7 Examples of use of $\bar{x}$, δ, *MR*, *MA*, *EWMA*

9.7.1 The data in Table 9.3 represent initial evaluation of a process for control chart application. They might be the proportion of an ingredient dosed into a mixer, or results from analysis of batches of material, etc. Only one result is

Table 9.3 Process evaluation data for x, δ, MA_5, MR_5, *EWMA*

Observation i	Value x_i	δ $x_i - x_{i-1}$	MA_5	MR_5	*EWMA* ($\alpha = 0.35$)
1	51.2	–	–	–	(51.20)
2	50.9	0.3	–	–	(51.10)
3	48.7	2.2	–	–	50.26
4	51.7	3.0	–	–	50.76
5	49.2	2.5	50.34	3.0	50.22
6	48.1	1.1	49.72	3.6	49.47
7	52.5	4.4	50.04	4.4	50.53
8	51.9	0.6	50.68	4.4	51.01
9	46.5	5.4	49.64	6.0	49.43
10	52.0	5.5	50.20	6.0	50.33
11	49.6	2.4	50.50	6.0	50.08
12	53.7	4.1	50.74	7.2	51.34
13	50.4	3.3	50.44	7.2	51.01
14	49.3	1.1	51.00	4.4	50.41
15	47.7	1.6	50.14	6.0	49.46
16	49.9	2.2	50.20	6.0	49.62
17	52.5	2.6	49.96	4.8	50.63
18	54.7	2.2	50.82	7.0	52.05
19	56.1	1.4	52.18	8.4	53.47
20	59.2	3.1	54.48	9.3	55.47
21	53.5	5.7	55.20	6.7	54.78
22	55.3	1.8	55.76	5.7	54.96
23	49.2	6.1	54.66	10.0	52.95
24	51.7	2.5	53.78	10.0	52.51
25	48.3	3.4	51.60	7.0	51.04
26	49.9	1.6	50.88	7.0	50.64
27	51.1	1.2	50.04	3.4	50.80
28	46.2	4.9	49.44	5.5	49.19
29	50.2	4.0	49.14	4.9	49.54
30	48.5	1.7	49.18	3.6	49.18

obtained at any one time, so $\bar{x}, R$ charting is not appropriate. We suppose the process is intended to run at a value of 50 units (e.g. percentage, kilogram, newtons or whatever is being measured), though initially it is necessary to assess whether the process is stable around its current average – even if that average is not precisely 50. Thus $\bar{x}$, the mean of the first 30 results, is used as an initial target, but later the limits will be set out around a target of 50 for on-going control.

The observed values are given in the x_i column of Table 9.2, and the statistics for the various charts are derived from them.

9.7.2 Summary statistics from Table 9.2 are required for setting up a trial control chart. The actual statistic will depend on which type of chart we adopt, and the following combinations are examined in this example.

(i) x, δ
Here we rquire, from all observations,
$\bar{x} = 50.99$, $\bar{\delta} = 2.824$.
From $\bar{\delta}$ we find $\hat{\sigma} = 2.824 \div 1.128 = 2.504$
(alternatively, we might have used $8\bar{\delta}/9 = 2.510$, or $\sqrt{(\frac{1}{2}MSSD)} = 2.287$).

(ii) MA_5, MR_5
We again require $\bar{x} = 50.99$ for use as a provisional centre-line, but here we need also $M\bar{R}_5 = 6.058$. Note that if we estimate $\hat{\sigma}$ from $M\bar{R}_5$, we have $6.058 \div 2.326 = 2.60$, a larger value somewhat inflated by a disturbance in the process around observations 18–22.

(iii) $EWMA$, δ
For this combination, we would need $\bar{x}$ and $\bar{\delta}$ as already determined.

9.7.3 Trial control limits for the charts can now be obtained as follows:

(i) x-chart. Fig. 9.1(a)
Using $\hat{\sigma} = 2.504$, we have (using 3σ and 2σ limits),
Action lines (or UCL, LCL) $50.99 \pm 3 \times 2.504, = 58.50, 43.48$,
Warning lines $50.99 \pm 2 \times 2.504, = 56.00, 45.98$.

(ii) δ-chart. Fig. 9.1(b)
UK practice would use $4.12\ \bar{\delta} = 11.63$,
US practice would use $3.27\ \bar{\delta} = 9.23$.
(both limits are shown in Fig. 9.1(b)).

(iii) MA chart. Fig. 9.2(a)
Taking this as part of the MA_5, MR_5 pairing, the measure of variation is $M\bar{R}_5 = 6.058$. Using the customary $\bar{x}, R$ chart factors for $n = 5$ then yields:
UK practice (but without warning lines):
$50.99 \pm 0.594 \times 6.058 = 54.59, 47.39$.
US practice:
$50.99 \pm 0.577 \times 6.058 = 54.49, 47.49$.
(the UK and US limits are almost indistinguishable on the chart)

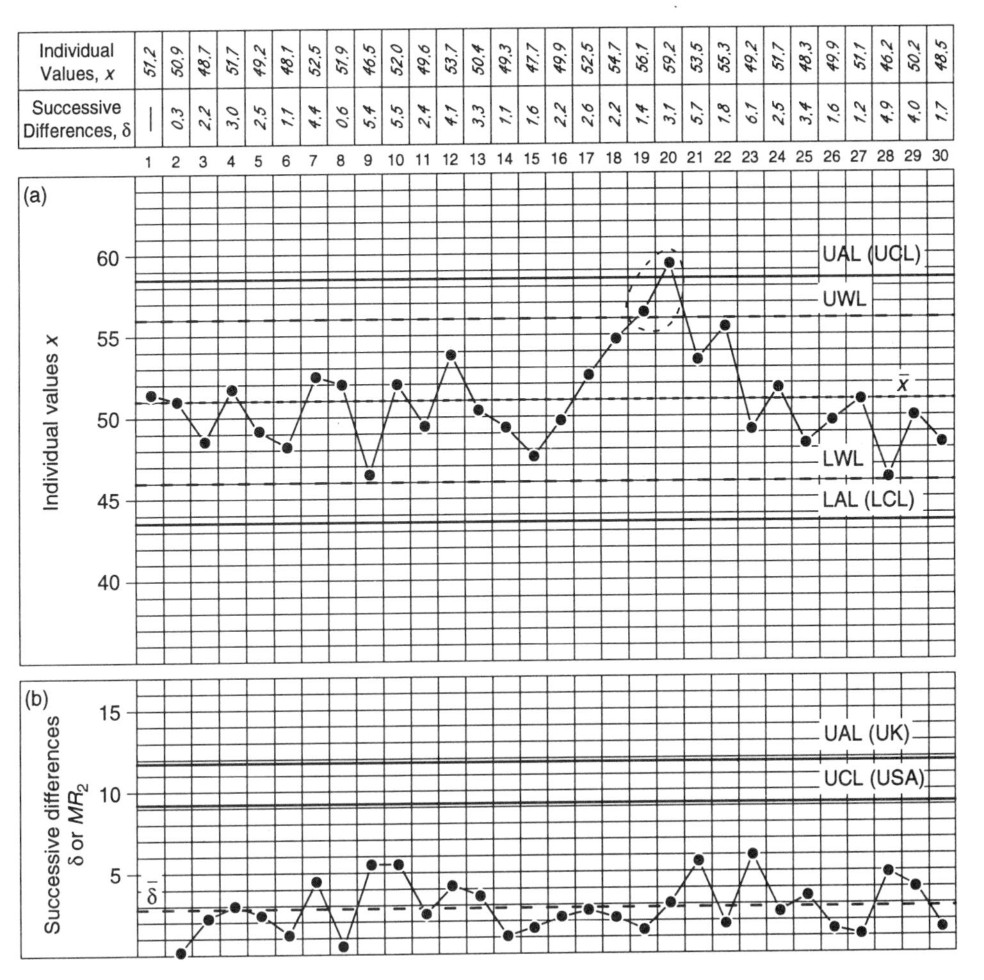

Fig. 9.1 Control chart for individuals and successive differences based on data of Table 9.2. (a) Individual values, x; (b) Successive differences, δ or MR_2.

(iv) MR chart, Fig. 9.2(b)
With $M\bar{R}_5 = 6.058$ and using the conventional factors for R charts with $n = 5$,

UK limit $= 2.34 \times 6.058 = 14.18$,
US limit $= 2.114 \times 6.058 = 12.81$.

(both are entered in Fig. 9.2(b))

(v) $EWMA$ chart, Fig. 9.3.
With $\alpha = 0.35$ to approximate the smoothing effect of MA_5, and using $\hat{\sigma} = 2.504$, the '3-sigma' limits are:

$50.99 \pm 3 \times 2.504 \times \sqrt{(0.35/1.65)} = 54.45, 47.53.$

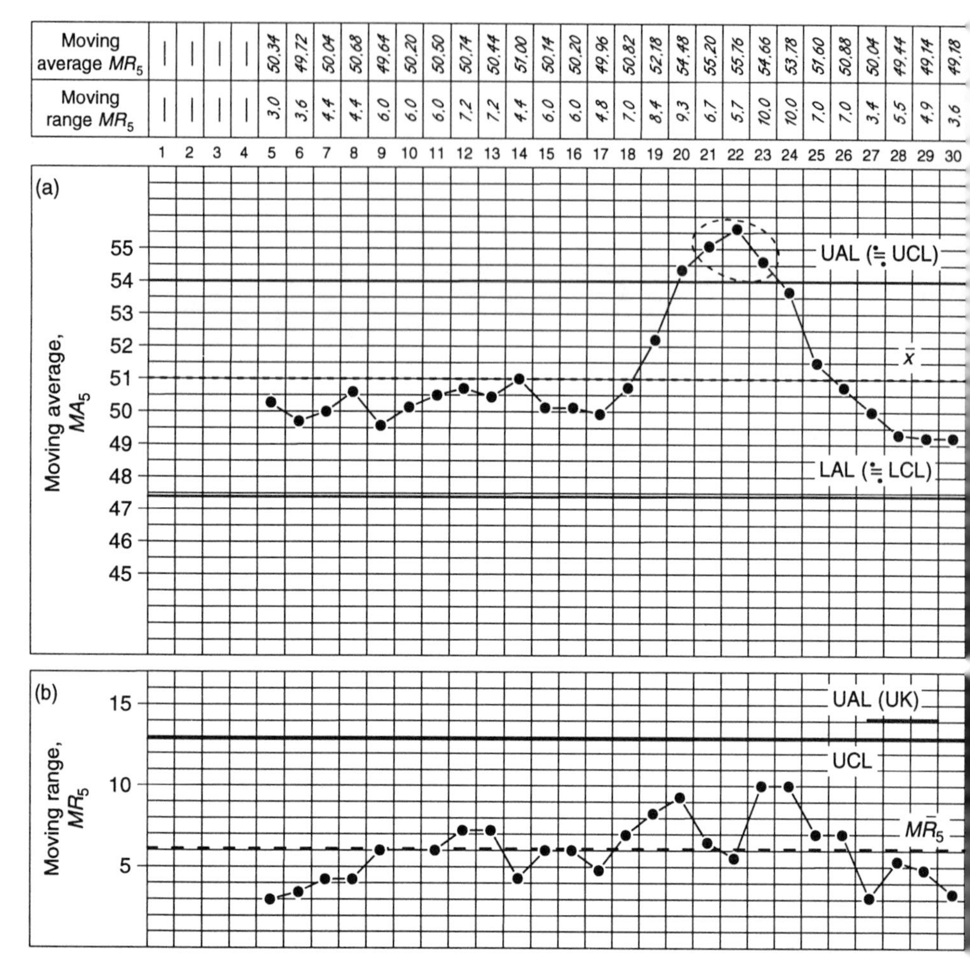

Fig. 9.2 Moving average (MR_5) and moving range (MR_5) charts based on data of Table 9.2. (a) Moving average, MA_5; (b) moving range, MR_5.

For tighter control, the alternative limits are:

$$50.99 \pm 2.58 \times 2.504 \times \sqrt{(0.35/1.65)} = 53.97,\ 48.01.$$

To illustrate the effect of closing the limits, the latter pair are used in Fig. 9.3.

9.7.4 The indications from the trial control charts are as follows:

(i) x-chart

There is one point (20) above the upper control limit, obviously part of a disturbance to the process around 18–22. This is confirmed, using UK

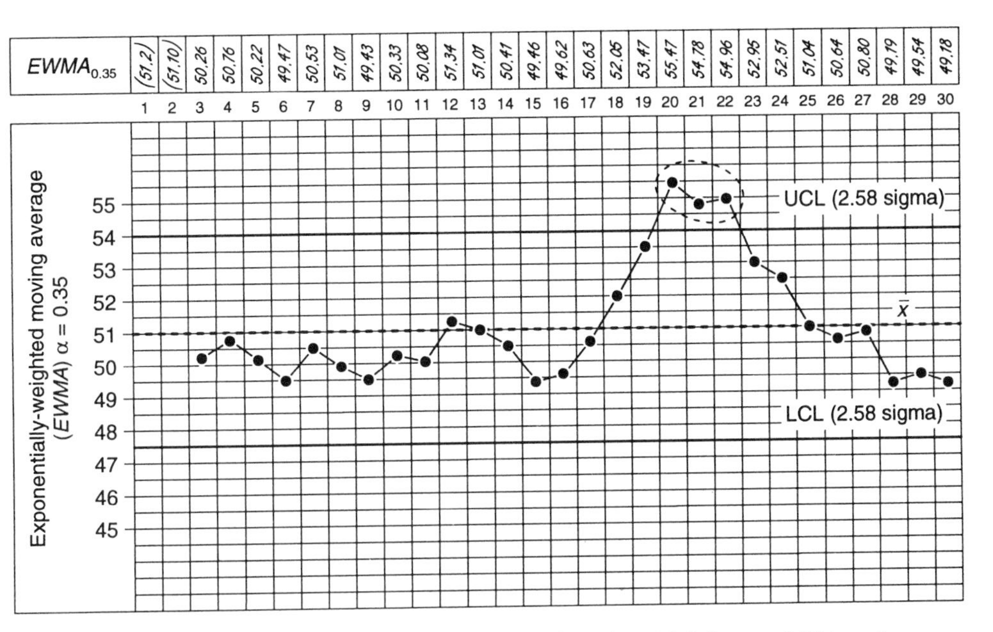

Fig. 9.3 Exponentially-weighted moving average chart ($EWMA$) for $\alpha = 0.35$ based on data of Table 9.2. (Tightened control limits: 2.58 sigma).

limits, by a warning value at 19 and a near miss at 22. There is a possible tendency towards values below the centre-line towards the end (23–30), but in fact they are close to the target value of 50; the centre-line has been displaced upward by the disturbance noted above.

(ii) δ-chart
This gives no rise for concern. A run of six values below the centre-line is typical of $\delta(MR_2)$ charts. No values violate either the US or UK control limits.

(iii) MA_5 chart
Irrespective of whether the US or UK practice is adopted, the MA chart shows the disturbance by continuous violation of UCL (or UAL) from points 20 to 23, with points 19 to 24 also associated. Although there is a run of nine initial points below the centre-line we require at least $7 + 5 = 12$ for an MA_5 chart; in any case, these values are close to the nominal value of 50, the displacement of the centre-line again producing this run.

(iv) MR_5 chart
No MR values violate (and only two approach) the control limit.

(v) $EWMA$ chart
This chart shows similar indications to MA_5, but it may be noted that the response to the disturbance is rather sharper – a clearer violation of UCL at point 20 and a quicker fall (towards 50) from points 23 to 25. This would

occur with either 3σ or 2.58σ limits. The latter, as used in Fig. 9.3, give a clear-cut rather than a marginal indication.

9.7.5 If we assume that the cause of the disturbance is identified and removed, we might delete values 18 to 22 and re-estimate the basic parameters. Bearing in mind that this also means re-calculating one of the δ's (the difference between x_{17} and x_{23} is required) and four MR_5 values, we have:

$\bar{x}$ (for 25 values) $= 50.036$;
$\bar{\delta} = 2.704$, whence $\hat{\sigma} = 2.397$; or $M\bar{R}_5 = 4.97$; $\hat{\sigma} = 2.137$.

(Note that $M\bar{R}_5$ now gives a **smaller** estimate (than $\bar{\delta}$) of $\hat{\sigma}$ after removal of the disturbance).

9.7.6 On-going control limits can now be calculated, and in many applications these could be centred on the nominal value of 50; $\bar{x}$ is in any case very close to this nominal. Proceeding as in section 9.7.3, but with the revised estimates, we have:

(i) x-chart	Action lines	57.19, 42.81
	Warning lines	54.79, 45.21
(ii) δ-chart	UK limit	11.14
	US limit	8.84
(iii) MA-chart (using revised $M\bar{R}_5$)		52.87, 47.13
(iv) MR-chart	UK limit	11.63
	US limit	10.51
(v) $EWMA$-chart	'3-sigma' limits	53.31, 46.69
	'2.58 sigma' limits	52.85, 47.15

Re-checking the trial control charts against these limits reveals no further irregularities.

9.7.7 Two further segments of data are shown plotted on the on-going charts to illustrate the response of the various procedures. The first twenty values incorporate a gradual downward drift. This is detected on the x-chart (Fig. 9.4(a)) by a run of seven values below 50 in samples 10 to 16. The MA_5 (Fig. 9.5(a)) and $EWMA$ (Fig. 9.6) charts produce runs **beginning** rather earlier, at sample 3 for MA_5 and sample 9 for $EWMA$. However, as we require $7 + 5 = 12$ points on MA_5, the signal appears at sample 14; for $EWMA$, a run of ten points signals at sample 17. With the tightened limit, a violation of LCL would occur at samples 18 and 20. Violation of a 3σ LCL would not quite occur on the $EWMA$ chart though it does so on MA_5 at sample 20. (We should not forget that a tighter limit can also be used with MA_5 if preferred.)

The second segment, after an imagined re-adjustment of the target value, includes an abrupt shift to a mean value of 52, i.e. a change of about one standard

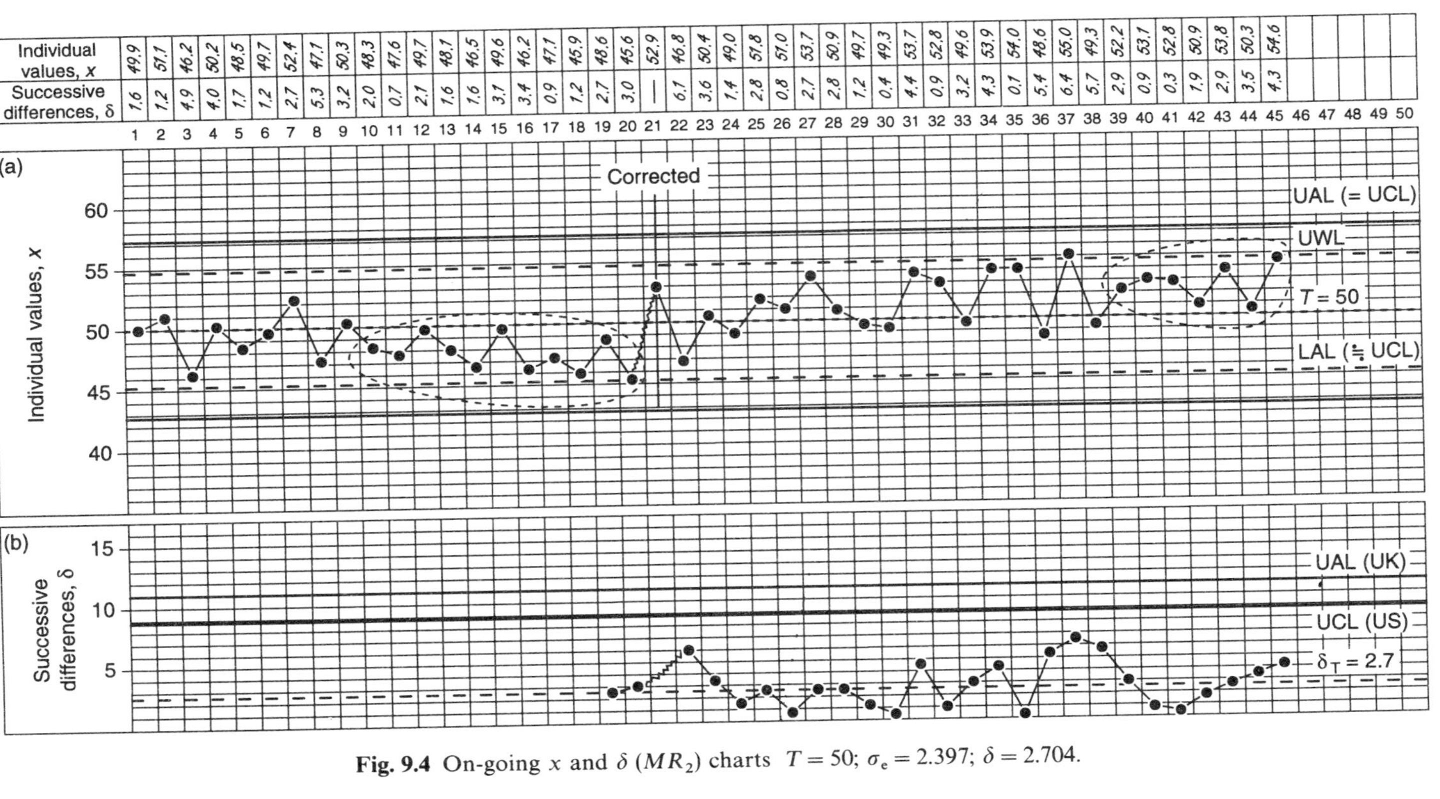

Fig. 9.4 On-going x and δ (MR_2) charts $T = 50$; $\sigma_e = 2.397$; $\delta = 2.704$.

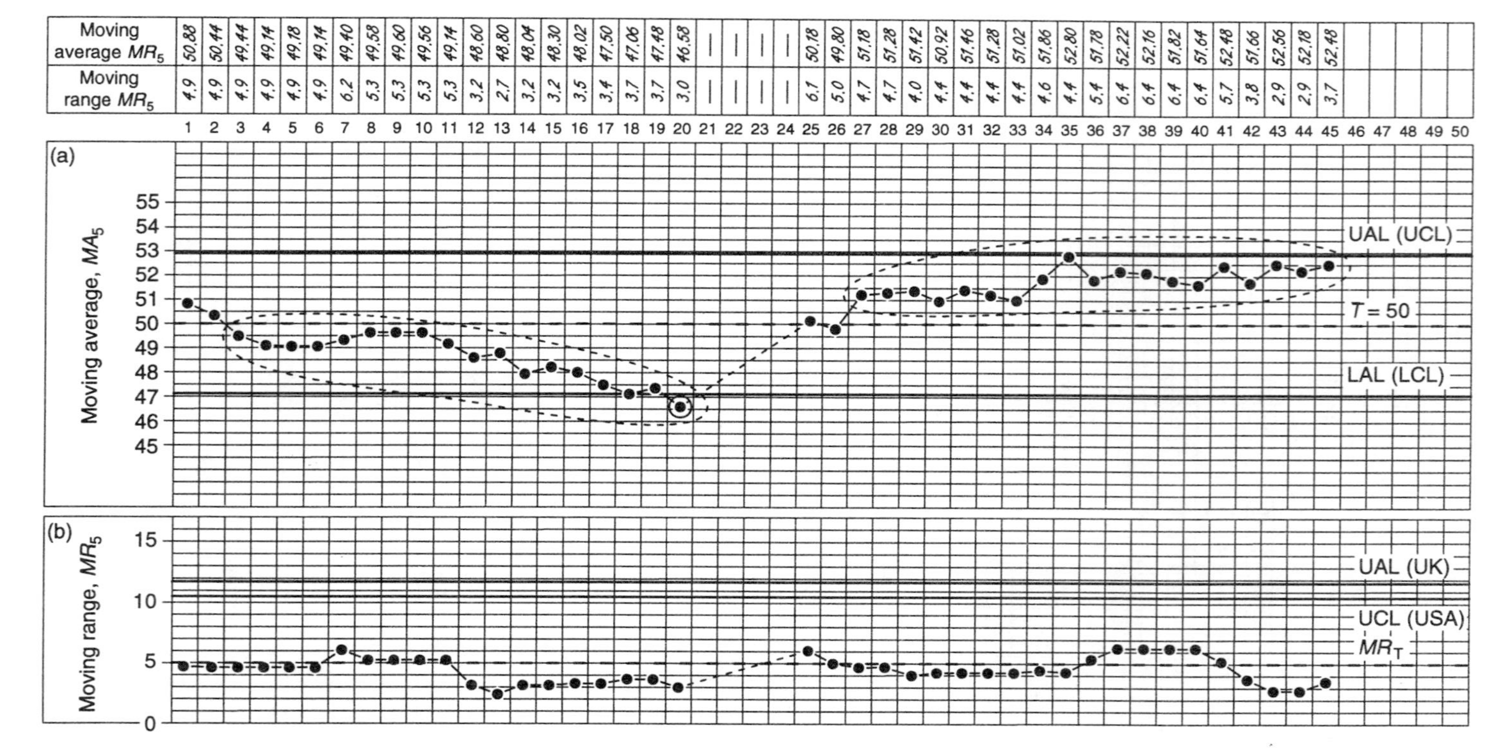

Fig. 9.5 On-going MA_5 and MR_5 charts. $T = 50$; $MR_5 = 4.971$.

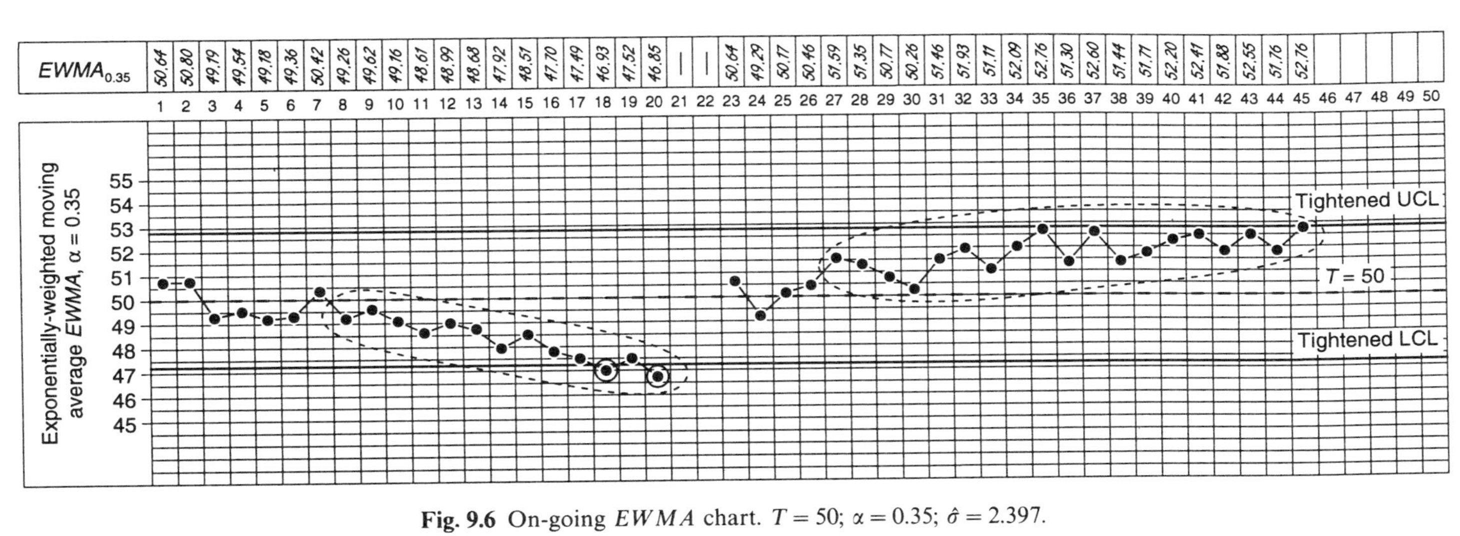

Fig. 9.6 On-going *EWMA* chart. $T = 50$; $\alpha = 0.35$; $\hat{\sigma} = 2.397$.

deviation. This is signalled as follows: x-chart by a run of seven at x_{45}; by the MA_5 at x_{38} with a run of 12, and an obvious drift towards the control limit; and by $EWMA$ at x_{34} (ten successive values above nominal), with the points lying consistently close to UCL from there onwards.

The δ (Fig. 9.4(b)) and MR_5 (Fig. 9.5(b)) charts indicate no irregularities, and are typical of their kind.

9.8 Summary of x, δ, MA, MR and $EWMA$ procedures

9.8.1 This section collects the main pairings of charts together, and lists the factors required for obtaining $\hat{\sigma}$ from $\bar{\delta}$ or $M\bar{R}$, and for calculating control limits.

First, the short term standard error, $\hat{\sigma}$, may be estimated in either of two ways from the succesive differences: Using the mean successive difference,

$$\hat{\sigma} = \bar{\delta}/1.128 \text{ (or } 8\bar{\delta}/9 \text{ approximately)}$$

Alternatively, and with greater efficiency,

$$\hat{\sigma} = \sqrt{(\tfrac{1}{2}MSSD)} = \sqrt{\left[\frac{1}{2(n-1)}\sum_{i=2}^{n}(x_{i+1}-x_i)^2\right]}$$

9.8.2 For individual value and successive difference charts (x, δ or x, MR_2) the

Table 9.4 Factors for *EWMA* charts

Weighting factor α	Nearest k-equivalent	Factors based on $\hat{\sigma}$		Factors based on $\bar{\sigma}$	
		'3 sigma'	'2.58 sigma'	'3 sigma'	'2.58 sigma'
0.6	2	1.96	1.69	1.74	1.50
0.5	3	1.73	1.49	1.54	1.32
0.45	3	1.62	1.39	1.43	1.23
0.4	4	1.50	1.29	1.33	1.14
0.35	5	1.38	1.19	1.225	1.05
0.3	6	1.26	1.08	1.12	0.961
0.25	7	1.13	0.975	1.005	0.864
0.2	9	1.00	0.860	0.887	0.762
0.175	10	0.929	0.799	0.824	0.708
0.15	12	0.854	0.735	0.757	0.651
0.125	15	0.775	0.666	0.687	0.591
0.1	19	0.688	0.592	0.610	0.525

limits are obtained as follows:

(a) For x, UCL, LCL or UAL, LAL $= T \pm 3\hat{\sigma}$
UWL, LWL (where required) $= T \pm 2\hat{\sigma}$
Alternatively, UCL, LCL (UAL, LAL) $= T \pm 2.66\bar{\delta}$
UWL, LWL $= T \pm 1.77\bar{\delta}$

(b) For δ, UK limit $= 4.12\bar{\delta}$
UK limit $= 3.27\bar{\delta}$

9.8.3 For *MA* and *EWMA* charts, control limit factors depending on k (for *MA*) and α (for *EWMA*) are listed in Tables 9.3 and 9.4.

10

Control charts for attributes and events

10.1 Introduction

10.1.1 This chapter deals with control charts for **counted** data. As indicated in Chapter 4, '**attributes**' implies the assigning of items into two or more categories. For control charts, the two categories are **conforming** and **non-conforming** to some standard or requirement. The terms effective and defective are occasionally used, but are best avoided especially when non-conforming items are not necessarily defective in the sense of failure to function (e.g. appearance or aesthetic imperfections, or perhaps even where the application involves a division into home or export orders, neither of which is in any way defective!).

10.1.2 '**Events**' are linked to occurrences such as faults, blemishes, accidents, stoppages. If these are regarded as non-conformities, some distinctions may be noted as compared with attributes (e.g. non-conforming **items**).

(i) A non-conforming item may have two or more non-conformities. Here, interest may centre on the rate of occurrence of the non-conformities, rather than the proportion of non-conforming items
(ii) 'Conformities' are often not observable as actual occurrences (i.e. what is a non-stoppage or a non-blemish?)
(iii) The subgroup for attributes necessarily comprises a number of 'items'; for events, the subgroup may be a quantity of material or a dimension such as time, length, etc. Examples are the number of faults in a given length of wire, the number of stoppages in an eight-hour production run, etc. The terms 'area of opportunity' or 'element size' are sometimes used to point up this distinction.

10.1.3 Both 'attribute' and 'event' situations may arise in conjunction with either constant or varying subgroup sizes. For attributes, the number of items may vary from one subgroup to another, especially where the subgroups represent complete production batches which are 100% inspected. In the case of events, the number of items on which non-conformities are counted may similarly vary, or the quantity of material (length, weight, area, volume, etc.) may fluctuate.

10.1.4 When subgroup sizes are constant, the actual **numbers** of non-conforming items or non-conformities provide the basis for charting. For varying subgroup sizes, it is necessary to derive the **proportion of items** non-conforming, or the **rate of occurrences** per given quantity, in order to provide valid assessment and comparisons.

Thus four main subdivisions are recognised, depending on whether attributes or events are counted, and whether subgroup sizes are constant or variable. When observations are based on 100% checking of variable production quantities, it is necessary to base comparisons on percentages, rates per unit quantity, etc., whereas for uniform subgroup sizes the numbers of non-conformities themselves provide an adequate and simple basis for comparison and control. We thus have the classification in Table 10.1.

10.1.5 The general principles of applying control charts to attributes or events are similar to those for measurements. A base level, representing satisfactory stable performance (with only inherent variation present and no effects from special causes) is established. When in statistical control, the average level and the variation between subgroups are closely related, and it is generally necessary to operate only one chart. Out-of-control conditions, whether due to change in average level of occurrence, to excess variability, or both, will be signalled by differing patterns on the same basic chart.

Once the base level has been estimated, control limits are calculated (although using different parameters from those applicable to measurements), and the limits are used both for checking the in-control state of the base data and for subsequent monitoring. Regular reporting, updating and the maintenance of the process log are as important for attributes applications as they are for measurements.

Table 10.1 Classification of control charts for attributes and events

	Subgroup size	
Type of data	Uniform	Variable
Attributes	'*np*' charts, using actual counts of non-conforming items	'*p*' charts. Divide number of non-conforming items by subgroup size to yield '*p*' value.
Events	'*c*' charts, using actual counts of events or non-conformities	'*u*' charts. Divide number of events by subgroup size to yield '*u*' value.

Examples:
np Number of test failures in regular samples of fifty on every occasion.
p Proportion of faulty items in production lots of varying sizes.
c Numbers of surface defects in units all of uniform size and/or quantity.
u Rate of occurrence (e.g. per running hour) of stoppages on a machine or in a system.

Table 10.2 Notation and terminology for attributes and events

	Attributes		Events	
Subgroup sizes	Constant	Varying	Constant	Varying
Chart identity	np	p	c	u
Subgroup nature (or area of opportunity)	n items	n items	n items or units of quantity	n items or units of quantity
Value plotted for each subgroup	np Number non-conforming	p Proportion non-conforming $= np \div n$	c Number of non-conformities	u Rate of non-conformities per unit $= c \div n$
Chart centreline	$n\bar{p} = \sum np \div k^*$	$\bar{p} = \sum np \div \sum n$	$\bar{c} = \sum c \div k^*$	$\bar{u} = \sum c \div \sum n$

$^*k$ = Number of subgroups included in evaluation study or completed control chart.

10.1.6 For measurements, the usual aim for process improvement is to reduce variation around an optimum or target level. In attribute/event applications concerned with product quality, the occurrences usually represent undesirable situations (rejection, impurity, disruption) and the object is to continuously reduce the rate of occurrences towards an ultimate zero.

For the four types of application the terminology and notation of Table 10.2 is used. In fact, it is possible to group the applications into two categories:

1. The number non-conforming (np) or number of non-conformities (c) charts.
2. The proportion non-conforming (p) or rate of non-conformities (u) charts.

The procedures for these two categories are now set out step-by-step, as for the $\bar{x}$, R and $\bar{x}$, s charts.

10.2 Practical guide to control charts for attributes and events

Step 1: Obtain evaluation data

This will comprise **counts** of the number of non-conforming items (np) or the number of non-conformities (c) in each of at least 20 subgroups. For np or c applications, *the subgroups must all be of the same size or quantity.*

The subgroup size (n) should be large enough to contain, on average, at least one and preferably two or more non-conforming items or non-conformities. The larger the subgroups, the more effective is the discrimination between good and off-standard performance, but one must note that practical constraints (cost, batch sizes, etc.) are often the limiting factor.

Step 2: Calculate sample statistics

For any subgroup, the 'statistic' is the number of non-conforming items or non-conformities, i.e. np or c, or the proportion or rate of occurrence,

$$p = \frac{np}{n} \quad \text{or} \quad u = \frac{c}{n}$$

For u charts, n may represent a number of quantity or time units rather than a number of items.

Step 3: Estimate parameters

The 'parameters' will be as follows for the various types of chart:

$$n\bar{p} = \frac{\text{Total non-conforming items}}{\text{No. of subgroups}} = \frac{\sum np}{k}$$

$$\bar{c} = \frac{\text{Total non-conformities}}{\text{No. of subgroups}} = \frac{\sum c}{k}$$

$$\bar{p} = \frac{\text{Total non-conforming items}}{\text{Total of subgroup sizes}} = \frac{\sum np}{\sum n}$$

$$\bar{u} = \frac{\text{Total non-conformities}}{\text{Total of subgroup sizes}} = \frac{\sum c}{\sum n}$$

Note that in UK literature, $\bar{c}$ is often denoted by m.

Step 4: Calculate trial control limits

We defer details of this step for a discussion of the statistical principles underlying the control limits.

Step 5: Scale the control chart

As for other charts, the scale must accommodate the limits with some room to spare. In most cases, it is reasonable to scale the lowest line on the chart as zero.

Step 6: Draw the control lines on the chart

The same convention as for other charts is preferred:

UAL and LAL	solid lines
UWL and LWL (where used)	broken lines
$n\bar{p}$, $\bar{c}$, $\bar{p}$, $\bar{u}$ (i.e. temporary target)	dotted line

Step 7: Plot the evaluation data

Plot the sample statistics for each subgroup, and join each point to its neighbours.

Step 8: Assess whether or not the process is in control

This step is similar to that for control charts for measurements, but we note some special points.

(i) Out-of-control values may arise from changes of standards of inspection or unintentional variations in subgroup size, as well as real changes in the process.

(ii) Violations of UAL, UWL or runs above the average level almost always indicate adverse effects, requiring prompt attention.

(iii) Violations of LAL, LWL or runs below the average, if real, generally indicate local 'good patches'. These should be investigated for their possible process improvement potential.

Step 9: Obtain new evaluation data if necessary

If step 8 reveals special causes, they must be investigated. There may be process problems, variations in assessment standards, or even some subgroups not being of the correct size.

After removal of special causes, and confirmation that the process is in-control, move on to step 10.

Step 10: Determine capability

Note that for attribute and event applications, performance has to be expressed in terms of proportions of non-conforming items, fault rates, breakdowns per unit time, or similar measures. Often there are no specified criteria, and the extent to which the process is judged capable may be a management decision based on circumstances. Chapter 13 deals with methods of capability evaluation.

Step 11: Consider adjustment of process setting

Unlike many measurement applications, the 'process setting' for *np* or *c* cannot be easily changed. However, there are circumstances where, for example, a change in sample size may be considered necessary; thus if the sample size is increased by 50%, $n\bar{p}$ or $\bar{c}$ will have to be scaled up by a factor of 1.5 if the non-conformity **rate** remains unchanged. Such considerations should be included when formulating the **target value** for on-going control.

Step 12: Set up on-going control limits

Using the value of $n\bar{p}$, $\bar{c}$, $\bar{p}$ or $\bar{u}$ selected as a target value in steps 1–11, recalculate the appropriate limits and draw them on the chart for on-going control, as for other charts.

Step 13: Run and monitor the process

Step 14: Periodically update the system

In attribute and event applications, this will require occasional recalculation of $n\bar{p}$, $\bar{c}$, $\bar{p}$ or $\bar{u}$ as performance summaries, and the revision of control limits when appropriate.

10.3 Statistical basis of control limits for attributes and events

10.3.1 We now take up the matter of control limits and their associated decision rules. The first point to consider is that of suitable statistical models for attribute and event situations. Under the assumption of a stable proportion, *p*, of observations in the 'non-conforming' category, and of random sampling from the process with constant sample size, *n*, a binomial distribution of observed numbers of non-conforming items would result for the *np* chart. Similarly, for events (non-conformities) occurring in a continuum, with constant average rate of occurrence, λ, and area of opportunity or sample size *n*, we have a Poisson distribution (with parameter λ).

Some features of the binomial and Poisson distributions were presented in Chapter 4.7. Except for fairly large values of *np* ($\geqslant 20$) or where *p* is close to 0.5 (an unusual value in SPC, though by no means impossible, especially in non-manufacturing applications not concerned with quality features), the binomial distribution is appreciably non-Normal. Similarly, except for large λ, say 20 or more, the Poisson distribution is non-Normal; it is worth stating again that the binomial distribution with mean *np* and the Poisson with $\lambda = np$ are very similar in shape provided that *p* is less than about 0.1

10.3.2 The UK practice for control limits is to find the value of *np* (for attributes) or *c* (for events) that would be exceeded, by chance, with specified low probability (α) for a binomial (n, p) or Poisson (λ) distribution. The calculations are tedious, but results can be tabulated or presented graphically, especially when the binomial can be approximated by a Poisson distribution, thus involving only one parameter.

British Standards 2564 and 5701 give details of the method and the control limits. Unfortunately lower control limits are not given, partly due to problems of presentation but also because these standards pre-date the SPC philosophy of seeking to identify opportunities for **improvement**. Additionally, the method of obtaining warning limits in BS 5701 does not lend itself to development for two-sided control. This problem is addressed in section 10.4.

10.3.3 US practice is to ignore the non-Normality of the binomial and Poisson distributions and to adopt 3-sigma control limits on either side of the target level of non-conforming items or non-conformities. This works reasonably well down to np or λ about 8 or 9 but then the lower control limit becomes negative and is discarded. The upper control limit cuts off rather more than 0.135% in the upper tail of the target distribution: as much as 1.5% if $\bar{c}$ or $n\bar{p}$ is just below 1.0, or even more for smaller $\bar{c}$ or $n\bar{p}$. This results in a tendency to false alarms and over-control.

The standard deviations of the binomial and Poisson distributions were noted in Chapter 4.7. To recapitulate, for binomial (n, p) we have

$$\text{Mean} = np, \quad \sigma = \sqrt{[np(1-p)]}$$

For Poisson (λ), the corresponding parameters are

$$\text{Mean} = \lambda, \quad \sigma = \sqrt{\lambda}$$

10.3.4 From the initial process study, $\bar{p}$ or $n\bar{p}$ are obtained from data with the process in-control, and for the US convention, n, $\bar{p}$ are then substituted into (10.1) to obtain the 3-sigma limits,

$$\text{UCL, LCL} = n\bar{p} \pm 3 \times \sqrt{[n\bar{p}(1-\bar{p})]} \qquad (10.1)$$

(discard LCL if negative). (If warning limits are required, the factor 2 is substituted for 3 in (10.1)).

Similarly for the Poisson distribution, the parameter λ is estimated from the available data by $\bar{c}$, and substituted for λ to give

$$\text{UCL, LCL} = \bar{c} \pm 3\sqrt{\bar{c}} \qquad (10.2)$$

Again for warning limits, use 2 in place of 3 in (10.2).

10.3.5 A similar approach is used for US practice in the case of p and u charts. For any particular sample size n, the standard error of $\hat{p}(=np/n)$ is given by

$$\sigma_{\hat{p}} = \sqrt{\left(\frac{p(1-p)}{n}\right)}$$

Here it is necessary not only to substitute the value of $\bar{p}$ but also an average, notional or target sample size. The approximate average, $\bar{n}$, of the sample sizes in the evaluation data is usually adopted, but any rounded or notional value close to what is likely to occur in routine operation may be used. The 3-sigma limits are then

$$\text{UCL, LCL} = \bar{p} \pm 3 \times \sqrt{\left(\frac{\bar{p}(1-\bar{p})}{\bar{n}}\right)} \tag{10.3}$$

The limits in (10.3) are usually regarded as being valid for sample sizes within about 25% either way from $\bar{n}$ (the limits are then about 12–13% too wide (or too narrow) for samples 25% smaller (or larger) than $\bar{n}$, respectively). The procedure for wider variations in n is to calculate special limits, using the same (target) value of $\bar{p}$ in (10.3) but inserting the actual value of n. An example is given at the end of this chapter. One may note that there are also techniques for either standardizing the values of p, using

$$z = (p - \bar{p}) \sqrt{\left(\frac{n}{p(1-\bar{p})}\right)}$$

or making small adjustments to the value of p to re-space it between $\bar{p}$ and the control limit. Both techniques have the disadvantage of losing direct contact with the observed p-values, and hence some loss of facility in interpreting the chart.

10.3.6 The corresponding approach for the u-chart is again to assume an average sample size, $\bar{n}$, and since the standard error of u is

$$\sigma_{\hat{u}} = \sqrt{\left(\frac{U}{n}\right)}$$

the expression for 3-sigma limits becomes

$$\text{UCL, LCL} = \bar{u} \pm 3 \times \sqrt{\left(\frac{\bar{u}}{\bar{n}}\right)} \tag{10.4}$$

If warning lines are required for either p or u charts, the factor 2 replaces 3 in (10.3) or (10.4).

10.4 'Exact' probability limits for *np*, *cp* and *u*

10.4.1 As noted in section 10.3.3 the normal approximations work reasonably well for mean numbers of occurrences of 10 or more per subgroup. For lower rates of occurrence, limits based on percentage points of the binomial or Poisson distributions may be preferred, especially if lower limits are considered useful. Figure 10.1 provides limits based on upper and lower 0.1%, 0.2%, 0.5%, 1%, 2.5% and 5% points. The method of using this chart is described in section 10.5.

Because the binomial and Poisson distributions are concerned only with integer values (i.e. counts of items or occurrences), it is rarely possible to define an exact limit which gives a precise probability of violation. The general practice is to use an upper limit, say x, such that the cumulative probability for all values up to and including x is at least $1-\alpha$. Often the limit is expressed with a decimal fraction, e.g. 4.5 (or in BS 5791, 4.7), to avoid ambiguity; in this example a count of four occurrences would lie inside the control limit, but five occurrences would constitute a violation and signal out-of-control. In Fig. 10.1 we use limits of the type $x.5$ for simplicity.

Similarly for lower limits, we require the smallest integer such that the cumulative lower tail probability does not exceed α, again using the style $x.5$ to avoid ambiguity.

10.4.2 By analogy with control charts for measurements and with US practice with 3-sigma limits for all control charts, it would seem that $\alpha=0.001$ would be a logical choice for action lines, and $\alpha=0.025$ for warning lines of required. However, it is widely recognized that counted data is less effective than measurements in detecting changes. To gain some sensitivity the in-control average run length is often reduced by using larger values of α. BS 2544 and 5701 use $\alpha=0.005$ (1 in 200) for action limits. Warning lines are not often used with such limits, though in BS 5701 they are, if required, set one unit inside the action line with an interpretation rule more complicated than a simple two-in-a-row or two-in-any-three.

In order to make a reasonable choice of α for action (A) and warning (W) lines, it is worth considering some *ARL* data. For single-sided control with action line only, the *ARL* is given simply by the reciprocal of the probability of violating the control limit, i.e. $1/P_A$. When in-control $P_A \doteqdot \alpha$, so $L_0 \doteqdot 1/\alpha$. For action and warning lines, with probabilities P_A and P_W of violation,

$$ARL = (1 + P_W - P_A)/\{P_A + P_W(P_W - P_A)\}$$

for the two-in-a-row warning rule. For a two-in-any-three rule, and especially when combined with runs rules, the expressions for *ARL* become more complicated, but Table 10.3 provides some approximate values of L_0 in typical cases.

Obviously the larger α-values combined with runs rules are likely to give unsatisfactory false alaram rates (overcontrol). We suggest the following:

(i) Where two sided control with A, W and runs rules is required (to approximate the performance of the US chart with its multiple rules):

use $\alpha = 0.001$ for A, $\alpha = 0.025$ for W, runs of 7 or 8.

(ii) For one-sided control (including the case where no lower control limits are possible) or where extra sensitivity is required:

use $\alpha = 0.005$ for A with *either* $\alpha = 0.05$ for W *or* a runs-of-7 rule (but not both).

(iii) A further useful possibility (with or without runs rules) is $\alpha = 0.002$ for A with 0.025 for W. This gives $L_0 = 400$ without runs rules for single-sided

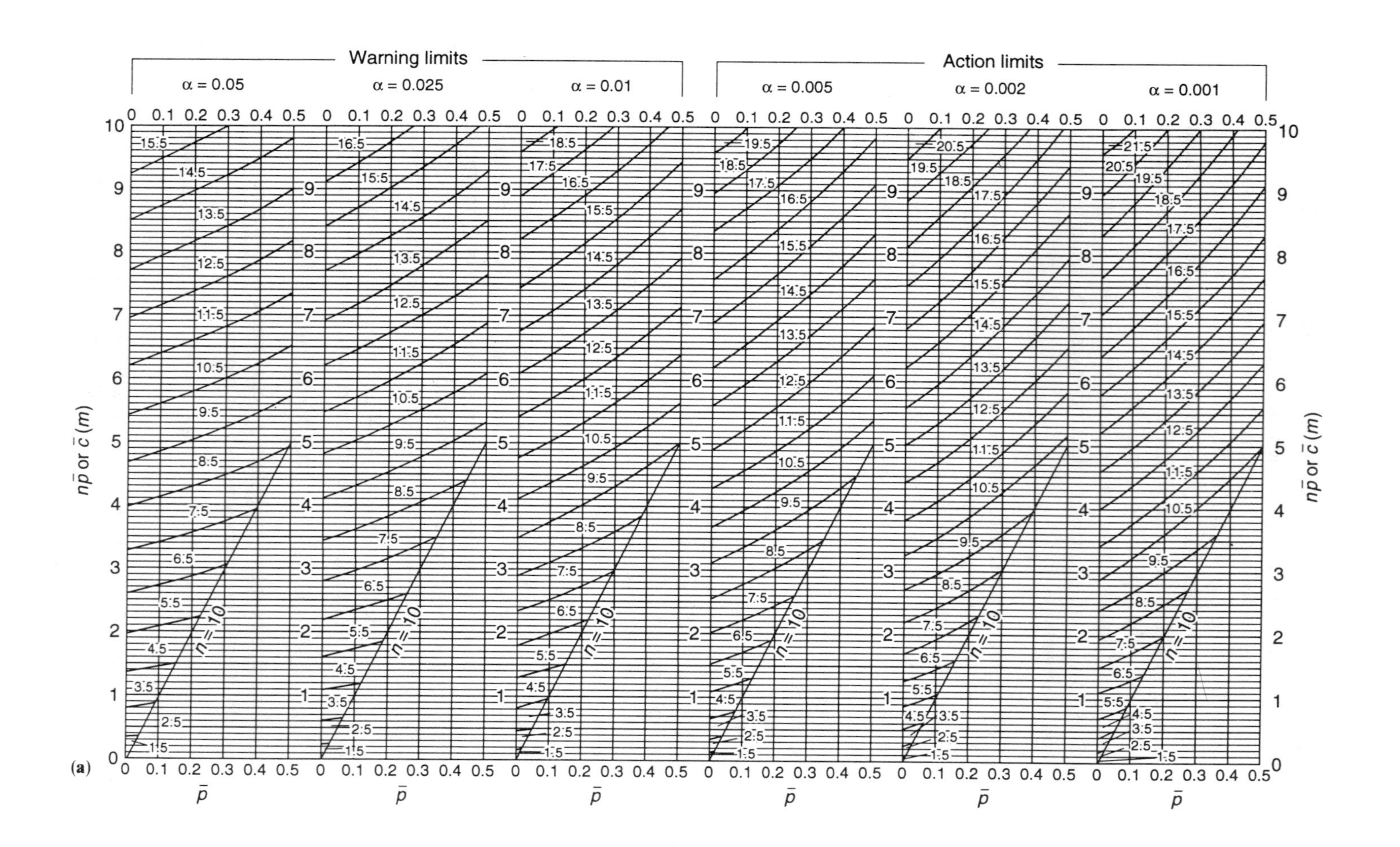

Warning limits
Action limits
α = 0.05
α = 0.025
α = 0.01
α = 0.005
α = 0.002
α = 0.001
$n\bar{p}$ or $\bar{c}$ (m)
$\bar{p}$
$n = 10$
(a)

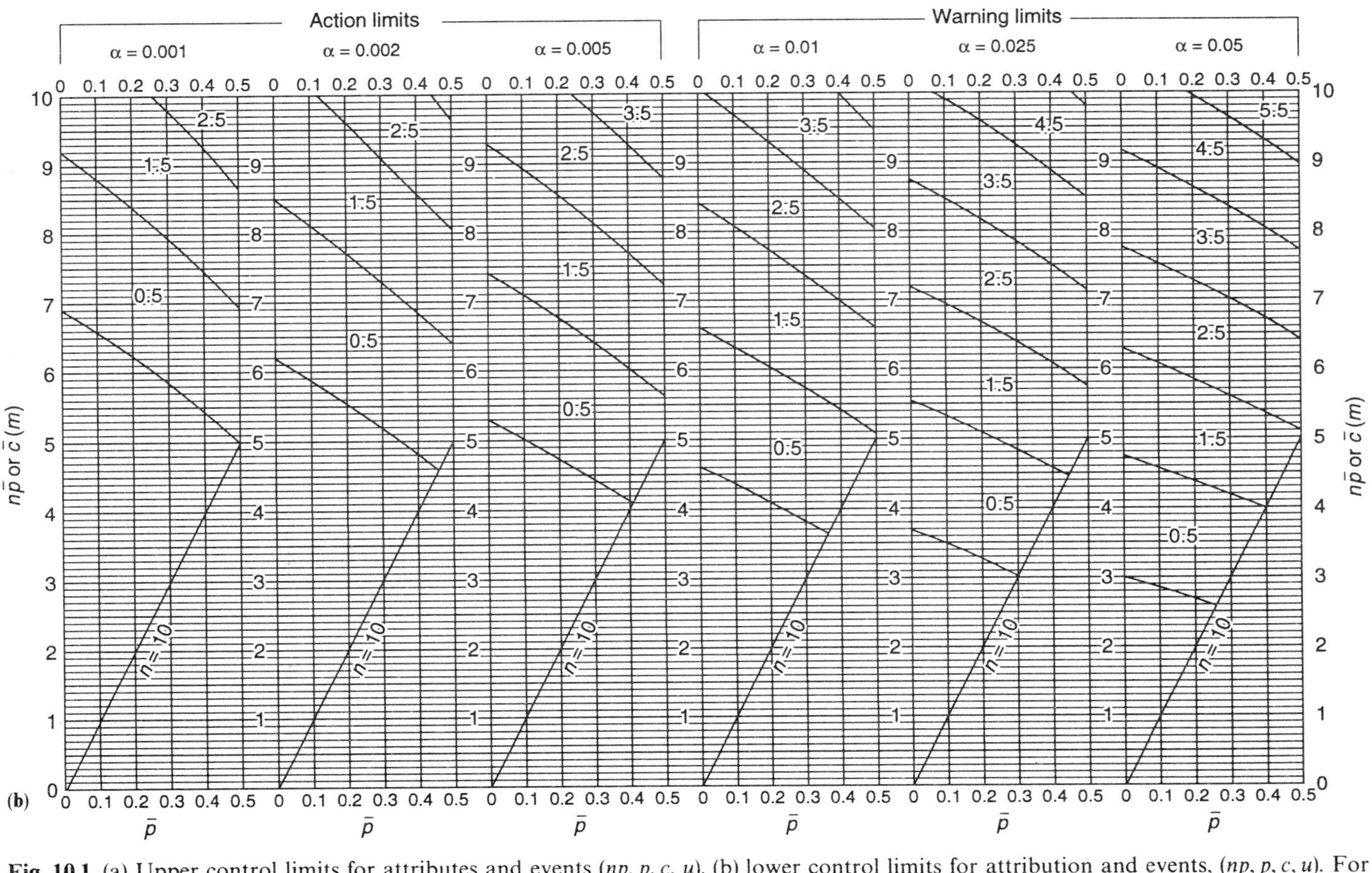

Fig. 10.1 (a) Upper control limits for attributes and events (np, p, c, u). (b) lower control limits for attribution and events, (np, p, c, u). For events (c, u) use $p = 0$; for $n < 10$ obtain limits from tables or by calculation.

Table 10.3 ARL implications of some control rules

Minimum values of L_0, average run length for in-control process

Control rules	One-sided control	Two-sided control
A only, $\alpha = 0.001$	1000	500
A, $\alpha = 0.001$; W, $\alpha = 0.025$	640	320
A, $\alpha = 0.001$; W, $\alpha = 0.025$; Runs of 7	≈ 200	≈ 100
A only, $\alpha = 0.005$	200	100
A, $\alpha = 0.005$; W, $\alpha = 0.05$	144	72
A, $\alpha = 0.005$; W, $\alpha = 0.05$; Runs of 7	≈ 100	≈ 50

control, and around 180 with a runs-of-7 rule. The values would be halved for two-sided control.

All of the above L_0 are minima, when the probabilities of the in-control process violating limits exactly correspond to the α-values. Because these probabilities are generally smaller than α (due to the discrete nature of the numbers of occurrences), L_0 will generally be larger. For example, with $\bar{c} = 2$ and combination (i) above, L_0 without a runs rule becomes about 2000 instead of 400; with runs rule it will reduce to about 220 instead of 180.

Figure 10.1 provides a range of control limits for various α, and we now turn to the use of this intercept chart.

10.5 Use of the intercept charts

10.5.1 The np chart

Values of $n\bar{p}$ and $\bar{p}$ are required. Locate $n\bar{p}$ on the vertical scales, find the intersection with the value of $\bar{p}$ in the appropriate section of Fig. 1, and simply read off LAL, LWL, UWL and LAL at the required percentage points.

10.5.2 The p chart

The values of $\bar{p}$ and a target or average sample size, say $\acute{n}$, are required.

Multiply $\dot{n}$ by $\bar{p}$ to obtain $\dot{n}\bar{p}$. Using this value, proceed as in section 10.5.1, but finally divide the resulting limits by $\dot{n}$ to obtain those relevant to the p chart.

For any unusally large or small values of n in individual subgroups or over a period, substitute the appropriate subgroup size, n for $\dot{n}$ in this procedure to obtain 'locally applicable' limits.

10.5.3 The c chart

Enter with $\bar{c}$ on the vertical scale, using $\bar{p}=0$ in Fig. 1. Read off the appropriate percentage points for LAL, LWL, UWL and UAL.

10.5.4 The u-chart

Values of $\bar{u}$ and the target sample size ($\dot{n}$) are required. Set $\dot{c}=\dot{n}\bar{u}$ and obtain limits as for c-charts in section 10.5.3. Finally divide each limit by $\dot{n}$ to obtain those relevant to the u-chart.

For sample sizes outside $\dot{n} \pm 25\%$, the procedure is to use the actual sample size, n, in this calculation in place of $\dot{n}$ (as in the case of p-charts).

10.6 Examples and illustrations

10.6.1 The first example (for a p-chart) comprises a complete set of evaluation data, with some subsequent re-assessment. Brief examples are then given of the calculation of control limits for np, c and u charts.

Table 10.4 lists the sample sizes and numbers of 'non-conforming' items for two years's monthly accounts data. The data comprise records of the numbers of invoices unpaid at the end of the second month after billing (the firm operates a 30 days grace policy). This data was intended to provide a base for subsequent monitoring of the effects of actions such as penalties, reminders, discounts, etc.

10.6.2 The initial estimates obtained from this data are:

$$\sum np = 1810, \quad \sum n = 9617, \quad k = 24;$$

whence $\bar{p}=0.1882$, $\bar{n}=400.71$.

With an average number of late payments (per month) of about 75, the 3-sigma technique for calculating control limits is appropriate, and we have **trial** limits:

UAL, LAL	0.247, 0.130;
UWL, LWL	0.227, 0.149.

Inspection of the trial chart, Fig. 10.2, reveals some interesting features. First, 17 points lie below $\bar{p}$ and only 7 above; of these 7, two lie above UAL and two more between UAL and UWL. One point lies between LWL and LAL. In fact the subgroup sizes for February '90 and '91 are usually large and those for

Table 10.4 Unpaid invoice data for p-chart

Invoice (month/year)	No. of invoices n	Late payments np	Proportion p	Above or below $\bar{p}$
4/89	419	71	0.169	−
5	392	73	0.186	+
6	406	72	0.177	+
7	443	86	0.194	+
8	417	74	0.177	+
9	396	71	0.179	+
10	410	81	0.198	+
11	374	51	0.136	−
12	322	82	0.255	(+)
1/90	385	66	0.171	−
2	511	123	0.241	(+)
3	368	64	0.174	−
4	424	76	0.179	+
5	421	85	0.202	+
6	406	75	0.185	+
7	393	65	0.165	−
8	367	60	0.163	−
9	421	76	0.181	+
10	438	74	0.169	−
11	407	75	0.184	+
12	279	75	0.269	(+)
1/91	374	64	0.171	−
2	498	117	0.235	(+)
3	346	54	0.156	−

The parenthetical (+)'s indicate control limit violations.
$\sum n = 9617$, $\sum np = 1810$, $\bar{p} = 0.1882$, $\bar{n} = 400.7$.

December '89 and '90 unusually small, so it is appropriate to check the control limits appropriate to these subgroup sizes. They are:

12/89	UAL = 0.254;	UWL = 0.232;
2/90	UAL = 0.240,	UWL = 0.222;
12/90	UAL = 0.258,	UWL = 0.235;
7/91	UAL = 0.241,	UWL = 0.223.

Thus the value of p for 12/89 still lies beyond UAL; that for 2/90 now lies just beyond UAL; 12/90 does not now reach UAL but is still beyond UWL; and 2/91 is now closer to, though still inside UAL. The overall picture is of some special cause operating in December and February of both years, but no particular reason was found for the warning value in 11/89.

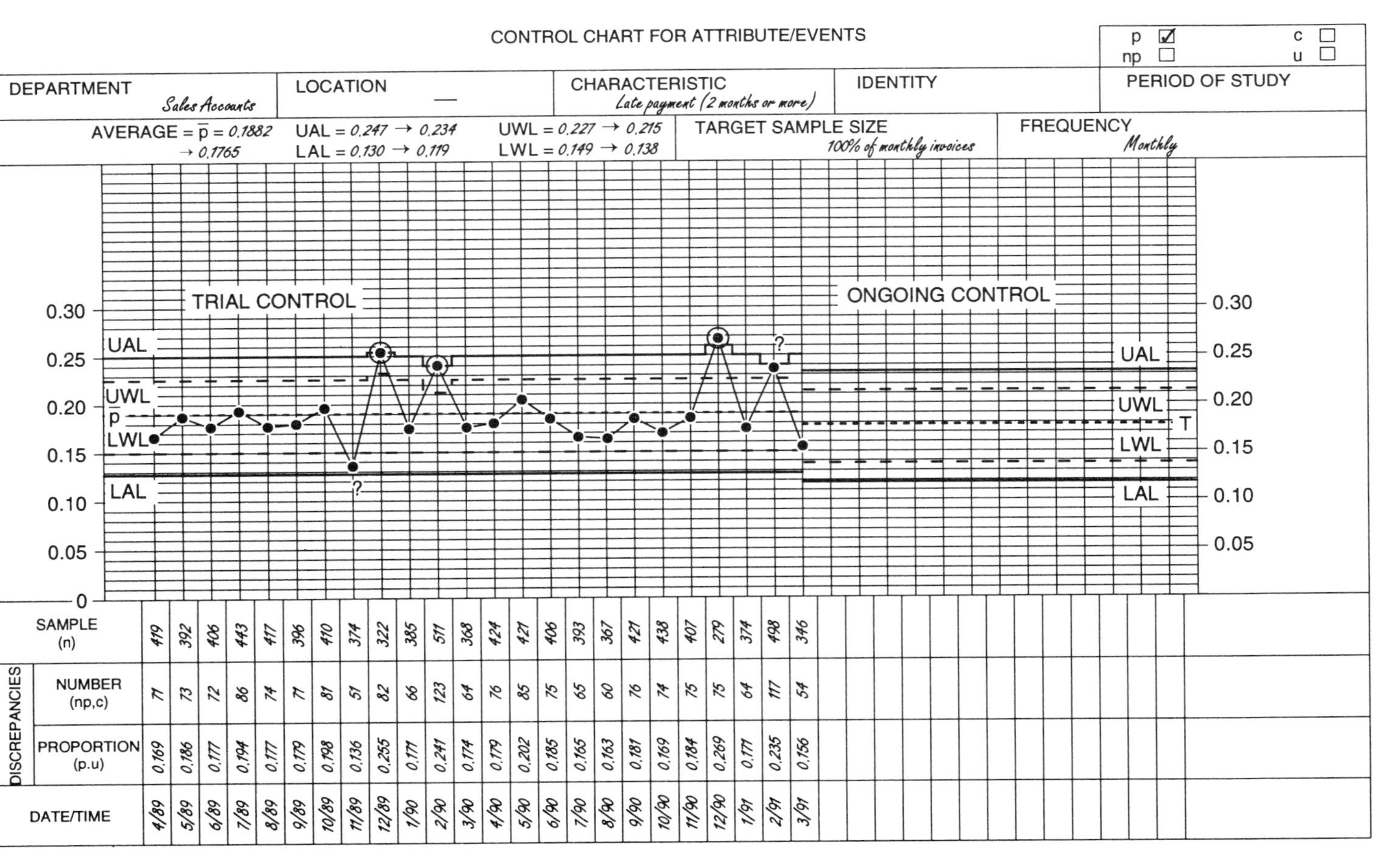

Fig. 10.2 Control chart (p) for invoice payment delays.

10.6.3 The nature of these special causes was not difficult to trace. December is a short working month due to Christmas and New Year holidays, giving rise to smaller numbers of invoices sent out. Postal delays and the aftermath of the holiday means that a proportion of invoices then miss qualifying dates for payment procedures. The large subgroups in each February arise from efforts to invoice as far up to date as possible to improve cash flow just before the end of the financial year, but this is partially thwarted by some customers deferring payment to improve *their* year-end finances! No immediate, short-term actions were proposed to deal with these phenomena, so it was considered appropriate to base the chart on the remaining data so as to monitor the general pattern of payment and identify any overall changes in proportion. This gives revised estimates:

$$\sum np = 1413, \quad \sum n = 8007;$$
$$\bar{p} = 0.1765, \quad \bar{n} = 400.35.$$

For $\dot{n} = 400$,

$$\text{UAL, LAL} = 0.234, 0.119;$$
$$\text{UWL, LWL} = 0.215, 0.138.$$

Re-appraisal of the 1989–1991 data against these limits reveals no outstanding patterns except for the December and February anomalies. There is a run of six values above the revised $\bar{p}$ from 5/'89 to 10/'89, but two of these are virtually on the centre-line. The low p for 11/'89 is now only just below LWL. The on-going control chart is now ready for further monitoring, and for assessing the effect of any policy changes.

10.6.4 An example of the *np*-chart arose in a programme to investigate and reduce packaging errors in a warehouse. Over a six month period, fifty packages per week were scrutinised for correctness of contents, labelling, documentation, etc. The average number of faulty packages per week was 3.8, so we have $n\bar{p} = 3.8$, $n = 50$, whence $\bar{p} = 0.076$.

In this case, $n\bar{p}$ is low enough to use Fig. 10.1. For these values of $n\bar{p}$ and $\bar{p}$, we find:

$$\text{UAL} = 11.5 \text{ for } \alpha = 0.001;$$
$$\text{UWL} = 8.5 \text{ for } \alpha = 0.025;$$
$$\text{LWL} = 0.5 \text{ for } \alpha = 0.025;$$
$$\text{LAL} = \text{Not applicable for } \alpha = 0.001.$$

Thus any sample with twelve or more faulty packages is a clear 'action' signal; two consecutive weeks with nine or more, or with none at all, would also prompt investigation. The runs rule can also be applied with this selection of α-values.

10.6.5 An application in accident reduction and prevention provides an example of the *c*-chart. Over an 18-month period in 1988/89, lost-time accidents averaged 22 per month. Although months vary slightly in length (thus perhaps suggesting a u-chart) the variations are small, and the c-chart was considered preferable on grounds of simplicity. Later, the presentation was changed to a cusum chart, which will be described in Chapter 12, along with some real data. Meanwhile, we note that the control limits, based on the 3-sigma, 2-sigma approach, are:

$$\text{UAL, LAL;}\quad 22 \pm 3\sqrt{22} = 36.1,\ 7.9;$$
$$\text{UWL, LWL;}\quad 22 + 2\sqrt{22} = 31.4,\ 12.6.$$

10.6.6 Final inspection of assembled chassis for electrical appliances revealed occasional faults, most of them fairly easily rectified. All chassis from production quantities of around 180 per day were checked, with a fault occurrence rate of approximately 4.7 per 100 chassis. Two or even more faults are sometimes found within one chassis, so that this is a u-chart rather than a p-chart application. Daily numbers of items vary, generally from 140 to 200, though material shortages, machine breakdowns and absence occasionally result in much smaller output, down to a little as 50 items. We note firstly that the system is in need of drastic improvement, with attention to elimination of the causes of the faults and to the problems yielding shortages and breakdowns.

Subject to these comments, u-chart control limits in the US style would be (with $\bar{u} = 0.047$ per **item**):

$$\begin{aligned} \text{UCL, LCL} &= \bar{u} \pm 3\sqrt{\left(\frac{\bar{u}}{\bar{n}}\right)} \\ &= 0.047 \pm 3\sqrt{\left(\frac{0.047}{180}\right)} \\ &= 0.0955,\ -0.0015 \end{aligned}$$

The lower limit is evidently unusable, so we consider the use of probability limits via Fig. 10.1.

With $\dot{n} = 180$, $\dot{n}\bar{u} = 8.46$, and suppose we choose to use only action lines based on $\alpha = 0.005$, along with a runs-of-seven rule as suggested in section 10.4.2(ii). Setting $\bar{p} = 0$ to obtain the limits, we have UAL $= 17.5$, LAL $= 1.5$. Dividing by $\dot{n}$ yields u-chart limits of 0.0872 (similar to UCL above), but a usable lower limit 0.0083. This means that if 0 or 1 fault is found in a typical day's output of 140–200 chassis, it could be worth investigating the cause.

For a low day's output of 50 items, use 50 in place of 180 for $\dot{n}$ to give $\bar{u}\dot{n} = 2.35$, and read off UAL (for $\alpha = 0.005$) as 7.5. In this case no lower limit is possible, and UAL $\div$ 50 to obtain a limit for u gives 0.15 as the local control limit for this subgroup size.

10.7 The multiple characteristic chart

10.7.1 In fact, section 10.6.6 is an example of the multiple characteristic chart, in that the chassis defects were of several types: loose or missing components, damage, incorrect assembly, etc. So, in fact are the accident data of section 10.6.5 if one notes that the accidents are of different kinds and arise from various causes. The packaging errors of section 10.6.4, yet again, involve several reasons for rejecting the packages, though in this case the chart was used to monitor faulty **packages** rather than packaging **faults**.

When any one item or unit of observation can contain several non-conformities rather than just one, and where the non-conformities can also be of different kinds, one has choices to make about methods for monitoring the system. An obvious but time-consuming possibility is to operate a separate control chart for each type of non-conformity, but a simpler method is usually preferable. The most popular choice is the multiple characteristic approach; all the non-conformities of different types are recorded (for reference and subsequent analysis), but the control chart is based either on the **principal** non-conformity, e.g. the one primarily responsible for rejection of the item, or on the total number of non-conformities of all types. The former gives rise to the *np* or *p*-chart (depending on whether subgroup sizes are constant or variable), the latter to the *c* or *u*-chart.

10.7.2 Some cautions about the use of multiple characteristic charts are worth noting. First, major and minor non-conformities should not be combined in one chart. Critical defects, faults which will cause malfunction, etc., should be handled separately from, say, blemishes which slightly mar appearance but cause no functional problems. Similarly in non-manufacturing applications, accidents causing death or disability should not be combined with other, especially minor, incidents.

Secondly, non-conformities or events which occur much more frequently than others should be separated for special attention, as they will otherwise mask changes in the rate of the less frequent occurrences. Thus if blistered surfaces occur ten times as frequently as end cracks, a doubling in the rate of end cracks may go unnoticed if both defects are combined, unless the subsequent analysis of chart data is scrupulously maintained.

Thirdly, a regular review of the balance of characteristics is essential, by lateral totalling of completed charts. It is not necessary to execute a fullscale Pareto analysis for every chart, but the pattern of relative frequencies of the various characteristics should be monitored, as well as the totals for each column representing a subgroup of observation.

Fourthly, signals from the chart need to be carefully interpreted. Is a change in the overall rate of occurrence of non-conforming items or non-conformities due to an outbreak of one or two characteristics, or to a factor that affects several characteristics? If the average rate *does* change, is the change consistent or are different characteristics prevalent on various occasions of observation?

Finally, when the data comprise varying subgroup sizes, the lateral summaries may require more detailed analysis to take account of these variations before drawing conclusions. Possible effects are, for example, that with large subgroups

CONTROL CHART

Plastic extrusion (1st data collection) 10m cuts

NO	CHARACTERISTIC OR OCCURRENCE	1	2	3	4	5	6	7	8	9	10	11	12	13	14	15	16	17	18	19	20	f	%	
1	Dirty surface	1	1	2	0	3	1	2	3	1	0	2	1	0	2	1	4	0	1	1	1	27	13.7	2
2	Scorch marks	1	1	1	0	0	0	1	1	1	0	1	0	0	0	0	0	0	0	0	1	8	4.1	9
3	Pinches	0	2	3	0	0	2	0	2	0	1	0	0	0	2	1	3	0	0	1	0	17	8.6	5
4	Discoloured	4	1	1	3	2	2	0	0	0	0	1	2	4	0	1	2	1	0	0	2	26	13.2	3
5	Splits	0	0	1	0	1	1	1	0	0	1	0	1	0	0	0	1	0	1	0	0	8	4.1	10
6	Blisters	0	0	0	0	0	0	0	0	0	0	1	0	0	0	0	0	0	0	0	1	2	1.0	11
7	Scores	1	0	0	0	1	0	0	0	1	0	0	1	1	0	2	1	0	3	3	1	15	7.6	6
8	Die lines	1	7	6	3	0	1	0	4	0	4	3	3	0	2	3	3	5	4	0	1	50	25.4	1
9	Dimples	1	0	0	2	1	1	1	1	0	1	1	1	0	0	0	3	0	0	1	0	14	7.1	7
10	Inclusions	0	0	1	3	0	1	1	1	3	1	2	0	0	0	1	1	1	2	0	0	18	9.1	4
11	Others	0	0	0	1	0	1	0	0	1	0	1	1	2	0	0	3	0	0	0	2	12	6.1	8
	Total	9	12	15	12	8	10	6	12	7	8	12	10	7	6	9	21	7	11	6	9	197	100.0	
SAMPLE SIZE	50 (cuts)																							
DATE/TIME		1	2	3	4	5	6	7	8	9	10	11	12	13	14	15	16	17	18	19	20			

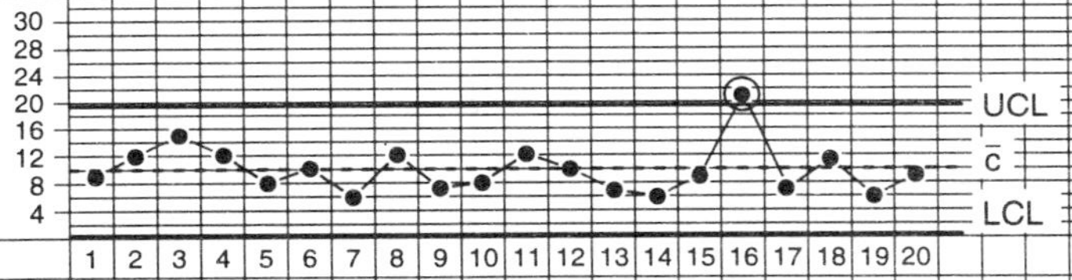

Fig. 10.3 Multiple characteristics: plastic extrusion faults. Excluding o.o.c. point $\bar{c} = 9.3$. Control limits for ongoing chart: US limits: UCL = 18.4, LCL = 0.1; UK limits: UAL = 20.5, UWL = 16.5, LWL = 3.5, LAL = 1.5.

more hurried working may increase the prevalence of some kinds of defects rather than others. Alternatively, with small subgroups, inspectors may have more time to spend on their task and detect faults that a cursory examination may not reveal.

10.7.3 Figure 10.3 illustrates the format of the multiple characteristic chart. This is of the more usual *c*-chart type, where subgroups of equal sizes (here 50 × 10 metre lengths of plastic extrusion) are examined, and all faults noted. If differing lengths were sampled on different occasions, a *u*-chart (defects per metre or per 100 metres) would be appropriate. For, say, an absence application with various reasons for absence, where only one reason is applicable to any one person, the total absences per day for all reasons would provide an *np* value, or data for a *p*-chart if *n* varies from day to day. Similar considerations would apply to accidents, and possibly to machine breakdowns (though if a machine can breakdown more than once in any period of observation a *c*-chart would become applicable). Each case must be considered on its merits.

10.7.4 The example in Fig. 10.3 lists data for 20 days from inspection of samples of plastic extrusion as a pilot study on the extrusion process. The trial control chart shows one out-of-control point (though it would be a Warning value rather than a control limit violation if the UK practice were followed). If it can be assumed that a cause for this high number of defects can be found, the on-going chart could be based on the mean of the remaining 19 days. However, close inspection of the data for day 16 does not indicate any particular type of defect to be unusually prevalent, and it might be difficult to identify a cause for this collection of 'odds and ends'.

By way of example, the columns at the end of the chart have been completed and the percentages ranked (from the highest downward) in preparation for possible Pareto analysis.

11

Control charts: problems and special cases

11.1 Introduction

11.1.1 Chapters 7–10 have dealt with the general principles underlying the control chart, and their application to a number of relatively uncomplicated situations. The common features of these situations, irrespective of whether attributes or variables are involved, or which particular type of chart ($\bar{x}, R$; individuals; moving averages, etc.) is adopted, may be summarized as follows:

(i) It is possible to divide the process time-scale into rational subgroups within which only the inherent (common cause) elements of variation operate; in cases where the subgroup size is one item, successive differences or moving ranges provide a valid measure of inherent variation.
(ii) It is reasonable to regard all other sources of variation (e.g. between subgroups) as special causes, or assignable, and to work on removing or reducing their effects.
(iii) The 'process' comprises a single homogeneous unit; the subgroups may be considered as representative samples of the whole process when it is in control.

These are the features of the type of process (often a simple single-machine engineering operation) to which Shewhart first applied his ideas: the term 'widget' process is often affectionately used in this connection to describe a stream of nominally identical items mass-produced by a single process unit.

11.1.2 Control chart techniques have been successfully used on many processes outside these 'widget' applications, for example in textiles, chemical production, steel-making and electronics. More recently, they have extended to non-production areas (stores, maintenance, sales, clerical) and indeed to service, as well as manufacturing industries. Sometimes, however, the basic methods need modification to cater for particular aspects of these processes. It is some of these special cases (and the problems that may arise when inappropriate rule-of-thumb procedures are applied indiscriminately) that form the subject of this Chapter.

11.2 Systematic shifts in process average

11.2.1 There are many types of process in which the average level may change in a systematic manner over time. One of the most frequently encountered is that due to tool wear in cutting or forming operations, where, for example, the diameters of holes produced by a slowly wearing drill may gradually reduce, or a grinding tool may progressively remove less material from the work-pieces so that they tend to increase in length, thickness, diameters, etc.

These are not the only examples of approximately linear trends over time. Chemical solutions, reagents or catalysts may gradually change in their composition or effectiveness; materials may slowly cool, or heat up due to exothermic reaction or by being mechanically worked; moulds for producing bottles or other items may gradually become larger due to abrasion. The common feature of such processes is a fairly predictable and roughly linear change in the average level over a period, with a sudden reverse change at the point where a tool is reset or replaced, when the solution is replaced or topped up, or when a new batch of material enters the process.

11.2.2 Another group of factors produces predictable effects with a cyclic (rather than linear) pattern. In the ultra-short term, the effect of AC electricity may produce cycles, though a production rate of many items per second would be required for this to show in the process output. However, such factors as wind-up mechanisms which yield differences in tension through reels of wire, textile fibres, etc. due to their traverse pattern; the minor oscillations of temperature from thermostatic controls; or the maintenance of pressure in a circulation system by cut-in/cut-out or header top-up devices, will produce cyclic patterns with periods from a few seconds to an hour or two. Longer-term cycles arise from diurnal patterns (e.g. changes in ambient temperature or humidity from day to night) and seasonal or annual effects.

11.2.3 As sources of, or contributions to, variation, these effects occupy a place intermediate between common and special causes. They do not form part of the variation present within subgroups when an $\bar{x}, R$ or $\bar{x}, s$-chart is used, nor do they contribute to successive differences between single values. The cause of variation is 'assignable' and understood, but it is usually necessary for both economic and technological reasons to **accommodate** at least part of this variation – it would be too expensive, troublesome or simply infeasible to 'control it out'. Long term aims may be to reduce its effects by process improvement or new technology, but at any particular state of the art some relaxation from strict control is likely to be appropriate to avoid a 'crying wolf' situation.

11.2.4 For a high capability process, the one-time solution was so-called 'modified control limits', sometimes envisaged as 'hanging the control limits from the edges of the specification'. Here, for upper and lower specification

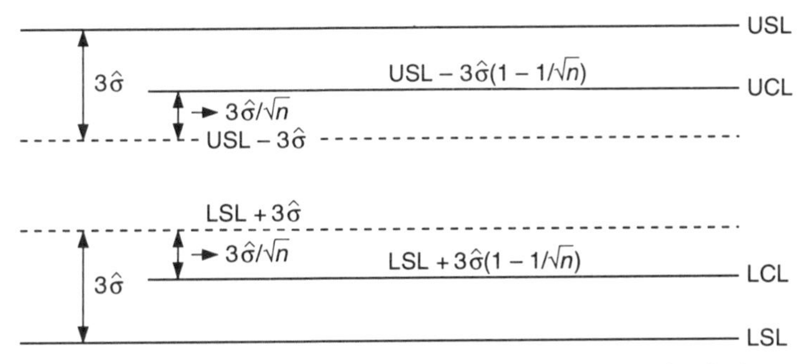

Fig. 11.1 Modified control limits. (Controlling to specification)

limits USL, LSL, upper and lower thesholds for process average (just compatible with the specification) are calculated, and the control limits then set outwards from these thresholds. For a 3σ interpretation of the specification requirement and $3\sigma_e$ control limits, we then have:

$$\left.\begin{aligned} \text{UCL} &= \text{USL} - 3\sigma + \frac{3\sigma}{\sqrt{n}} = \text{USL} - 3\sigma\left(1 - \frac{1}{\sqrt{n}}\right) \\ \text{LCL} &= \text{LSL} + 3\sigma - \frac{3\sigma}{\sqrt{n}} = \text{LSL} + 3\sigma\left(1 - \frac{1}{\sqrt{n}}\right) \end{aligned}\right\} \tag{11.1}$$

for subgroup size n. Appropriate factors based on average range are available in various national and international standards (e.g. BS 2564), sometimes based on a 0.1% (3.09σ) interpretation for both control limits and capability to specification. Figure 11.1 illustrates the principle.

11.2.5 This approach is both contrary to the philosophy of continuous improvement (it uses up the whole specification width with no incentive for reduction in variation) and also begs some key questions. Among these are:

(a) What if the process is not of high capability? In this case, the use of specification limits as part of the control mechanism will generate over-control and thereby actually **increase** the variability of the process.
(b) In many SPC applications, there are no specification limits: the object is to maintain process stability and identify opportunities for improvement. How should one proceed when unavoidable drift or cycles are present?
(c) Should one not distinguish between situations where common cause variation plus drift (or cycle amplitude) represent a large proportion of the specification and those where it is only a minor or moderate contribution?
(d) How does one monitor against real special causes (other than the drift or cyclic element)?
(e) There is finally the strange anomaly that, from (11.1), the larger the value of n the further the control limits move **inwards** from the specification limits,

although with larger n any adverse change will be detected more quickly. This arises from the flaw in philosophy which implies that the process is satisfactory unless there is strong evidence that it violates the specification.

The intelligent and cautious use of **relaxed** control will cover these points. Where there is a specification, one must decide in the light of the desirability for better-than-basic capability, and technical/economic factors, how much of the specification width should be allocated to the systematic effect(s). One possible way is to specify a target Cp index (see Chapter 13), and provided that the observed Cp based on within subgroup variation exceeds this target, calculate a relaxation zone based on:

$$\text{Zone width} = 6(\text{Target } Cp - \text{Actual } Cp)$$

Note that where target $Cp = 1$ and actual Cp is more than 1, this amounts to the use of modified limits. Where target Cp exceeds 1, it represents a compromise between strict (but unrealisitc) control and modified limits. Figure 11.2 illustrates the principle.

Obviously, if target Cp exceeds actual Cp, no relaxation is possible and the process is in urgent need of improvement action before any systematic effects can be accommodated.

Where no specification exists, a similar approach can be adopted, but the estimation of the relaxation allowance will be based entirely on economic and technical evaluation. If the systematic pattern is cyclic, the allowance will often correspond to the cycle amplitude; for linear drift, aspects concerned with the cost or inconvenience of frequent adjustment, replacement or intervention will predominate.

11.2.6 Once the allowance has been derived, the control limits are set out from the upper and lower edges of the resulting target zone. For a central nominal or target value, we thus have:

$$\text{UCL} = \text{Nominal (or target)} + \tfrac{1}{2}\text{relaxation allowance} + 3\sigma/\sqrt{n};$$
$$\text{LCL} = \text{Nominal (or target)} - \tfrac{1}{2}\text{allowance} - 3\sigma/\sqrt{n}.$$

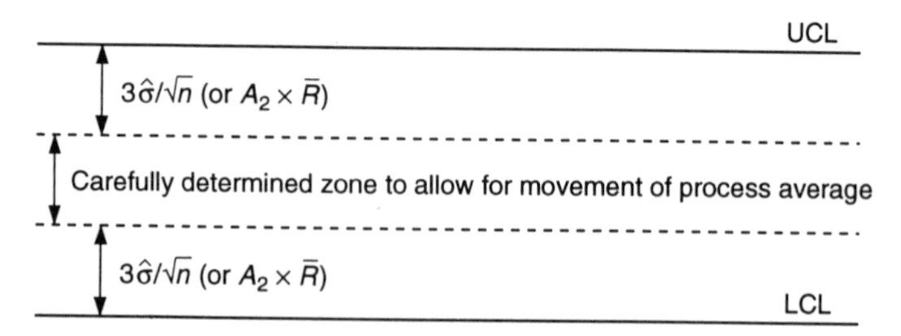

Fig. 11.2 Relaxed control for high precision process. (Often with systematic movement of process average). Note: zone width may be determined on technical/economical grounds, or from minimum required capability.

Equivalents in terms of $\bar{R}, \bar{s}$, etc. may be obtained via standard control chart factors in place of $3\sigma/\sqrt{n}$. Warning lines may be added if desired, as for conventional strict control charts.

11.2.7 There remains the need to check for special or assignable causes (other than the systematic effect that led to the relaxation). Apart from gross anomalies signalled by points beyond control limits, this can only be achieved by retrospective analysis of the control chart. For example, with a linear drift a line may be estimated and drawn on the chart, and sloping control lines drawn at $\pm 3\sigma/\sqrt{n}$ (and $\pm 2\sigma/\sqrt{n}$ for warning lines) on either side of the trend, as in Fig. 11.3. In rare instances, it may be possible to draw these lines in advance where the pattern is fully predictable. For cyclic patterns, analysis is a little more difficult, as cycles may not be totally regular, but a moving average and 'wavy' control limits drawn retrospectively, may be a possible approach. Figures 11.3 and 11.4 illustrate the approach for a linear trend and a cycle respectively.

Proper summaries of performance become more important with the special cases for this Chapter, and the methods of Chapter 16 are likely to prove useful.

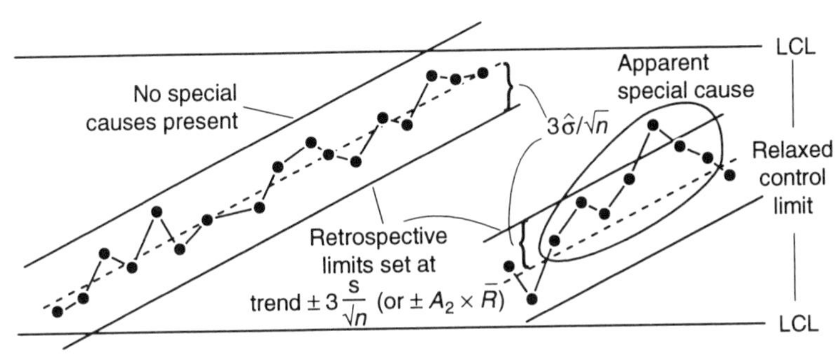

Fig. 11.3 Retrospective check for special causes: linear trend. Dashed diagonal line indicates estimated or predicted trend.

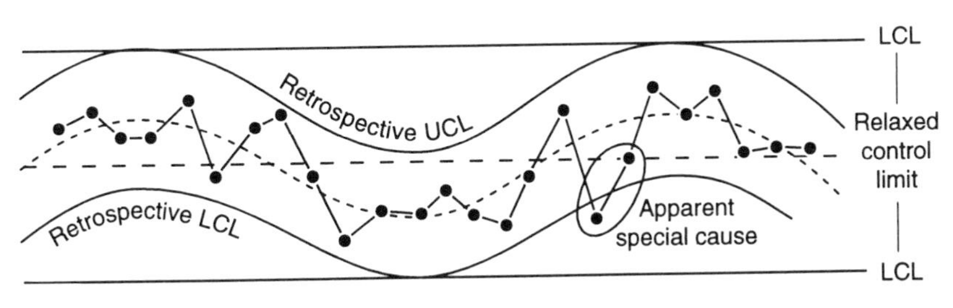

Fig. 11.4 Retrospective check for special causes: cyclic component. Wavy dashed line indicates predicted or estimated cycle (e.g. via moving average). Retrospective limits at $\pm 3\sigma/\sqrt{n}$ (or $A_2 \times R$) around cycle.

11.3 Between-sample common causes

11.3.1 The previous section dealt with between-sample effects of a systematic nature. We now consider another type of between-sample effect which is (to a large extent) random, in that *no* systematic pattern is discernible. Such effects often manifest themselves in the form of control charts which are rather 'noisy', sometimes yielding points beyond control limits for which no assignable cause can be found, or fairly frequent warning values (when warning lines are used), or appreciably more than one-third of the points in the outer thirds of the distance between control limits.

The problem here is often that the within-sample variation (usually measured by $\bar{R}$ or $\bar{s}$), on which the control limits are based, does not include all of the real inherent variation in the process. A couple of examples will illustrate the point. In a forging process, samples of five consecutive forgings were measured and monitored via an $\bar{x}$, R chart (Fig. 11.5). The sets of five items showed very little within-sample variation, and thus $\bar{R}$ (used for calculating the control limits) was small, and capability **appeared** to be very high. These samples of five items were taken every 15 minutes, and using a conventional control chart nearly half of the $\bar{x}$ points lay outside the control limits. The reason lies a little further back in the process – in the furnace. A stream of metal blanks passed through a gas-fired thermostatically controlled furnace on a conveyor. The five items within a sample obviously experienced virtually identical furnace conditions, and with little variation added at the forging stage, were very consistent. Fifteen minutes later, the next set of items has experienced a somewhat different heating history due to the 'wobble' of thermostatic control, slight variations in draught

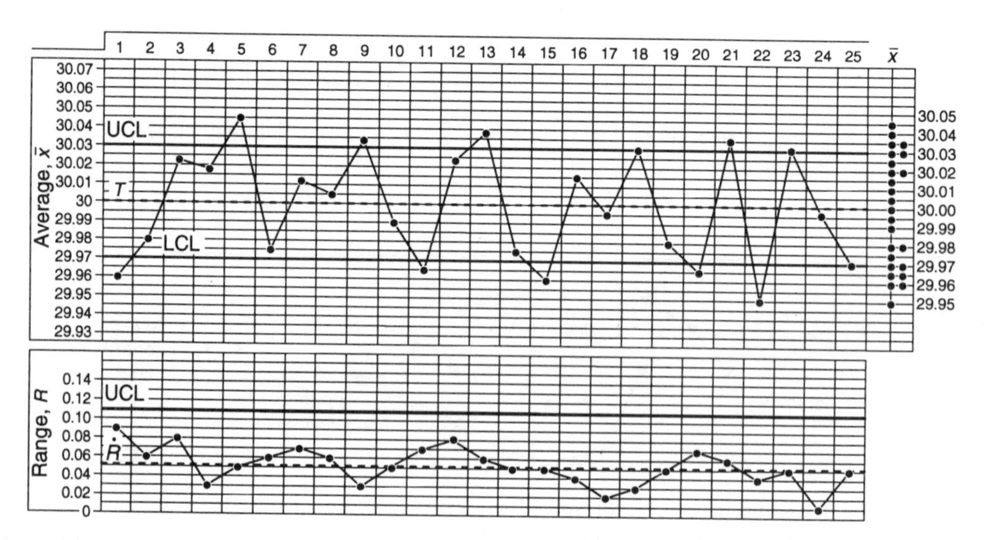

Fig. 11.5 Control chart for forged components showing between-sample common causes.

and possible ambient effects, as well as factors from earlier elements in the process. As no tighter control of these conditions was technically feasible, the between-sample **common** causes need to be measured and allowed for in control limit evaluation, so that genuine special causes (thermostat malfunction, furnace overload, changes in quality of blanks) can be properly signalled.

The second example concerns a filling line for cans of a food product. Samples of a few consecutive items again reveal very little within-sample variation in weight, but sample means for groups taken at intervals of 30 minutes or more show frequent 'out of control' to conventional control limits. In this case the source of additional variation is the make-up of batches of ingredients, temperature/pressure conditions in the sterilizer and small fluctuations between consecutive batches of cans. None of these variations is such as to produce inferior product (e.g. over or under fill), and again they constitute part of the inherent variation at the current state of the art.

11.3.2 Where some elements of common-cause variation do not appear in the time span between consecutive items, or within the time that repeat measurements can be made on a continuous process, there are two main methods of allowing for the medium term elements.

First, the formation of rational subgroups may be modified so that opportunity arises for all elements of true inherent variation to appear **within** subgroups. Care is necessary to guard against the inclusion of assignable causes like changes in conditions, shift change-over, etc. In the forging example, subgroups of five items taken at a rate of one item every two or three minutes would provide an opportunity for the inherent variation in furnace conditions to exert an influence, while still providing a 15 minute between-samples interval.

In the filling line example, it is evidently necessary to obtain some information on variation within ingredient batches, but equally (if not more) important to quantify the acceptable extent of between-batch variation. In this case, because batches vary, each from its predecessor and successor, at discrete points in time rather than continuously (like the furnace conditions), two or more items from each batch will provide a suitable sampling plan. The within-batch variation might be further resolved if two or more subgroups, each comprising two items, were sampled from each batch.

11.3.4 In both examples, and in many industrial (and other) processes, contributions to inherent variation may be resolved into two or more components. For the forgings, one, albeit small, contribution is the variation between items produced within a short time span when furnace conditions have not appreciably changed. This contribution can be measured from within-sample variation if two or more items are sampled on any one occasion. The more important contribution in this example arises from common-cause, but between-sample, fluctuations in process contribution. These can be evaluated using successive differences or moving ranges between sample means, in effect treating the $\bar{x}$

values in the same way as individual values (Chapter 9). A control chart may have three elements in such applications:

(a) A chart for $\bar{x}$, with control limits based on the output of the second element, (b), as a measure of common cause variation of the sample means;
(b) A chart for the successive differences or moving ranges, δ or MR, between the $\bar{x}$ values. This provides a measure of variation for the control limits (a), and a means of monitoring the between-sample variation. The present author has a strong preference for successive differences in this context, as they provide the basis of useful further analysis described in Chapter 16;
(c) A conventional range chart to monitor the short-term within-sample variation.

It may be found that when between-sample effects are relatively large, samples of only two items are preferable, with effort concentrated on more frequent sampling; say $n=2$ every 15 or 20 minutes rather than $n=5$ every half-hour.

11.3.5 The $\bar{x}$ and δ parts of the chart, in effect, deal with the sample means in the manner of single values, as in section 9.3. Following the same principle, and combining the range chart into the format, we proceed as follows.
For each sample, calculate:
$\bar{x}$, the sample mean,
R, the sample range,
δ, the (absolute) difference between the current and immediately preceding values of $\bar{x}$.

As summary statistics, obtain:

$\bar{\bar{x}}$ (or use a suitable target value T),
$\bar{R}$, the average range,
and either $\bar{\delta}$, or $\hat{\sigma}_{\bar{x}} = \sqrt{(\frac{1}{2}MSSD)}$ as defined in (9.2).

Control limits are then obtained using the appropriate UK or USA convention, for example:

$T \pm 3\hat{\sigma}_{\bar{x}}$ or $T \pm 2.66\bar{\delta}$ for action lines or US control limits,
$T \pm 2\hat{\sigma}_{\bar{x}}$ or $T \pm 1.77\bar{\delta}$ for warning lines.

The control limits for R are obtained via the factors in Table 8.1, and those for δ are either $4.12\bar{\delta}$ for UK practice or $3.27\bar{\delta}$ for US practice.

It would, of course, be appropriate to use the $\bar{x}, s, \delta$ combination if this is preferred in a particular application.

Interpretation of the chart follows the usual rules, noting that warning lines are not suitable for the δ-chart, and runs of ten or more (rather than seven or more) are needed to signal significant changes.

Finally, it must be noted that where capability indices are required, allowance must be made for the additional variation contributed by sample-to-sample fluctuations. As will be pointed out in Chapters 13 and 16, the overall standard deviation must be estimated as

$$\left.\begin{aligned}\hat{\sigma} &= \sqrt{\left\{\left(\frac{\bar{\delta}}{1.128}\right)^2 + \frac{n-1}{n}\left(\frac{\bar{R}}{d_n}\right)^2\right\}} \\ \text{or} \qquad \hat{\sigma} &= \sqrt{\left\{\tfrac{1}{2}MSSD + \frac{n-1}{n}\left(\frac{\bar{R}}{d_n}\right)^2\right\}}\end{aligned}\right\} \tag{11.2}$$

not simply as $\bar{R}/d_n$.

11.3.6 As an example of the $\bar{x}$, R and δ-chart, we take up the data of Table 8.2. Some instability in the $\bar{x}$ values was noted in the analysis of section 8.6.1 (step 8). The use of the triple chart might well be wise in these circumstances. For the trial data, excluding sample 23, we had $\bar{\bar{x}} = 239.53$, $\bar{R} = 2.421$. We now additionally require the δ values (shown in Fig. 11.6) and $\bar{\delta}$, which is found to be 0.833 (alternatively, $\sqrt{(\frac{1}{2}MSSD)} = 0.739$).

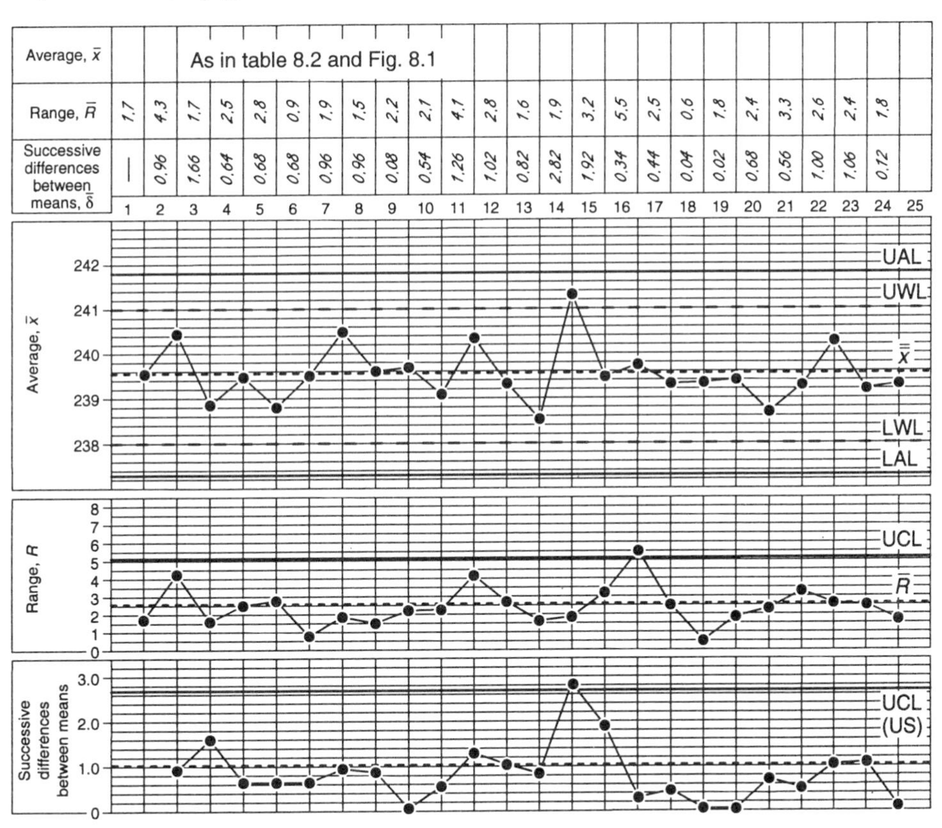

Fig. 11.6 Example of $\bar{x}$, R, δ-chart. (Trial control for data of Table 8.2).

For illustration, the US-style limits have been used in Fig. 11.6, giving:

For $\bar{x}$, $239.53 \pm 2.66\bar{\delta} = 241.75, 237.31$ for UCL, LCL

(with warning lines at $\pm 1.77\bar{\delta}$ if required); compare the US limits of 240.93 and 238.13 in Table 8.4.

For R, $\text{UCL} = 2.114 \times 2.421 = 5.12$;
For δ, $\text{UCL} = 3.27 \times 0.833 = 2.72$.

Some features to note in Fig. 11.6 include:

(i) The $\bar{x}$ of 241.34 for sample 14 is no longer a violation of the upper control limit (if warning lines are used, it is a warning signal, but this is discounted by the next sample);
(ii) The R of 5.5 in sample 16 is an out-of-control point when using US control limits, but not with the UAL of the UK chart (it then becomes a warning value). The US chart has a tendency towards over-control;
(iii) Similarly, the δ of 2.82 (produced by the $\bar{x}$ values for samples 13 and 14 well-scattered on each side of $\bar{\bar{x}}$) is out-of-control with the rather severe US limit of $3.27\bar{\delta}$. It would not yield a signal when using the UK limit of $4.12\,\bar{\delta} = 3.43$.

It is of course important that the $\bar{x}$, R, δ approach should be reserved for cases where there are real between-sample common causes, and not regarded simply as a relaxation of control limits to 'make life easier' when processes are genuinely prone to special causes.

11.3.7 The situation of the forging process (Fig. 11.5) could be tackled similarly. Here the control limits based on $\bar{\delta}$ become 30.0 ± 0.096 (in place of 30.0 ± 0.030 in the original chart), but it unfortunately turns out that allowing for the furnace fluctuations, capability is inadequate and some means of reducing this variation is urgently needed. Simply narrowing the control limits would induce over-control at the forging press, causing unnecessary interruptions and spurious corrections which would need to be reversed soon after they were made. Note also that this is a situation where small but more frequent samples (say $n = 2$ every 10 min) would be both more efficient and more economical.

11.4 Within-sample special causes

11.4.1 The effects considered here are, in a way, the converse of those in the previous section. Where a sample or subgroup comprises items from various subdivisions of a process, it may happen that one or more of these subdivisions differs consistently from the others. Sometimes, for example, a subgroup is made up of one item from each position of a machine tool, or from each head of a filling machine. If one or two of these units are incorrectly set up (compared with the remainder) or if there is differential drifting from an initially correct setting, then the within-sample variation will contain this **assignable** effect as well as the inherent variation in the system.

Strictly, of course, such samples do not fall under Shewhart's original definition of 'rational subgroups'; nevertheless such sampling strategies are used, and it is advisable to note the problems they may introduce and some possible ways of dealing with them.

To measure the true inherent variation, it is necessary to identify the subunits and relate them to the observed values in the samples. If this is impossible (e.g. random mixing occurs between the subunits themselves and the point at which samples are taken) it may be necessary to carry out a special study to determine the within-subunit variation.

Assuming that over a series of samples, each comprising one item from each subunit, the individual values relating to each subunit are available, the most effective means of estimating the inherent variation is to treat each subunit as a process. The successive differences between items within each subunit provide a measure of variation, and if required any subunits showing anomalously large or small variation may be investigated. Alternatively, the successive differences between $\bar{x}$ values may be used as the basis for setting control limits (in the manner of the $\bar{x}, R, \delta$ chart of section 11.3) as these will be less influenced by subunit anomalies than the subgroup ranges or standard deviations.

In the absence of significant discrepancies in variation between the sub-units, a pooled estimate of within-subunit variation can be calculated, and this will provide a basis for control limits for an $\bar{x}$, R-chart (see example in section 11.4.3). In operating the R-chart, a note should be made of the subunits yielding the highest and lowest values in each sample, so that action can be taken if one subunit drifts away from the general average. In cases of severe instability of the subunits, or where it is impossible to bring outlying subunits into line with the others, the R-chart will be of little value, and it may be necessary to treat each subunit effectively as a separate process, though obviously this is tedious and time consuming.

Periodically, the within-subunit successive differences should be used to up-date the measure of inherent variation and to revise control limits where appropriate.

11.4.2 Finally, some of the ways in which misleading indications may arise from within-sample special causes should be noted:

(i) Because a persistent anomaly between one (or a few) subunits and the remainder will inflate the apparent commoncause variation, control limits for $\bar{x}$ charts will be too widely spaced and hence ineffective in signalling other aspects of special cause variation;
(ii) The increase in apparent standard deviation may yield poor estimates of capability whilst failing to indicate its real cause; the true capability may, in fact, be appreciably *better* than indicated by conventional indices;

(iii) Because the random element of inherent variation is, at least partially, smoothered by the more regular effects between subunits, sample ranges may vary appreciably *less* than is general for an *R*-chart for an in-control process. Chapter 16 indicates a method of testing for this effect.

(iv) Following from the preceding points, it is even possible that an $\bar{x}$ chart may appear to be rather too well in-control – a phenomenon sometimes called 'centre-line hugging', though the actual $\bar{x}$ values may in fact be off-target due to the effect of the anomalous subunit(s).

11.4.3 The example shown in Fig. 11.7 is from the field of quantity control. One package from each of four filler heads was sampled on each occasion (approximately at 15 minute intervals). The weighing was performed on an automatic balance with microprocessor which operated a conventional $\bar{x}$, *s*-chart. Earlier data had indicated a standard deviation of 1.3 for individual packages,

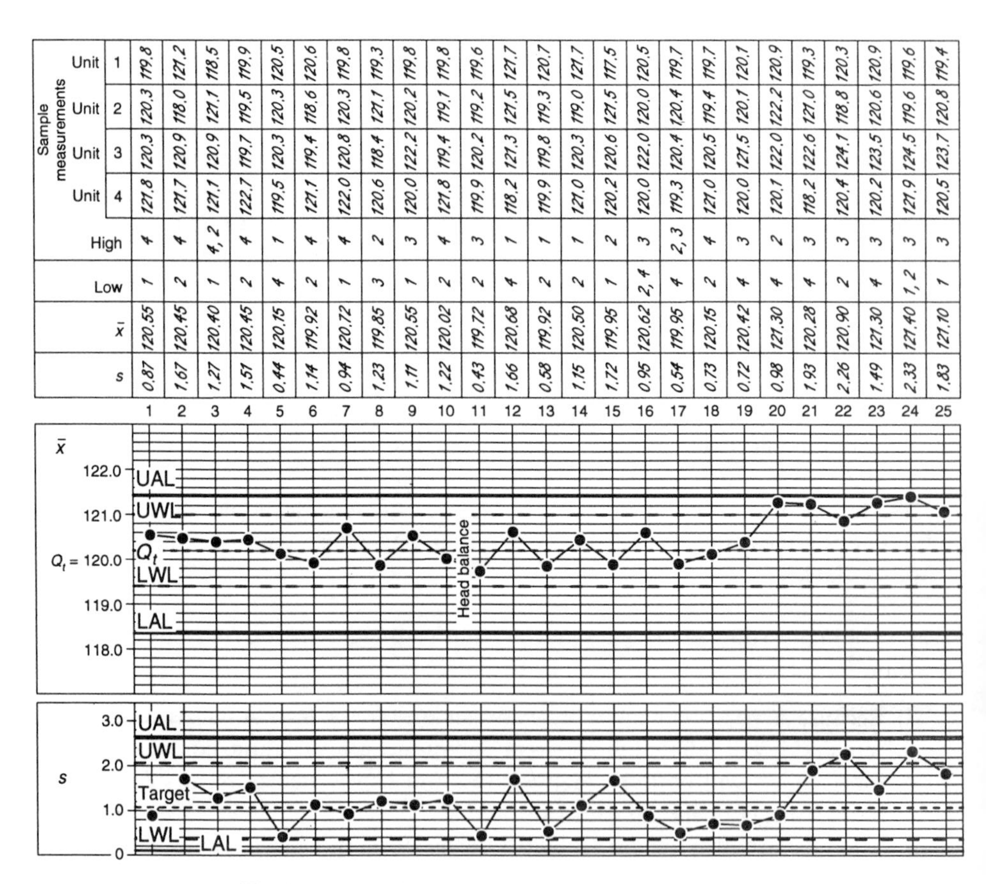

Sample measurements	1	2	3	4	5	6	7	8	9	10	11	12	13	14	15	16	17	18	19	20	21	22	23	24	25
Unit 1	119.8	121.2	118.5	119.9	120.5	120.6	119.8	119.3	119.8	119.8	119.6	121.7	120.7	121.7	117.5	120.5	119.7	119.7	120.1	120.9	119.3	120.3	120.9	119.6	119.4
Unit 2	120.3	118.0	121.1	119.5	120.3	118.6	120.3	121.1	120.2	119.1	119.2	121.5	119.3	119.0	121.5	120.0	120.4	119.4	120.1	122.2	121.0	118.8	120.6	119.6	120.8
Unit 3	120.3	120.9	120.9	119.7	120.3	119.4	120.8	118.4	122.2	119.4	120.2	121.3	119.8	120.3	120.6	122.0	120.4	120.5	121.5	122.0	122.6	124.1	123.5	124.5	123.7
Unit 4	121.8	121.7	121.1	122.7	119.5	121.1	122.0	120.6	120.0	121.8	119.9	118.2	119.9	121.0	120.2	120.0	119.3	121.0	120.0	120.1	118.2	120.4	120.2	121.9	120.5
High	4	4	4, 2	4	1	4	4	2	3	4	3	1	1	1	2	3	2, 3	4	3	2	3	3	3	3	3
Low	1	2	1	2	4	2	1	3	1	2	2	4	2	2	1	2, 4	4	2	4	4	4	2	4	1, 2	1
$\bar{x}$	120.55	120.45	120.40	120.45	120.15	119.92	120.72	119.85	120.55	120.02	119.72	120.68	119.92	120.50	119.95	120.62	119.95	120.15	120.42	121.30	120.28	120.90	121.30	121.40	121.10
s	0.87	1.67	1.27	1.51	0.44	1.14	0.94	1.23	1.11	1.22	0.43	1.66	0.58	1.15	1.72	0.95	0.54	0.73	0.72	0.98	1.93	2.26	1.49	2.33	1.83

Fig. 11.7 Quantity control: 4-head filling machine.

Table 11.1 Control limits for package weight control

Target weight 120.2 g

		Based on apparent variation, including sub-unit imbalance	Based on true common cause variation
$\bar{x}$	UAL	122.21	121.46
	UWL	121.47	121.00
	LWL	118.93	119.40
	LAL	118.19	118.94
$\bar{s}$	UAL	3.29	2.68
	UWL	2.49	2.04
	LWL	0.38	0.31
	LAL	0.21	0.10

estimated from pooled sample values of s (via $\sqrt{\text{Av}(s^2)}$). Because of occasional head imbalance, this is larger than the true common-cause variation, which is of the order of $s = 1.0$. In the data of Fig. 11.7, the combined estimate via successive differences from all four heads is 1.06, and the standard error of the $\bar{x}$ values (again via differences) is 0.41. A UK-practice control system was in use (though without any actual charting) and the control limits for the original system are compared with those based on the true common cause variation in Table 11.1.

The wider limits originally in use masked some appreciable changes shown in Fig. 11.7. The use of the correct limits, along with identification of the highest and lowest head values in each sample, gives much clearer indications of these changes. In fact a run of more than three successive occurrences of the same head number as either highest or lowest should draw attention to a possible anomaly. The use of a chart with all values plotted, and lines of different colours identifying each subunit, would give even clearer indications.

In this example, the persistence of head 4 as the largest value between samples 1 and 10 (7 of the 10 maxima involve head 4) is a clear signal. From samples 17 to 25, head 3 is increasingly frequent as the largest value, and the warning values for both $\bar{x}$ and s are entirely due to the drift in setting on head 3.

11.5 Short-run processes

11.5.1 To a large extent, the key in dealing with short production runs is to identify those elements in the process itself that can be controlled and monitored, rather than elements peculiar to the different production runs. This is not always easy, though in surprisingly many cases control of input materials, methods and process conditions can be achieved by the usual 'continuous' methods of SPC.

It is worth the effort to identify similarities between the different production runs. In mechanical engineering processes (machining, stamping, drilling, etc.). whilst product dimensions may differ, process performance around the nominal may be similar for the various jobs. In such cases, a straightforward control chart for deviation from nominal will probably provide a suitable means of control. A few particular points should be noted.

(a) The change-over from one job to the next should be noted on the control chart.
(b) Care is needed to ensure that appropriate start-up, set-up or first-off checks are carried out at the start of each job or run.
(c) The estimate of variation ($\bar{R}$ or standard deviation) should be obtained within each run. Where variation for the various runs is of similar magnitude, the estimates can be pooled to give greater precision.
(d) Data relating to start-up or initial adjustment for a run may need to be excluded from general summaries (though it should not be destroyed as it may contain useful information about initial problems on changing from one job, or run, to another.
(e) Where the product differs appreciably in size from job to job (dimensions for machining, volume or weight for filling, etc), one should bear in mind the possibility that variability may be proportional to nominal size, i.e. that coefficient of variation may be a more appropriate measure than standard deviation (but note that this must be based on actual data values, not on deviations from nominal – which can result in infinite coefficient of variation ($s/\bar{x}$) where the mean is almost exactly at the nominal value).

11.5.2 Sometimes the differences between runs, jobs or products are too great to permit the above approach, and each run must be treated on its own merits. An efficient record system is needed here, so that data from previous runs of the same type can be called up to provide a basis for targets and control limits. This can be a considerable problem where many different products are produced, and an approach based on grouping products into types sufficiently similar to permit data pooling (as above) may prove useful. Each product can be assigned an identifying code to faciliate calling up the appropriate data bases for control limits.

There is no doubt that, except where differences between runs are sufficiently small to permit the 'deviation from nominal' approach based on pooled data, control of short runs by product measurement can be tedious and requires ingenuity and dedication. These difficulties emphasize the desirability of 'upstream' control of the process itself, reducing the reliance on information which, as well as being more difficult to gather, identifies problems rather than preventing them from occurring.

The use of cusum methods is particularly beneficial for short runs, where the prompt indication of off-target performance by a change in slope of the cusum chart provides not only a signal but an estimate of the correction needed, either to restore the current mean level to the target or to achieve a prescribed overall average for the complete run. These points will be further developed in Chapter 12.

11.6 Non-Normality

11.6.1 The most widely used control chart limits are based on symmetrically disposed lines at 3 standard errors on either side of a target value. For a Normally distributed variable, when the process is in-control each line will be violated by about one sample in 740, (i.e. about one 'false alarm' per 370 samples for a two-sided chart) when 3-sigma limits are used. Further false alarms may arise from randomly occurring runs of points above or below the centre line, an excess of points in the outer thirds of the chart, pairs of consecutive values beyond warning lines, etc., so that the chart with these multiple rules tends to have a false alarm rate of around 1 in 150–200 samples.

When a real shift in process mean of, say, one standard error occurs, such charts will detect this situation in an average of around 10 samples; for a two standard error shift, the average run length is about samples. Obviously these average response times will depend on the precise combination of rules used to interpret the chart. Broadly, the more rules that are included, the quicker the response to real changes but the greater is the risk of false alarms.

11.6.2 The same collections of rules are frequently used with variables that are not Normally distributed. For $\bar{x}$-charts, the tendency for sample means to be more Normal than the individual values (see Chapter 5) is often sufficient to render the use of conventional limits satisfactory. For other sample statistics, and for charting individual values from a non-Normal distribution, the conventional limits and rules may be quite unsuitable. We take, as a well-known example, the case for the range in samples of $n = 2$ where individual values *are* Normally distributed. The difference between random pairs of values then has a Normal distribution with standard error of $\sigma\sqrt{2}$, where σ represents the variation in individual values. However, when forming the range, the sign of $x_1 - x_2$ is ignored, so that half of this Normal distribution is folded on to the other. This yields a distribution of ranges (R_2) with mean $\sigma\sqrt{(4/\pi)} = 1.128\sigma$, and with standard error 0.8525σ or 0.7555σ times the mean range (Fig. 11.8). The mode is, of course, at zero (because of the half-Normal shape) and the median does not coincide with the mean but is at $0.854\mu_R$. The mean in fact divides the R_2 distribution in the proportions 57.5% below μ_R, 42.5% above μ_R, and to obtain the same false alarm risks as for the Normal distribution, the runs rule should be modified to nine or more below μ_R but six or more above μ_R.

For US-practice range charts, control limits remain (as for $\bar{x}$) at ± 3 sigma.

For the range distribution, this yields

$$\mu_R \pm 3 \times 0.7555\mu_R, \quad \text{i.e. } 3.267\mu_R, \ -1.267\mu_R$$

The lower limit is obviously unusable and is ignored. However, the problem with the upper limit is that it cuts off much more than the intended 0.135% in the distribution tail when the process variation is stable: in fact just under 1%. This means that when in control, the range chart will, on average, contain 1% of values above UCL. The effect is only partially offset by the reduced sensitivity of the runs-of-seven rule. Overall, the US control limit tends to increase the risk of over-control at the upper end and decrease it at the lower–although with the compensating feature that real changes will be signalled slightly more rapidly (but improvements more slowly) than the UK chart with limits based on 0.1% and 2.5% tail probabilities.

Figure 11.8 illustrates the distributions of R_2 and the positions of the US and UK control limits.

The UK limits, as noted in Chapter 8, are set to cut off 0.1% of the target distribution at the action line and 2.5% at the warning line. For R_2, this approach yields UAL $= 4.12\bar{R}$, UWL $= 2.81\bar{R}$, LWL $= 0.04\bar{R}$, LAL $= 0$. The latter limit is due to the presence of the mode at $R = 0$, and in practice it is also found

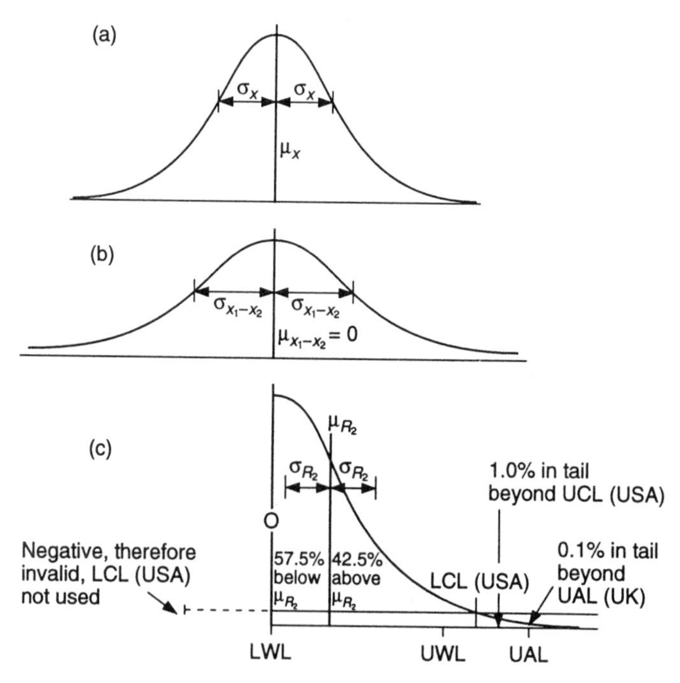

Fig. 11.8 Non-Normal distribution, R_2. (a) Normal distribution of x; (b) Normal distribution, $x_1 - x_2$; distribution of R_2.

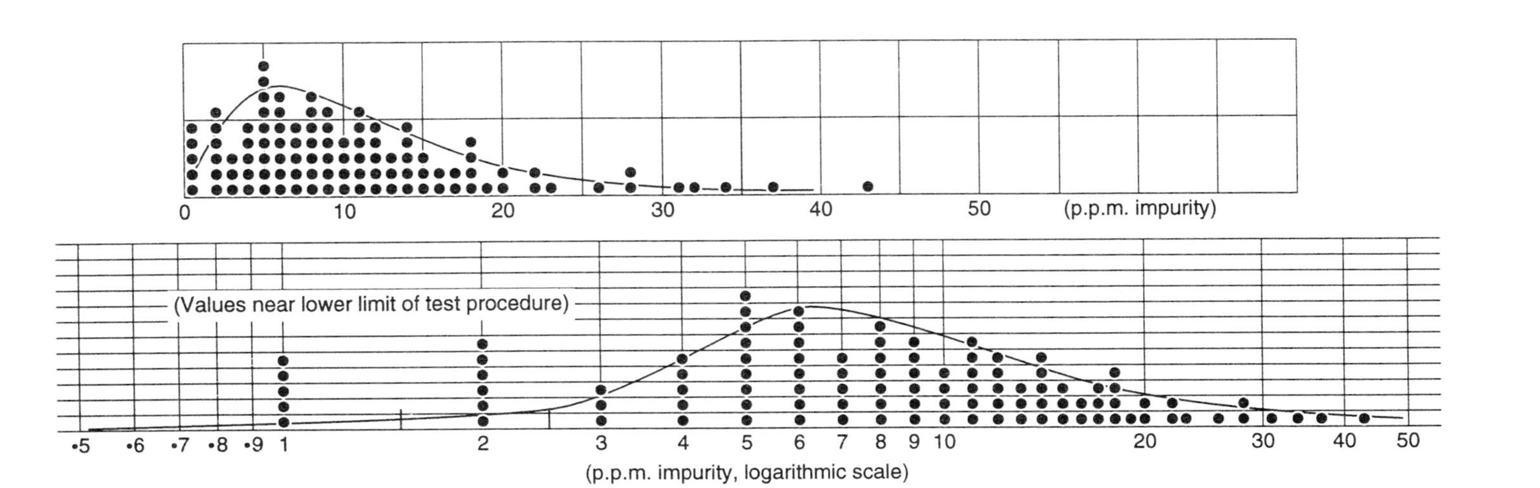

Fig. 11.9 Data for metallic impurity (p.p.m.) in surgical components.

that LWL often lies below the limit of discrimination of the measurement system. The lower limits may thus become zero, but for logical reasons, rather than the failure of a Normal approximation when circumstances lead to non-Normality.

11.6.3 Similar problems occur with the binomial and Poisson distributions for attributes and events. Suitable probability-based limits are given in Fig. 10.1. These examples, as well as indicating the problems of non-Normality, outline the approach for dealing with it: that is, to use limits based on percentage points of a suitable distribution model, or obtained empirically from an observed distribution. Transformations, including $\log(x)$, $\sqrt{x}$ and others (even x^2 or $1/x$) may sometimes be useful, and probability plotting on suitable grids (as noted in Chapter 14) may also provide estimates of percentage points.

11.6.4 Figure 11.9 shows, in the form of a blob diagram, the results from 100 determinations of the amount of metallic impurity in plastic components for surgical use. The values were obtained by occasional though unsystematic sampling during production trials, and it was then decided to monitor regular production using samples of two items per batch. This rather low sampling rate was unavoidable because of the destructive nature of the test and an expensive analytical procedure, but an alternative electronic test was under development. Quality control is exercised by strict observance of process settings and input material certification, and the control chart with $n = 2$ will provide an assurance for these procedures and a more extensive data base for development and improvement.

The distribution of results is highly skewed, but in fact log (impurity) is reasonably Normally distributed. In order to gain interpretability, it was agreed to use a procedure based on ln(p.p.m), but transformed back to ppm for plotting. The antilog of the mean of log-transformed values is the geometric mean, and for $n = 2$ it corresponds to $\sqrt{x_1 x_2}$, so that the calculations are in fact fairly straightforward. Similarly for the range chart, it turns out that the ratio of the larger to the smaller of the two sample values is required, deriving from antilog $\{\log(x_1) - \log(x_2)\} = x_1/x_2$.

For the data of Fig. 11.9, $\bar{x} = 10.89$, $s = 8.143$, but if we set $y = \ln(x)$, we have $\bar{y} = 2.092$, $s_y = 0.8421$. Now for samples of $n = 2$, the standard error of means of 2 will be about $0.8421/\sqrt{2} = 0.6$ approximately, and the mean range will be $1.128 \times 0.8421 = 0.950$. A mean for ln(p.p.m.) of 2.092 corresponds to a geometric mean of 8.1 p.p.m., and the mean range of 0.95 for ln(p.p.m.) corresponds to a value of about 2.6 for the **ratio** of the larger to the smaller value in a sample of 2. These statistics and the control limits derived from them are listed in Table 11.2.

Since the geometric mean for $n = 2$ is simply $\sqrt{(x_1 x_2)}$, and the antilog of the range of two log-transformed values is $x_{\text{larger}}/x_{\text{smaller}}$, the chart can be readily operated in a 'root and ratio' format.

For this application, the control chart could alternatively be drawn up on a

Table 11.2 Data summary and control limits for metallic impurity data

ln(p.p.m.) (y)		p.p.m. (x)	
Arithmetic mean	2.092	Geometric mean	8.1
Standard deviation	0.8421	–	
Mean range	0.95	Typical $x_{larger}/x_{smaller}$	2.6
Control limits ($n=2$)			
UK $UAL_{\bar{y}}$	3.946	UAL $\sqrt{(x_1, x_2)}$	51.7
$UWL_{\bar{y}}$	3.268	UWL $\sqrt{(x_1, x_2)}$	26.3
$LWL_{\bar{y}}$	0.916	LWL $\sqrt{(x_1, x_2)}$	2.5
$LAL_{\bar{y}}$	0.238	LAL $\sqrt{(x_1, x_2)}$	1.3
UAL_R	3.914	UAL $(x_{larger}/x_{smaller})$	50.1
UWL_R	2.670	UWL $(x_{larger}/x_{smaller})$	14.4
(no lower limits for R)			
US $UCL_{\bar{y}}$	3.878	UCL $\sqrt{(x_1, x_2)}$	48.3
$LCL_{\bar{y}}$	0.306	LCL $\sqrt{(x_1, x_2)}$	1.36
UCL_R	3.106	UCL $(x_{larger}/x_{smaller})$	22.3

Table 11.3 Operations of control chart for impurity (p.p.m.)

Sample	Impurity (p.p.m.)	Geometric mean $\sqrt{(x_1, x_2)}$	Range ratio x_1/x_2 or x_2/x_1
1	2, 5	3.2	2.5
2	9, 16	12.0	1.8
3	17, 5	9.2	3.4
4	10, 12	11.0	1.2
5	4, 6	4.9	1.5
6	20, 7	11.8	2.9
7	16, 11	13.3	1.5
8	12, 16	13.9	1.3
9	7, 7	7.0	1.0
10	6, 2	3.5	3.0
11*	38, 20	27.6 (warning)	1.9
12*	32, 2	8.0	16, 0 (warning)
etc.			

*(Samples 11 and 12 for illustration).

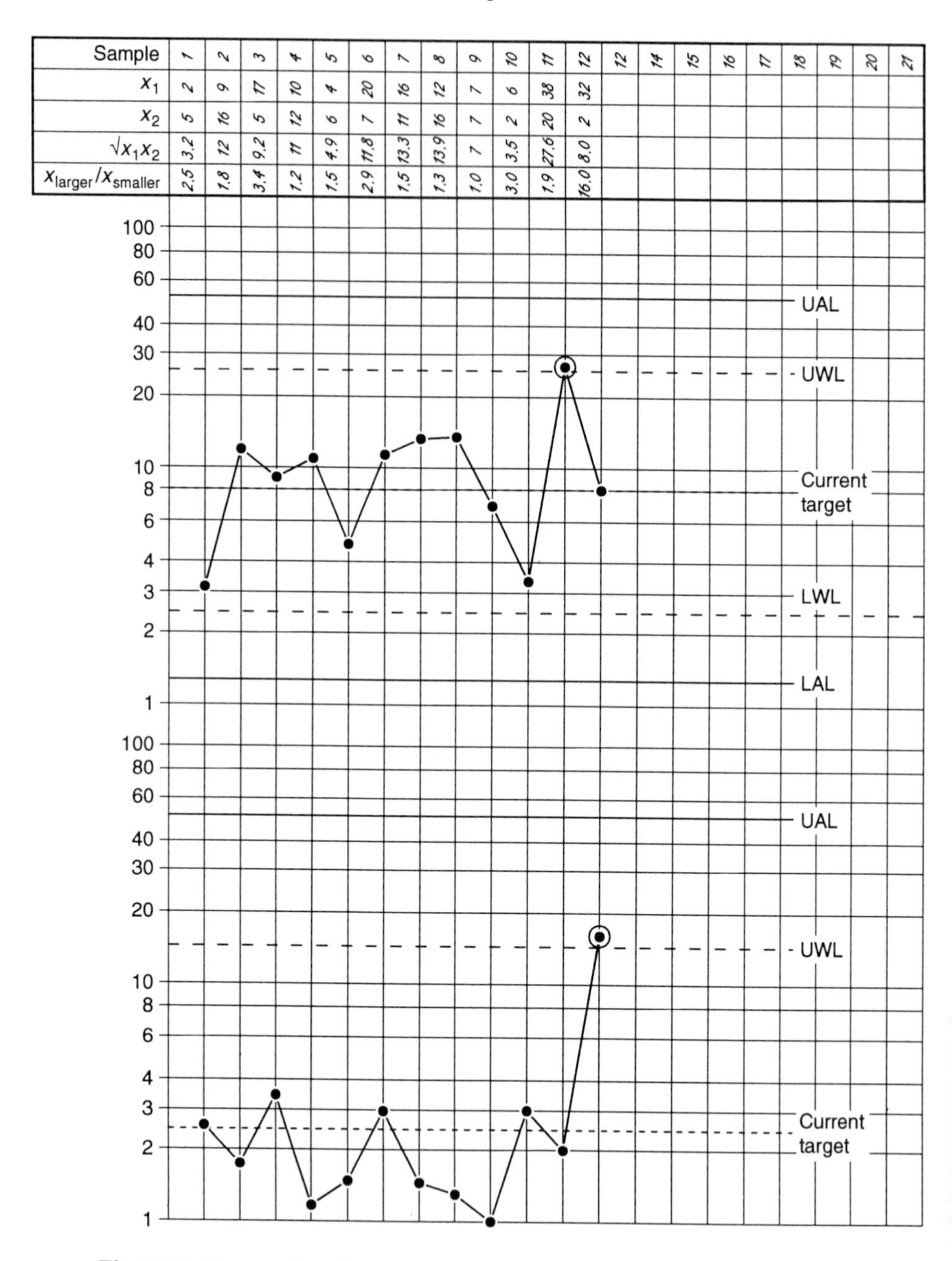

Sample	1	2	3	4	5	6	7	8	9	10	11	12	12	14	15	16	17	18	19	20	21
x_1	2	9	17	10	4	20	16	12	7	6	38	32									
x_2	5	16	5	12	6	7	11	16	7	2	20	2									
$\sqrt{x_1 x_2}$	3.2	12	9.2	11	4.9	11.8	13.3	13.9	7	3.5	27.6	8.0									
$x_{larger}/x_{smaller}$	2.5	1.8	3.4	1.2	1.5	2.9	1.5	1.3	1.0	3.0	1.9	16.0									

Fig. 11.10 Control chart for metallic impurity (based on log transformation).

log-scale (one-way) graph grid. An example of the calculations is shown in Table 11.3, and of a suitable control chart in Fig. 11.10.

11.6.5 The problems of non-Normality and the reasons for it are very varied, and methods for handling the situations need to be adaptable. Variables which frequently give rise to non-Normal distributions include measures of distortion (warp, eccentricity, 'out-of-square'), surface finish, angular displacements and many chemical, physical and electrical test characteristics. For minor deviations from Normality, samples of four or five items may be sufficient to Normalize the distribution of $\bar{x}$ and to leave the performance of R or s-charts little affected. Even here, however, care is necessary in interpreting for capability, as this depends on the distribution of individual values, not of sample means.

Chapter 14, dealing in general with capability measurement for non-Normal situations, provides techniques for assessing the Normality or otherwise of data, and gives suggestions on alternative distribution models or transformation methods. A full treatment of this topic lies beyond the scope of this book.

12

Cusum methods

12.1 Introduction

12.1.1 Cumulative sum or 'cusum' techniques were developed in the late 1950's, initially in a search for greater efficiency (than conventional control charts) in detecting changes in the mean of a process parameter. It turns out that they are an excellent data presentation tool, with applications far beyond those of process control: for example, in retrospective analysis, literary criticism, medical research and administrative systems.

They depend on a simple summation of deviations from a target value – precisely in the manner of scoring cumulatively in golf by summing the deviations (in number of strokes at each hole) from 'par'. When these cumulative deviations are plotted, differences in performance from par (i.e. from a target value) appear as upward or downward **slopes** on the chart. On-target performance gives a predominantly horizontal plot, and changes from the target average can be clearly and quickly recognized.

12.1.2 The cusum mode of plotting has considerable advantages as compared with conventional (Cartesian) presentation of series data.

(i) Any indication of a change in mean level is given by the slope of the chart over a number of points, and is thus less prone to influence by occasional 'wild' values (which can in any case be clearly seen against the general background).

(ii) Local average levels, either for sequences of particular interest or for routine reporting, may be directly estimated from the chart without recourse to the individual data values.

(iii) It is nearly always easier to locate approximate change points on a cusum chart than on a conventional graph. Collating change points with other information can yield useful diagnostics.

(iv) When decision rules are applied, cusum methods are usually more efficient than control charts with limit lines, in the sense that equally reliable decisions can be based on less sample information. In fact, simple control charts are effectively discretised cusums, this discretization resulting in some loss of information.

(v) With proper scaling, features such as trend, changes in variability, cycles, are more clearly recognized from plots in the cusum mode.

There are two disadvantages to be noted:

(i) The slightly greater complexity involved in preparing and operating the charts may be a drawback if unskilled or untrained operators are required to maintain them.
(ii) The interpretation of slope (rather than ordinate) as indicating mean level may cause conceptual difficulty to the uninitiated.

12.1.3 As for conventional methods of plotting series data, cusum charts may be used for many kinds of numerical information. Some examples are:

(i) individual values of a series, such as sales or production figures, index numbers, physical, chemical or other measurements;
(ii) sample statistics e.g. mean, range, counts of defectives or defects in samples taken at regular intervals, such as for quality monitoring;
(iii) Binary data, recordings such as yes/no, success/failure, in sequences of trials;
(iv) Discrepancies between predicted and observed values, such as (performance minus forecast) in time series modelling, or (observation minus estimate) in evaluating the adequacy of a regression function;
(v) measurements or counts arising from sample elements of varying sizes, where the object is to monitor the mean level or rate per unit of observation;
(vi) accumulated or historical data, plotted retrospectively, from which much may be learned about past behaviour of the system. This in turn may aid the formulation of hypotheses or indicate avenues for further investigation;
(vii) data from different sources, where a cusum plot provides an effective means of demonstrating differences (e.g. in average, proportion, rate of use, etc.) between the sources.

12.1.4 Some further points to be noted are that the greater sensitivity of the cusum method can be exploited either to achieve closer control (i.e. more rapid detection of changes) or to reduce sample size/frequency – or a combination of both. Cusum methods lend themselves particularly well to applications involving computers, programmable calculators or dedicated microprocessors. Finally, the generally clearer presentation obtained via the cusum plot, achieved by smoothing the common cause variation to enhance the 'signals', does not require any decision on grouping or weighting (as do the moving average or EWMA) methods), and any rogue data values can be identified against the general random noise of the cusum path.

12.1.5 A number of derivations of the basic cusum idea have received attention in statistical and quality technology journals. Generally known as **cu-scores**, they depend on dividing the observed values into zones which are scored as 0, ± 1, ± 2 etc., and these scores are then treated as data for cusum analysis. These methods achieve virtually the same efficiency as cusums, with some gain in simplicity, but they may sacrifice some of the other cusum features connected with estimating local averages and process corrections.

12.2 Preliminary requirements

12.2.1 As for other charting or analytical methods, some preparation for the use of cusums is necessary. The points in sections 7.1 and 7.2 are applicable (objectives, priorities, deciding what to measure, setting up a data collecting/handling procedure and considering sources of variation). Cusums, like control charts, depend on rational subgrouping, the formulation of a target value (further considered in section 12.2.2) and consideration of risks of both over-control, i.e. false alarms, and delayed response to real changes.

Because of the additional arithmetic involved and the change to representing averages by slopes rather than ordinates (i.e. height on the y-axis of a chart), cusum methods may not always be appropriate for 'on the bench' applications where sheer simplicity is the major concern. They are particularly useful in connection with management information, situations where data or testing is expensive and may involve samples of one item at a time, and also where sample or element sizes vary much more than the $\pm 25\%$ usually regarded as reasonable for methods such as p and u-charts. They can be readily linked to data held on computer files, and routines can be very simply added to spreadsheet operations.

12.2.2 It is necessary to formulate a **target value**, to be subtracted from each observation. This target corresponds to the centre-line on a conventional chart, and we pause to consider the choice of an appropriate value. Among the possibilities are:

(i) the midpoint or nominal of a specification, where it is intended to keep the process average at or close to the nominal level;
(ii) an 'aim' value or intended level. Where no specification exists, or where a process (rather than product) parameter is involved, there is often good technical sense in aiming the average level at some value known to give good performance;
(iii) sometimes a recent (and satisfactory) level of performance will serve as a suitable target, e.g. $\bar{\bar{x}}$, $n\bar{p}$, etc. from earlier data;
(iv) where a one-sided tolerance is involved, the mean level may be targeted at a safe threshold such as $3\sigma, 4\sigma$, even 5σ, away from the limit for individual values, so that high capability is obtained, and a signal will be generated if the capability is jeopardized;
(v) in management applications, agreed budgets, rates of use of materials, sickness or accident rates, will provide the target value;
(vi) in some rather specialized applications for binary values (0, 1 scores for good/bad, pass/fail characteristics) where single values are obtained for

reasons of expense, time, complexity of test procedure, a target proportion p (which is a decimal fraction between 0 and 1) constitutes a target level.

(vii) In evaluating the performance of prediction or forecasting systems, the target may be zero when the observation is of the kind

$$\text{Observed} - \text{Predicted}$$

on the grounds that if the system is operating correctly, the average discrepancy should be zero. Usually, of course, there is a time lag between the prediction and the availability of the corresponding observed value, e.g. actual sales minus sales predicted by a forecasting model;

(viii) finally, when plotting a retrospective cusum, a suitable target is provided by the mean value of the whole data series. This mean value may be slightly rounded to simplify subsequent arithmetic.

12.2.3 For reasons connected with scaling (section 12.3.3), as well as for decision criteria, a measure of variation of the plotted values is required. Because these values are often means, proportions, rates, etc., we shall refer to this measure as the standard error, σ_e. When individual values are plotted, this becomes simply their standard deviation.

Sometimes the standard error will be known either from previous records or from the nature of the problem. Thus if the values are means in samples of size n from a stable process with standard deviation σ, then the standard error will be $\sigma/\sqrt{n}$. However, one should beware of any between-samples component of variation, and it is generally a wise precaution to check by the estimation method given below.

Another situation is that of sampling from a binomial or Poisson distribution, where (in conventional notation) the standard error will be either $\sqrt{[np(1-p)]}$ for the binomial or $\sqrt{c}$ where c is the mean, for the Poisson distribution.

For empirical estimation, the most useful method of measuring variation is via the differences, δ_i, between successive values of the plotted variable. Two methods of using these differences were introduced in (9.1) and (9.2). For convenience, we repeat them here:

$$\hat{\sigma}_e = \sqrt{\left(\frac{1}{2}MSSD\right)} = \sqrt{\left[\frac{1}{2(k-1)}\sum \delta_1^2\right]} \qquad (12.1)$$

the square root of half the mean square successive difference;

$$\hat{\sigma}_e = \frac{1}{1.128}\bar{\delta} \doteq \frac{8}{9}\bar{\delta} \qquad (12.2)$$

Note that (12.1) is preferable for counted data, especially if containing numerous 0's and 1's, or if only a small number of differing values occurs, as with small numbers of non-conforming items or non-conformities.

Methods of estimating variation from varying element sizes (weighted estimation) are introduced in connection with weighted cusums in section 12.9.

12.3 Cusum procedure and introductory example

12.3.1 By way of introducing some important aspects of cusum charting, we take the set of 32 observations listed in Table 12.1 in the column headed 'Observation, x'. The remaining columns will be used as this section progresses.

Table 12.1 Data for cusum analysis

Observation number, i	Observation, x	Cumulation, $\sum x$	Cusum of $\sum(x-\bar{x})$	Cusum of $\sum(x-14)$	Successive differences, Δ
1	13.6	13.6	−1.23	−0.4	4.8
2	18.4	32.0	2.34	4.0	5.2
3	13.2	45.2	0.71	3.2 etc.	2.2
4	11.0	56.2	−3.12	0.2	6.5
5	17.5	73.7	−0.46	3.7	2.2
6	15.3	89.0	0.01	5.0	4.8
7	10.5	99.5	−4.32	1.5	1.5
8	12.0	111.5	−7.15	−0.5	3.7
9	15.7	127.2	−6.28	1.2	1.0
10	16.7	143.9	−4.41	3.9	1.6
11	15.1	159.0	−4.14	5.0	5.3
12	9.8	168.8	−9.18	0.8	5.9
13	15.7	184.5	−8.31	2.5	1.9
14	13.8	198.3	−9.34	2.3	1.2
15	15.0	213.3	−9.17	3.3	1.7
16	13.0	226.6	−10.70	2.6	2.1
17	11.2	237.8	−14.33	−0.2	0.7
18	11.9	249.7	−17.26	−2.3	3.0
19	14.9	264.6	−17.19	−1.4	4.1
20	10.8	275.4	−21.22	−4.6	0.8
21	11.6	287.0	−24.46	−7.0	2.0
22	13.6	300.6	−25.69	−7.4	3.3
23	16.9	317.5	−23.62	−4.5	1.3
24	18.2	335.7	−20.25	−0.3	5.1
25	13.1	348.8	−21.98	−1.2	7.4
26	20.5	369.3	−16.31	5.3	6.3
27	14.2	383.5	−16.94	5.5	5.1
28	19.3	402.8	−12.48	10.8	0.5
29	18.8	421.6	−8.51	15.6	5.6
30	13.2	434.8	−10.14	14.8	5.5
31	18.7	453.5	−6.27	19.5	2.4
32	21.1	474.6	0	26.6	

$\bar{x} = 14.83125$; $s = 3.076$; $\bar{\delta} = 3.3774$
$\bar{\delta}/1.128 = 2.994$; $\sqrt{\frac{1}{2}MSSD} = 2.775$;
$(8/9\bar{\delta} = 3.002)$.

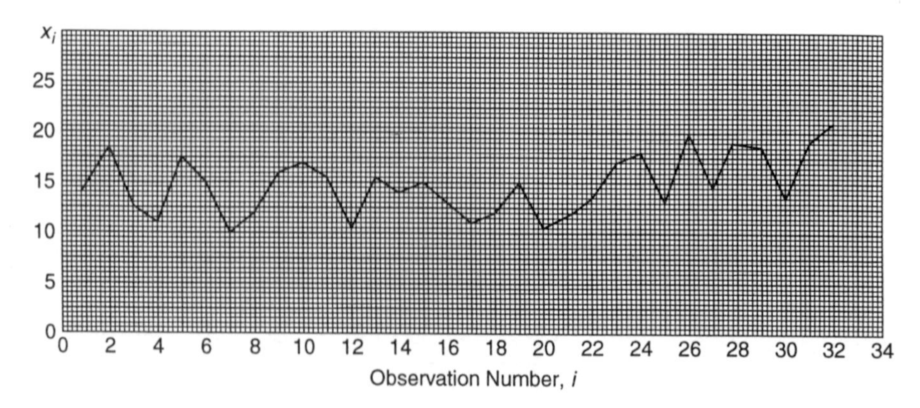

Fig. 12.1 Run chart for data of Table 12.1.

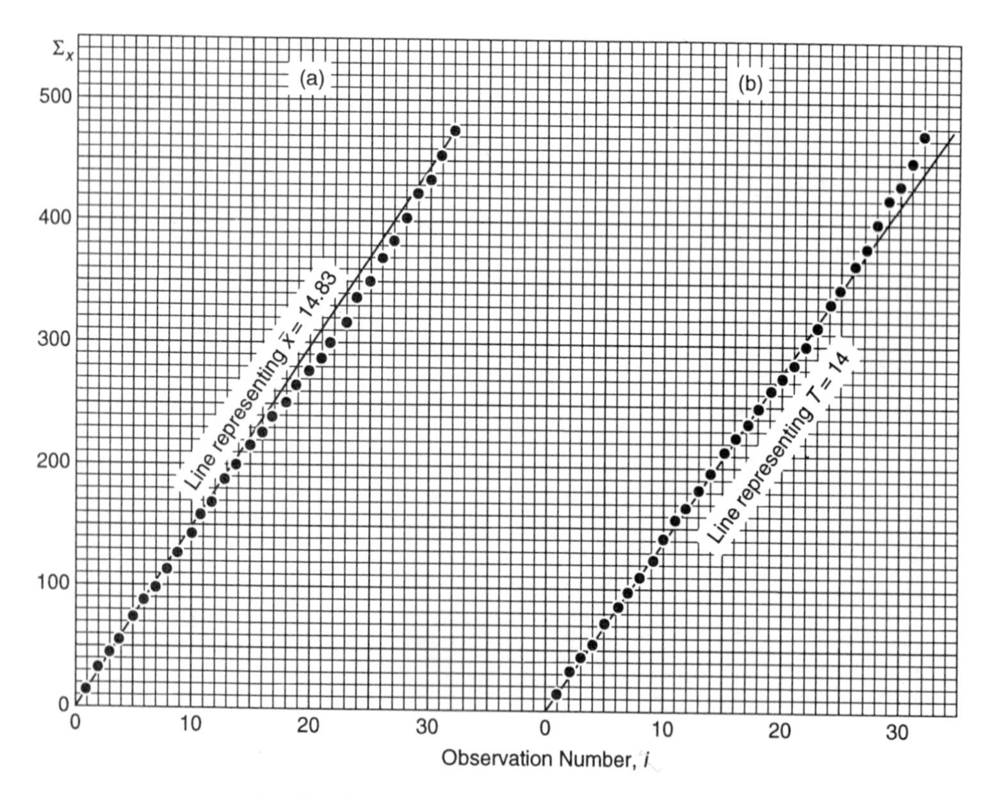

Fig. 12.2 Cumulative charts for data of Table 12.1.

First, we plot the data on a run chart, Fig. 12.1. The plot is not remarkable excepting for an indication that the average of the last seven points may be slightly higher than the earlier part of the sequence. Possibly the average of the values 12–22 may be a little lower than the first ten or eleven.

A device often used in production, financial or sales data is the **cumulative** chart (this must be distinguished from the **cusum**). The Σx column of Table 12.1 develops the running total of the observations, and these totals are plotted as the cumulative chart of Fig. 12.2 Part (a) of this graph shows a sloping line representing the overall average of the observations, and the main feature is a curved deviation from this line centred around observation 22 – a 'dip' below the overall average followed by a 'catching up' with the $\bar{x}$ line.

Part (b) of Fig. 12.2 shows the same cumulative totals, but with a line representing a target value of 14 – perhaps a monthly budget, a nominal dimension or a sales target. Here there is little evidence of a fall below the target, but a clear rise above the target line from around observation 28 onward.

12.3.2 The cumulative plots provide some indications of changes, but the change points, or regions where the mean shifted from one value to another, are not clear. The cusum mode of plotting offers an advantage here. By rotating the plot until the path is approximately horizontal, deviations from either the overall $\bar{x}$ or a target value, T, become more apparent. The vertical scale can also be more generous, as the cumulated deviations cover a much smaller range than the running total of original values.

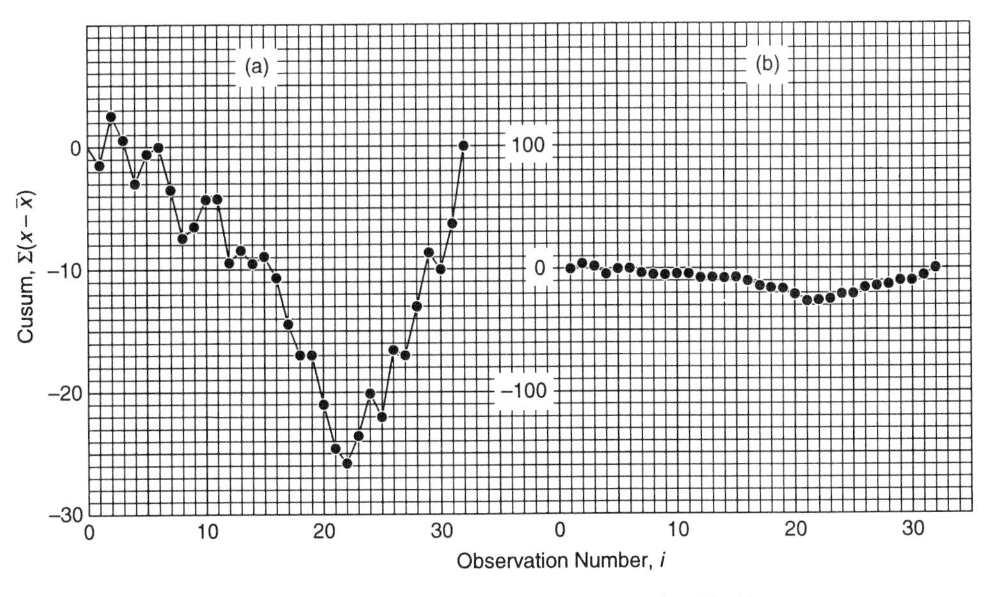

Fig. 12.3 Cusum plots, $\sum(x - \bar{x})$, for data of Table 12.1.

Nevertheless, the scale should not be too wide, otherwise the inherent common cause variation (or 'noise') can obscure other effects. Unfortunately some computer software incorporating cusums operates on the principle of filling the available screen space. Figure 12.3(a) illustrates this for the data of Table 12.1, using $\bar{x}$ as a target value for the cusum. While the three possibly different averages show as differing slopes ($i=0$ to 15, 16 to 22 and 23 to 32 approximately) the plot is rather noisy, and some local peaks and troughs attract undeserved attention (e.g. at $i=2$, 4, 10–11, etc.). Damping the scale too much makes it difficult to read anything into the plot at all, as in Fig. 12.3(b), where the scale of the cumulative chart (Fig. 12.2) has been used.

12.3.3 Scaling the cusum axis evidently deserves some attention. A useful general convention is first to select a convenient distance on the i-axis for the interval between successive samples: this may vary from 0.1 inch (or 2 mm) for desk use to 1 inch (or 20–25 mm) for wall charts. The same distance on the cusum axis is then taken to represent approximately two standard errors ($2\sigma_e$). The actual value of $2\sigma_e$ is rounded to a convenient scale factor: for example with $\sigma_e=2.3$ the exact value of $2\sigma_e=4.6$ would be rounded to 4 or 5 as a convenient scale factor. Generally, numbers such as 1, 2, 2.5, 4, 5, 10 form convenient factors.

Special rules may be necessary for some applications (e.g. when the data comprise strings of 0's and 1's with a target value between 0 and 1). BS 5703, especially Part 4, gives advice on this point.

For the data in the present example, the standard error is around 3 (the various methods of estimation give slightly differing results), but $2\sigma_e=6$ is not a convenient number for scaling. Five units, or possibly 10, would be preferred. Figure 12.4 shows a chart with a vertical scale of 5 units corresponding to one sample interval, with the cusum based on $\Sigma(x-\bar{x})$.

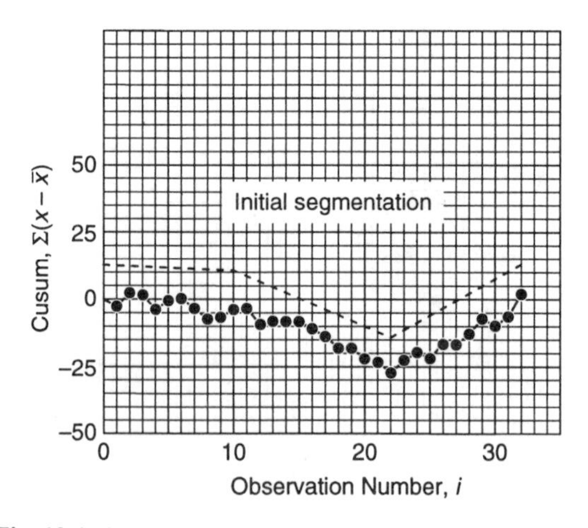

Fig. 12.4 Correctly scaled cusum for data of Table 12.1.

12.3.4 With less distraction from the common cause variation, it is now more apparent that three segments occur within the whole series, as indicated by the sloping lines above the plotted points in Fig. 12.4. Evidently one would wish to assess whether these are **significant** changes or special causes, and the subject of decision rules is taken up in section 12.5.

In most SPC applications, one would require not so much a retrospective analysis of $\Sigma(x-\bar{x})$ but a procedure for monitoring against a target value. Suppose that $T = 14$ as for the second part of Fig. 12.2(b). The resulting cusum

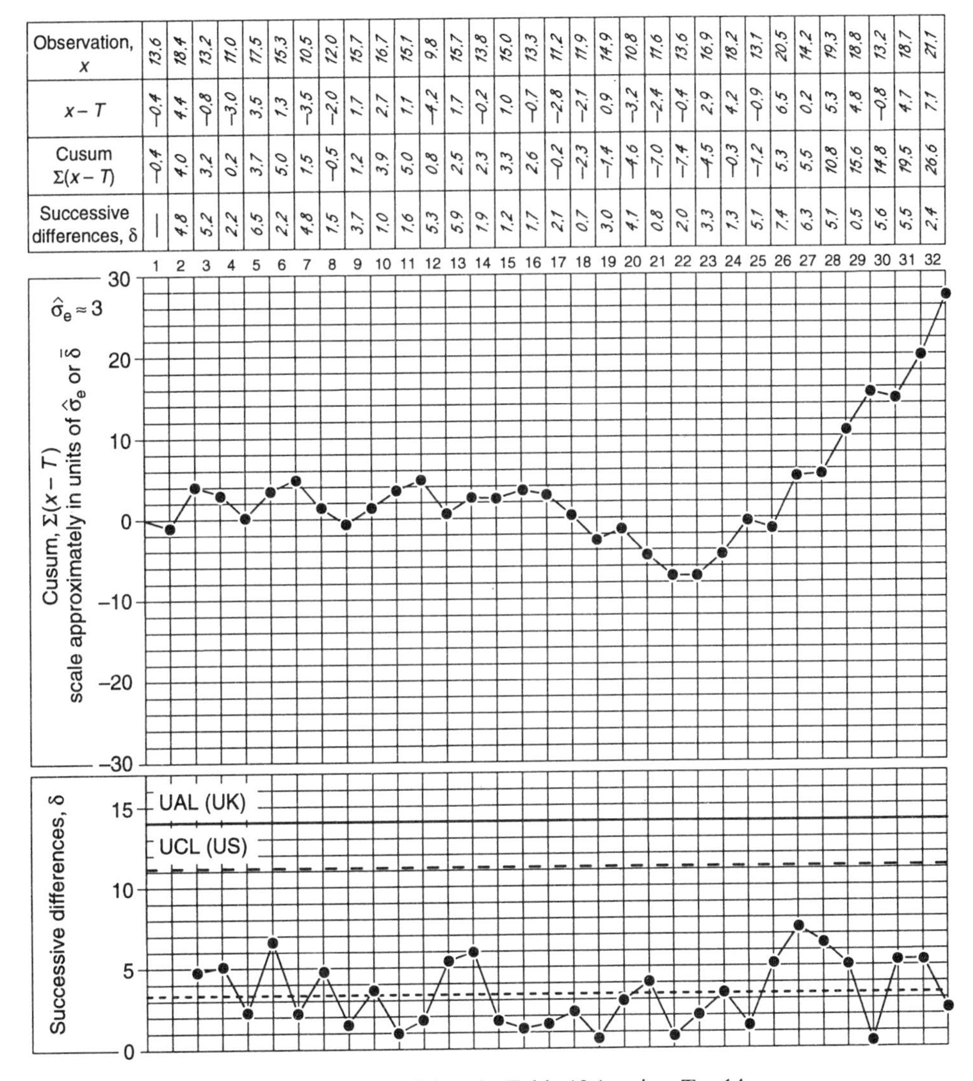

	1	2	3	4	5	6	7	8	9	10	11	12	13	14	15	16
Observation, x	13.6	18.4	13.2	11.0	17.5	15.3	10.5	12.0	15.7	16.7	15.1	9.8	15.7	13.8	15.0	13.3
$x-T$	−0.4	4.4	−0.8	−3.0	3.5	1.3	−3.5	−2.0	1.7	2.7	1.1	−4.2	1.7	−0.2	1.0	−0.7
Cusum $\Sigma(x-T)$	−0.4	4.0	3.2	0.2	3.7	5.0	1.5	−0.5	1.2	3.9	5.0	0.8	2.5	2.3	3.3	2.6
Successive differences, δ	—	4.8	5.2	2.2	6.5	2.2	4.8	1.5	3.7	1.0	1.6	5.3	5.9	1.9	1.2	1.7

	17	18	19	20	21	22	23	24	25	26	27	28	29	30	31	32
Observation, x	11.2	11.9	14.9	10.8	11.6	13.6	16.9	18.2	13.1	20.5	14.2	19.3	18.8	13.2	18.7	21.1
$x-T$	−2.8	−2.1	0.9	−3.2	−2.4	−0.4	2.9	4.2	−0.9	6.5	0.2	5.3	4.8	−0.8	4.7	7.1
Cusum $\Sigma(x-T)$	−0.2	−2.3	−1.4	−4.6	−7.0	−7.4	−4.5	−0.3	−1.2	5.3	5.5	10.8	15.6	14.8	19.5	26.6
Successive differences, δ	2.1	0.7	3.0	4.1	0.8	2.0	3.3	1.3	5.1	7.4	6.3	5.1	0.5	5.6	5.5	2.4

Fig. 12.5 Cusum of data in Table 12.1, using $T = 14$.

(the penultimate column of Table 12.1) is shown in figure 12.5 on a standardized cusum grid. By arranging the vertical scale divisions as half the width of the horizontal divisions, the cusum axis may simply be scaled in units of approximately one standard error – or even of $\bar{\delta}$, since there is only a discrepancy of about 11% between $\bar{\delta}$ and $\hat{\sigma}_e$. This standard chart contains a lower section which can be used either to plot successive differences or as a run chart of the x-values to assist in interpreting the cusum.

The interpretation of Fig. 12.5 is evidently similar to that of Fig. 12.4. The use of $T = 14$ in place of $\bar{x} = 14.83$ as the basis of the cusum has made some slopes steeper, some less so. The 'process' (in this case a simulation) is on target at least to observation 16, as indicated by the horizontal path of the cusum. A slight downslope (mean perhaps below 14) is followed by a dramatic upslope, suggesting a sudden slippage to a mean that is some way from the target. That this latter is a significant change or special cause will be demonstrated in section 12.6.

12.3.5 We now need to formalize the cusum procedure. After choosing an appropriate target value, T, the target is subtracted from each observation, x_i, and the deviations from target, $x_i - T$, are cumulated to form the cusum. At the ith data point, the cusum, C_i, may be denoted by

$$C_i = \sum_{r=1}^{i} (x_r - T) \qquad (12.3)$$

where r is a dummy variable for summation purposes.

The cusum chart is then formed by plotting C_i on the usual vertical axis (the y-axis or ordinate) against i on the horizontal axis (x-axis or abscissa).

Because any slope on the chart corresponds to an average value for the x-variable, the calculation of average slope across any segment of the chart provides an estimate, **identical to** $\bar{x}$, of the average for the segment. Of course, some precision may be lost if slopes are estimated from the graph rather than by calculation from the actual cumulations, but even graphical estimates will correspond closely to the calculated $\bar{x}$ for any segment. The details of these estimation procedures are now considered.

12.4 Calculation of averages from a cusum chart

12.4.1 As noted in section 12.1.2, local averages may be calculated from the cusum values over any section of the chart. Using the notation of section 12.3.4, the arithmetic mean of the x-values from the $(j + 1)$th to the kth inclusive is given by

$$\bar{x}_{j+1,k} = T + \frac{C_k - C_j}{k - j} \qquad (12.4)$$

If a graphical representation of this average is required, the procedure is to join the cusum point at observation j to that at observation k by a straight line. Note that it is *not* necessary (indeed, it is incorrect) to draw a 'line of best fit' through the intermediate cusum points.

12.4.2 Exploiting the relationship between cusum slope and the mean of the observations, it is also possible to prepare a slope guide or protractor for approximate interpretation of the chart. When the process is on target, the mean is represented by a horizontal line. For a mean deviation from target of, say, g units, the corresponding line will slope away from the horizontal by g units per sample – often more conveniently measured as $10g$ units per 10 sample intervals. The procedure is illustrated in Fig. 12.6, producing a fan of lines that provide quick estimates of local averages by eye or using a parallel or rolling ruler.

12.4.3 We now illustrate these estimation methods using the same example. Figure 12.6 reproduces the plot of section 12.5, with lines representing the mean values in the following segments:

Segment A $j+1=1$, $k=16$ (whence $j=0$)
Segment B $j+1=17$, $k=22$ (giving $j=16$)
Segment C $j+1=23$, $k=32$ ($j=22$).

The number of points contributing to each mean is $k-j$, viz 16 in segment A, 6 in segment B and 10 in segment C.

12.4.4 The calculation of these arithmetic means is via (12.4):

Segment A. $j=0, k=16;$ $C_0=0,$ $C_{16}=2.6;$

$$\bar{x}_{1,16}=14+\frac{2.6-0}{16-0}=14.1625:$$

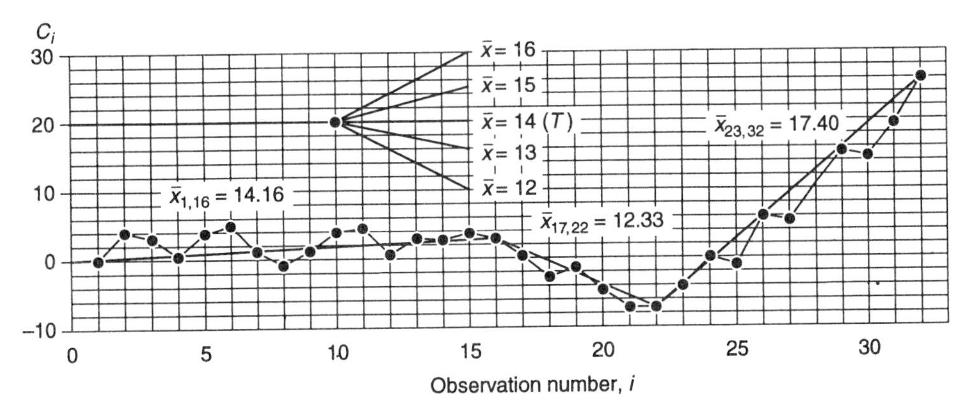

Fig. 12.6 Cusum chart with slope guide and average lines.

Segment B. $j = 16, \quad k = 22; \quad C_{16} = 2.6, \quad C_{22} = -7.4;$

$$\bar{x}_{17,22} = 14 + \frac{-7.4 - 2.6}{22 - 16} = 12.\dot{3}:$$

Segment C. $j = 22, \quad k = 32; \quad C_{22} = -7.4, \quad C_{32} = 26.6;$

$$\bar{x}_{23,32} = 14 + \frac{26.6 - (-7.4)}{32 - 22} = 17.40.$$

The slope guide (used with a parallel ruler) suggests for segment A an average nearer to 14 than to 15; for segment B a mean between 12 and 13; and for segment C a mean considerably in excess of 16 (the protractor could include a wider spread of averages if desired).

12.5 Estimation of trend from a cusum chart

12.5.1 The calculation of $\bar{x}$ for a segment presumes that the average was fairly stable throughout that segment. Often, a process or system characteristic is subject to a gradual drift or trend away from a target or previous average. Trend or drift is indicated on a cusum chart by a curved path of the plotted points. It is not always easy to distinguish between a 'step' change in mean and a drift (it is even more difficult to do so via a run or control chart) and the example illustrates this point. For segments A and B combined, is there (subject to the vagaries of the background variation) a gradual curvature rather than a sudden change? The same may be true of segment C, where there is a suspicion (no more) that the path may be steepening towards the extreme right end. When a trend is very marked, or persists for a long period, it becomes easier to detect.

12.5.2 Using *three*, preferably equally spaced, cusum values, it is possible to estimate both the rate of change and the transient mean at the centre of a segment. Consider segment AB. Here we have

$$C_0 = 0; \quad C_k = -7.4 \text{ at } k = 22$$

We require a further cusum approximately half way along the segment, say $C_{11} = 5.0$. The corresponding observation numbers may be denoted by i, j, k, thus:

$$\begin{aligned} i &= 0 & C_0 &= 0; \\ j &= 11, & C_{11} &= 5.0; \\ k &= 22, & C_{22} &= -7.4. \end{aligned}$$

Then the rate of change, say Δ units per sample interval, is estimated as

$$\hat{\Delta} = 2\left\{\frac{C_k - C_i + (k-i)T}{(k-j)(k-i)} - \frac{C_j - C_i + (j-i)T}{(k-j)(j-i)}\right\} \tag{12.5}$$

The mean at the midpoint of the segment, corresponding to observation $(k + i)/2$, is obtained via (12.4), but substituting i for j, as

$$\bar{x}_{i+1,k} = T + \frac{C_k - C_i}{k - i}$$

For the present example, these estimates become

$$\hat{\Delta} = 2\left\{\frac{-7.4-0+(22-0)\times 14}{(22-11)(22-0)} - \frac{5.0-0+(11-0)\times 14}{(22-11)(11-0)}\right\} = -0.1438;$$

$$\bar{x}_{1,22} = 14 + \frac{-7.4-0}{22-0} = 13.66$$

Thus the mean around observation 11 is estimated as 13.66, but appears to be falling by about 0.144 units per observation over the whole segment. This would imply a mean of

$$13.66 + 11 \times (-0.1438) = 12.08$$

at the end.

Cusums are in fact very efficient at detecting or giving early warning of trends, another of their advantages over conventional control charts.

12.6 Decision rules for monitoring applications

12.6.1 To make an intelligent choice among the many possible alternative decision rules, we take up the subject of **average run length** (*ARL*) introduced in section 10.4.2. The *ARL* is defined as the average number of samples taken, with a stable process average, until a signal is generated by the control scheme. For a conventional control chart, such a signal may comprise a value beyond a control limit, or two successive warning values, or a run of seven or more points on one side of the centre-line, etc. When the process is on target, such signals are false alarms, and the average number of samples between such false alarms is usually denoted by L_0.

A control chart using *only* a control (or action) limit rule has L_0 approximately 700–1000, depending on the positioning of the limit. On adding a warning rule, this falls to about 500–650, and further rules (seven-in-a-row, etc.) may bring L_0 down to around 200. For two-sided control, the risks of false alarms are doubled, so L_0 is halved.

The *ARL* also provides a means of measuring the effectiveness of a control scheme in detecting real changes. For a step change of one or two standard errors, the *ARL*'s are shown in Table 12.2 for some typical sets of decision rules.

12.6.2 The earliest cusum rule to achieve widespread use was designed to match the performance of the conventional chart, (with warning rule but not runs rules)

Table 12.2 Average run length data for control charts

Decision rules	Deviation of mean from target value $0 \times (L_0)$*	$1.0 \times \sigma_e$	$2.0 \times \sigma_e$
UCL or UAL only	700–100	45–55	~7
Additional warning rule	500–600	~20	~4
Additional runs-of-7 rule	~200	~12	~3.5

*Values of L_0 are for one-sided control, and are halved for two-sided control. Other ARL's (for 1 and $2\sigma_e$ shift) are little affected.

at zero and $2\sigma_e$ shifts (this was around 1960, and runs-of-n rules were not then in general use). The rule is implemented by constructing the mask shown in Fig. 12.7, and placing it over any point of interest on the chart (often the most recently plotted point when there is a suspicion of a change) as in Fig. 12.8(a).

The parameters of this mask are the **decision interval** of $5\sigma_e$, representing the half-width at the narrow end, and the slope, $0.5\sigma_e$. Because of the simplicity of the construction illustrated, it is widely known as the 5, 10, 10 mask. Its *ARL* performance is shown in Table 12.3 along with that of some alternatives.

One of the criticisms of the 5, 10, 10 mask is that it does not generate a signal when an action violation or two warning values occur: it is more effective at detecting sustained changes in average of 0.5 to 2 standard errors. The conventional chart detects such slippages via the various runs rules, largely developed after the 5, 10, 10 mask was designed. As illustrated in Table 12.2, additional rules reduce the value of L_0, but if occasional false alarms are not a problem,

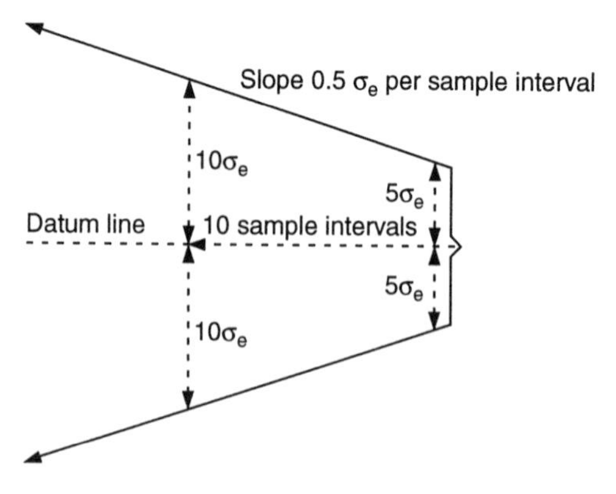

Fig. 12.7 Construction of 5σ, 0.5σ mask (5, 10, 10). The vertical distances for $5\sigma_e$ and $10\sigma_e$ are measured from the vertical scale of the cusum chart to which the mask refers.

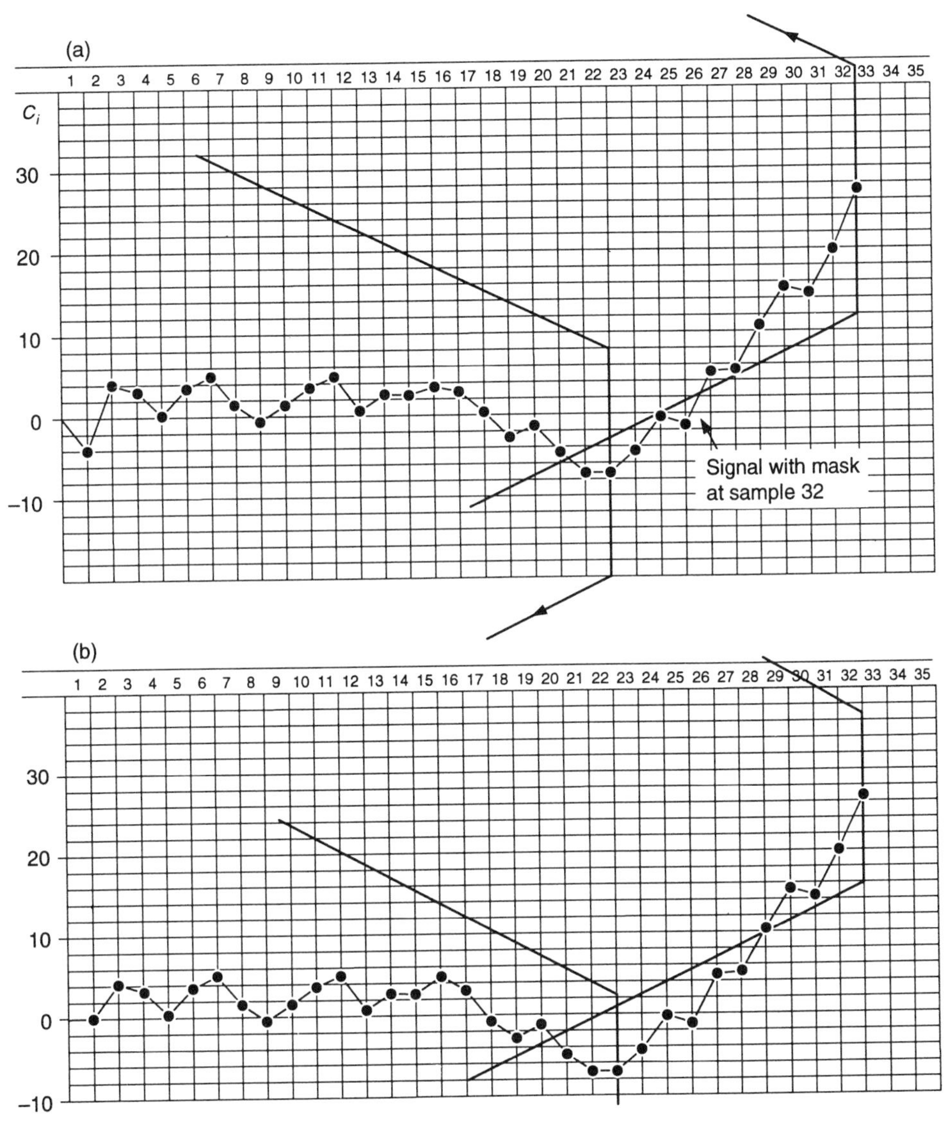

Fig. 12.8 Use of truncated mask on cusum chart. (a) Mask of Fig. 12.7; (b) mask of Fig. 12.9(c).

alternative cusum masks can be designed to combine the advantages of the cusum on small sustained shifts with the signalling of occasional large individual deviations.

In all cases, the datum of the mask is placed over the most recently plotted point (or the point at which a significant change may be signalled). The central line must be parallel to the sample axis, because this decision procedure is

Table 12.3 ARL performance of various control methods* (single sided L_0)

Shift from target ($X\sigma_e$)	Control charts: Control limit only	Control charts: A and W†	Control charts: A, W† + runs of 7	Cusums: 5, 10, 10	Cusums: 3.5, 10, 8.5	Cusums: SPB‡ or snub-nosed
0	700–1000	500–650	~200	930	200	440
0.25	400–450	~180	~75	140	55	110
0.5	~200	50–70	~35	38	22	35
0.75	~100	30–40	~20	17	11.5	16.5
1.0	~50	~20	~12	10.5	7.4	10
1.5	~18	~7.5	6	5.8	4.3	5.3
2.0	~7	~4	~3.5	4.1	3.0	3.4
2.5	~4	~3	2.3	3.2	2.4	2.3
3.0	~2.2	~2	1.7	2.6	2.0	1.7
4.0	~1.2	~1.2	~1.1	1.9	1.5	1.2

*ARL's for control charts are approximate, and depend on whether control limits are at 3.09 or $3.0\sigma_e$, and warning lines at 2.0 or $1.96\sigma_e$.
†A = action; W = warning
‡SPB = semi-parabolic mask of Fig. 12.9(a).

intended to signal any significant departure from the **target value**, which is of course represented by any horizontal line across the chart. Criteria for detecting changes *between* segments (rather than deviations from target) are discussed in 12.7.

Figure 12.9 illustrates some alternative masks and gives details for their construction. Further information, and a wider choice of mask shapes, is given in BS 5703, especially in Part 3.

12.6.3 The masks of Fig. 12.9(a) and (b) would give identical indications, for the data of the example, to those already obtained from the 5, 10, 10 mask. Their main advantage is that single points on the chart that would violate a control limit, or pairs of successive points that would give warning signals, will trigger an out-of-control response. Their general performance is very similar to those of control charts with additional rules, but with an appreciable reduction in false alarms and the benefit of the additional cusum features listed in section 12.1.2.

For tighter control (in the sense of even more rapid response to real deviations from target, with no higher risk of false alarms than the control chart with additional rules) the mask of Fig. 12.9(c) is preferable. Use of this mask in Fig. 12.8(b) would have yielded a signal at observation 29, but the small downslope over observations 16–22 would not have yielded an alarm. This mask is recommended for general use as a possible replacement for the control chart.

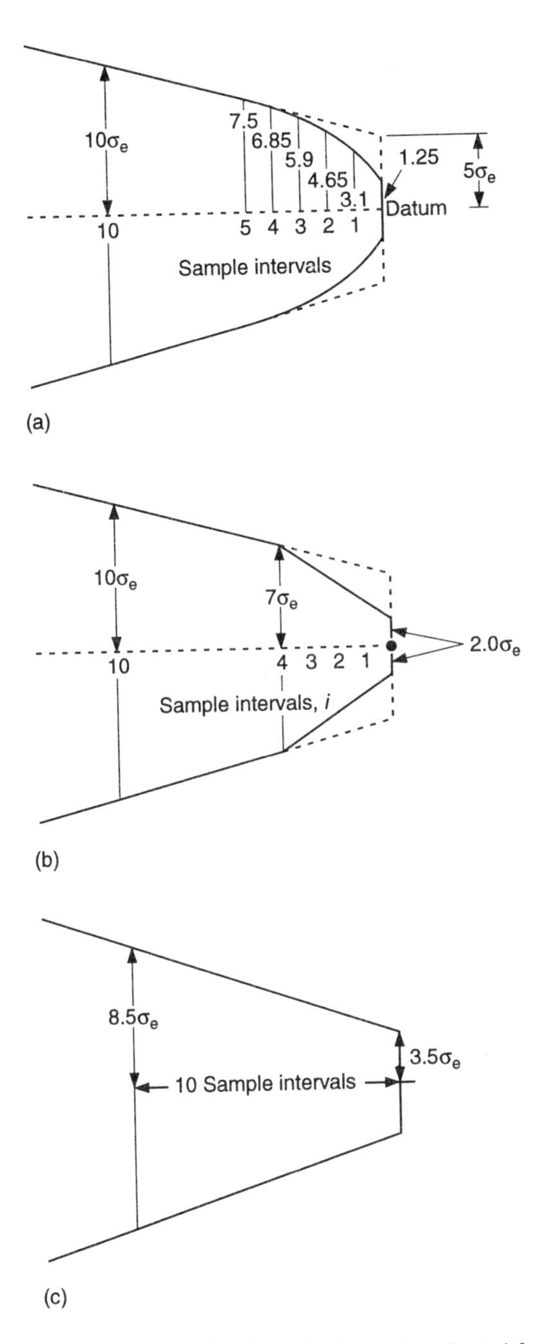

Fig. 12.9 Further cusum masks. (a) Semi-parabolic mask, adapted from 5, 10, 10 mask. Basic construction as Fig. 12.7 with curved section over first five sample intervals. Ordinates in units of standard error (σ_e). (b) Snub-nosed mask, adapted from 5, 10, 10 mask: Basic construction as Fig. 12.7 with inner line joining basic mask at 7σ, $i=4$ to 2σ, $i=0$. (c) Straight-sided mask for tightened control; 3.5, 10, 8.5. Similar to mask of Fig. 12.7 but with $h=3.5$, $f=0.5$, giving construction as shown above.

12.6.4 The decision rules described above are, like those of the control chart, intended for use where the plotted variable is approximately Normally distributed. However, the cusum decision criteria (implemented via the masks as illustrated) are rather more robust to non-Normality than control chart limits, largely because cusum decisions are generally based on the average indication over several individual points. For marked non-Normality as occurs for attribute charts with fewer than about 5 non-conformities per subgroup, or with small-sample ranges or standard deviations, alternative masks and rules are provided in BS 5703 (Part 3 for R or s, Part 4 for attributes and events).

It should be noted also that cusums lose some of their advantage in efficiency when monitoring for an increase in a positively skewed distribution, though they retain the benefits of average estimation and change-point indication. For this reason, a cusum for sample means is often accompanied by a conventional R, s or δ chart to monitor the level of common cause variation.

12.6.5 We conclude this section with some practical aspects of cusum charting, and especially of mask construction.

Scaling is important, and the use of a standard format like that of Fig. 12.5 offers considerable advantages. As in the example, space can be provided for the cusum calculations, but these are simplified by the use of a calculator with an M+, MR function, or with program/learn facility.

For manual use, a mask is best drawn on tracing paper, or better still, acetate sheet as used for overhead projectors. The construction can then be made with the transparent material placed over the actual cusum scale on the chart, improving the accuracy of shape as well as simplifying the construction.

Occasionally, when the average level of the system has been displaced from target for some time, a cusum path may run off the upper or lower bounds of the chart. It is here that one must remember that the actual magnitude of the cusum is of little importance – it is **changes** in the cusum that matter. Thus at any time, the cusum may be re-centred either by deducting a suitable number from C_i for a particular i, or by re-marking a cusum scale suitably displaced to bring the path of the points back to the chart centre. There are some applications where a cusum slope is desirable: e.g. a downslope on a chart for attributes indicating a reduction in non-conformities. In such cases, rather than re-centre, the chart can be displaced to re-start the cusum so that a further slope in the same sense can be accommodated. Figure 12.10 gives a slightly exaggerated example for purpose of illustration.

In connection with cusum slopes, it is worth noting that following a signal, it is *not* usually desirable to reverse the slope of the cusum by an over-adjustment so as to restore the cusum to zero. A proper adjustment to target will simply give a horizontal cusum from the adjustment point onward. There are some less common applications (in budgetary control, or progressive sampling of sub-batches to obtain an intended overall batch average) where 'slope reversal' to achieve a zero cusum is appropriate, but this is not the general case. Figure 12.11 includes an illustration of this point.

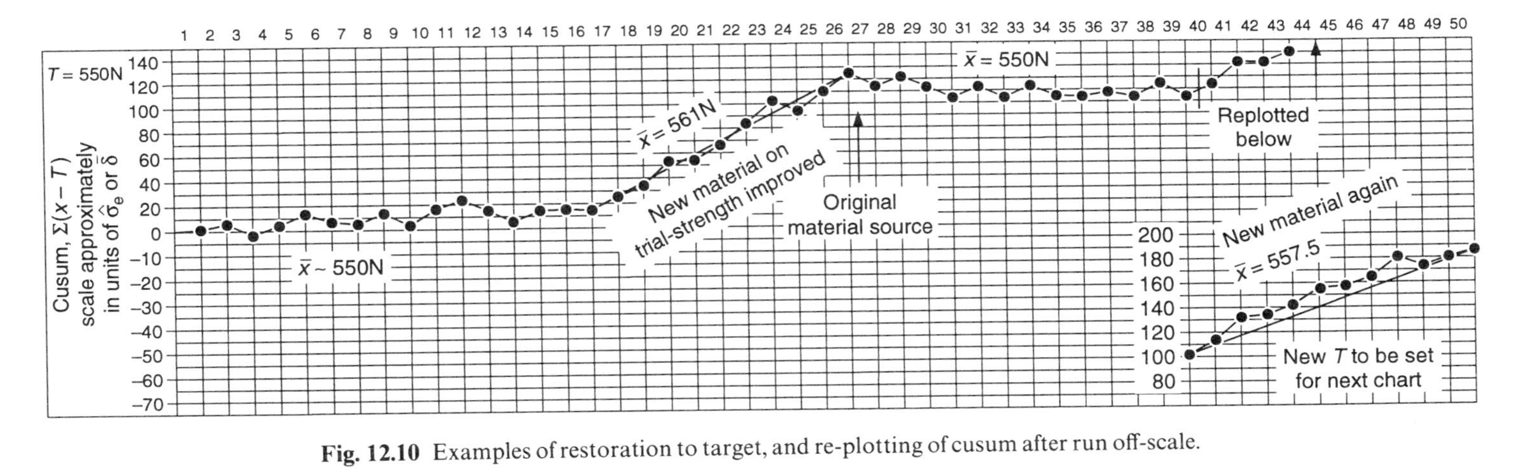

Fig. 12.10 Examples of restoration to target, and re-plotting of cusum after run off-scale.

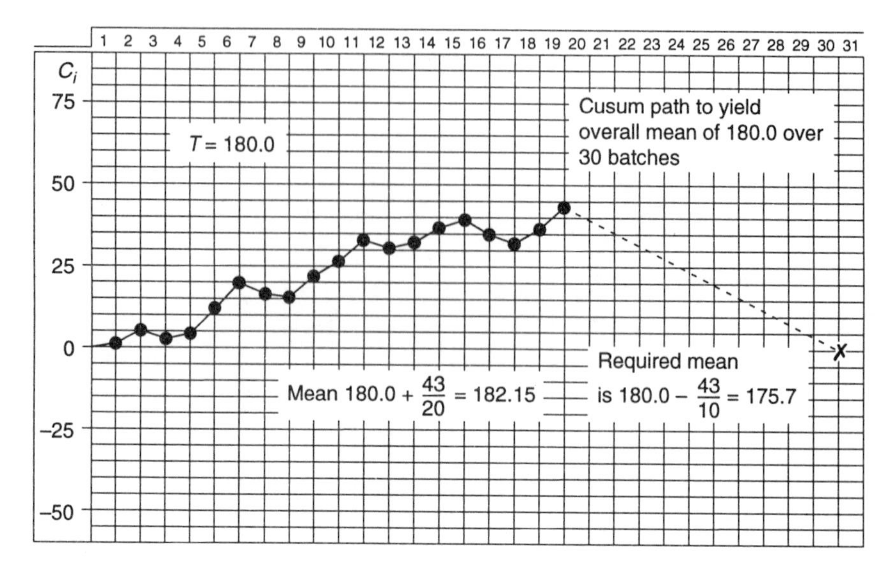

Fig. 12.11 Use of cusum to correct overall run average. Batches 1–20: average 182.15; target for 30 batches: 180.0. Revised target for batches 21–30: 175.7.

12.6.6 Cusums can, in fact, be operated in tabular mode without the use of charts. Such procedures are mainly useful in implementing computer algorithms for cusum control, but with present-day availability of screen display and graphical output, the tabulation format is now of limited use. The interested user may refer to BS 5703 Parts 2 and 3 for details, or the IOS *Introduction to Cusum charts* (1989) listed in the bibliography. Much of the advantage of the cusum technique is in its visual impact, and this should be exploited wherever possible.

12.6.7 Figure 12.12 gives an example of a cusum chart used for monitoring batch production of a chemical compound. Control is exerted on process conditions, material supply, formulation and mixing, but a final chemical analysis is the only available measure of the effectiveness of these procedures. It is required to signal any significant departure from a nominal value of 14.4, but with a standard deviation of 0.2 for individual values and a specification of ±0.5 there is no scope for any relaxation of control (though batches can be rectified by further processing or blending with other batches having opposite deviations from nominal).

The procedure was operated via a software package with user options, and a decision rule equivalent to a 3.5, 10, 8.5 mask was adopted. An updated cusum plot was displayed with each entry of a new observation, along with calculation of mean in the event of any significant change. The standard error (for single observations) had been derived from earlier data.

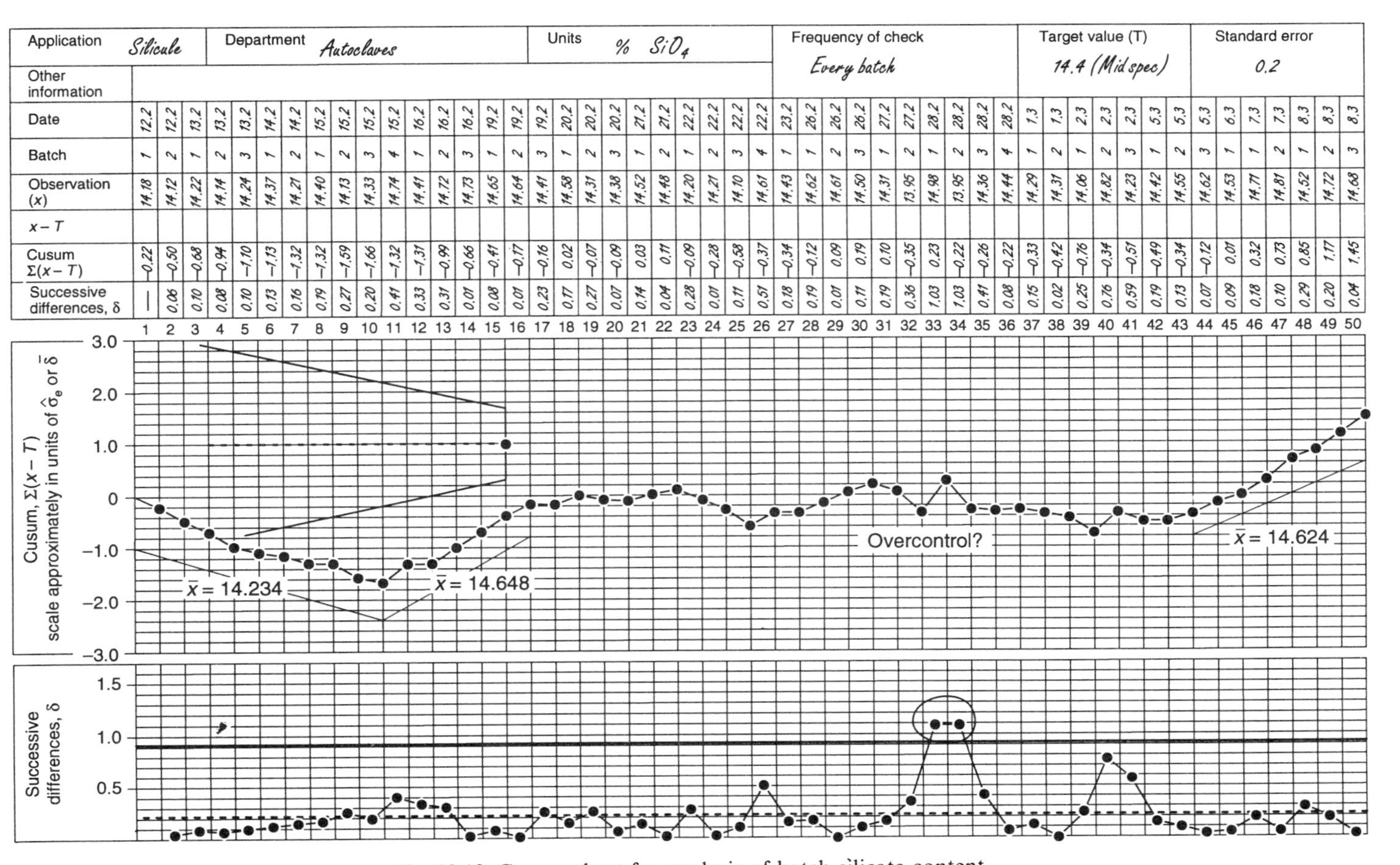

	1	2	3	4	5	6	7	8	9	10
Date	12.2	12.2	13.2	13.2	13.2	14.2	14.2	15.2	15.2	15.2
Batch	1	2	1	2	3	1	2	1	2	3
Observation (x)	14.18	14.12	14.22	14.14	14.24	14.37	14.27	14.40	14.13	14.33
x − T										
Cusum Σ(x − T)	−0.22	−0.50	−0.68	−0.94	−1.10	−1.13	−1.32	−1.32	−1.59	−1.66
Successive differences, δ	—	0.06	0.10	0.08	0.10	0.13	0.16	0.19	0.27	0.20

	11	12	13	14	15	16	17	18	19	20
Date	15.2	16.2	16.2	16.2	19.2	19.2	19.2	20.2	20.2	20.2
Batch	4	1	2	3	1	2	3	1	2	3
Observation (x)	14.74	14.41	14.72	14.73	14.65	14.64	14.41	14.58	14.31	14.38
x − T										
Cusum Σ(x − T)	−1.32	−1.31	−0.99	−0.66	−0.41	−0.17	−0.16	0.02	−0.07	−0.09
Successive differences, δ	0.41	0.33	0.31	0.01	0.08	0.01	0.23	0.17	0.27	0.07

	21	22	23	24	25	26	27	28	29	30
Date	21.2	21.2	22.2	22.2	22.2	22.2	23.2	26.2	26.2	26.2
Batch	1	2	1	2	3	4	1	1	2	3
Observation (x)	14.52	14.48	14.20	14.21	14.10	14.61	14.43	14.62	14.61	14.50
x − T										
Cusum Σ(x − T)	0.03	0.11	−0.09	−0.28	−0.58	−0.37	−0.34	−0.12	0.09	0.19
Successive differences, δ	0.14	0.04	0.28	0.01	0.11	0.51	0.18	0.19	0.01	0.11

	31	32	33	34	35	36	37	38	39	40
Date	27.2	27.2	28.2	28.2	28.2	28.2	1.3	1.3	2.3	2.3
Batch	1	2	1	2	3	4	1	2	1	2
Observation (x)	14.31	13.95	14.98	13.95	14.36	14.44	14.29	14.31	14.06	14.82
x − T										
Cusum Σ(x − T)	0.10	−0.35	0.23	−0.22	−0.26	−0.22	−0.33	−0.42	−0.76	−0.34
Successive differences, δ	0.19	0.36	1.03	1.03	0.41	0.08	0.15	0.02	0.25	0.76

	41	42	43	44	45	46	47	48	49	50
Date	2.3	5.3	5.3	5.3	6.3	7.3	7.3	8.3	8.3	8.3
Batch	3	1	2	3	1	1	2	1	2	3
Observation (x)	14.23	14.42	14.55	14.62	14.53	14.71	14.81	14.52	14.72	14.68
x − T										
Cusum Σ(x − T)	−0.51	−0.49	−0.34	−0.12	0.01	0.32	0.73	0.85	1.17	1.45
Successive differences, δ	0.59	0.19	0.13	0.07	0.09	0.18	0.10	0.29	0.20	0.04

Fig. 12.12 Cusum chart for analysis of batch silicate content.

Although no formal signal had occurred, it was decided (from the appearance of the displayed cusum) to make an upward adjustment after batch 3 of 15 February. Either this was a false alarm (though one more result below 14.3 *would* have given a signal) or there was an over-response. It was assumed that the raw material was low on silicate, and quantities were therefore reformulated. The opposite decision was triggered at batch 1 of 19 February, and a smaller (downward) adjustment was made on completion of batch 2. A fairly stable period then prevailed, except for some over-enthusiasm by a deputy shift foreman, unused to the cusum, around the end of February. This yielded the two violations of the δ-chart control limit, as well as some wobbles in the cusum path. Finally a signal after batch 2 on 7 March could not be traced to materials, and several further batches were made up before the problem was ovecome. These had to be stored pending make-up of a few special batches with low SiO_4 for blending off the highest silicate batches (although none were actually beyond the upper specification limit).

At the end of this period it was noted that, if the over-control period is discounted, a decrease of around 15% in standard deviation appeared to have occurred. If maintained, this would permit narrowing of the decision rule and a corresponding improvement in sensitivity.

A year later, the process had improved to the extent that some relaxation of strict control was possible, with standard deviation a little over 0.1%. A target zone of 14.35 to 14.45 was then specified for the process mean to further reduce the risk of over-control. This also reduced reformulation activity (a time-consuming task involving one or more trial batches) and simplified technical management of the process.

12.6.8 It is appropriate here to compare the performance of the cusum chart with that of the various methods of dealing with individual values developed in Chapter 9. Such comparison depends on the particular collection of rules used for control charts or the shape of mask used with a cusum chart. For example, if runs rules are not used, the individuals chart of Fig. 9.4 would not have signalled either of the changes. Again, if 3-sigma rather than 2.58-sigma limits were used on the *EWMA* chart of Fig. 9.6, the violations at samples 18 and 20 would not have occurred, though a run of 10 or 12 points below target could have been considered significant.

Table 12.4 summarizes the comparisons, in each case considering two variations of the rules or decision procedures, excepting for the individual values chart. The data is shown in cusum format in Fig. 12.13(a) and (b).

12.6.9 A final example from the sphere of TQM concerns industrial accidents. Figure 12.14 gives the data, cusum and run chart for accidents in a steel works from 1986 to 1989. The first 18 months, with a mean of 22 accidents per month, was used as a base period, and the subsequent data plotted with 22 as a target value. A deterioration in the latter part of 1987 was attributed to lapse of safety

Table 12.4 Comparison of control procedures

Individual values: Figs 9.1, 9.4; MA_5: Figs 9.2, 9.5; *EWMA*: Figs 9.3, 9.6; Cusum: Fig. 12.13(a) and (b).

	Sample number giving signal for various charts						
Sequence	Individuals $3\sigma_e$ (plus runs of 7)	MA_5* $3\sigma_e$	MA_5* $2.58\sigma_e$	*EWMA** $3\sigma_e$	*EWMA** $2.58\sigma_e$	Cusum 5, 10, 0 (Fig. 12.7)	Cusum 3.5, 10, 8.5 Fig. 12.9(c)
Initial data, disturbance 18–22	20	20	19	20	19	20	19
Downward trend, ongoing data 1–20	15	20	18	20	18	20	18
Run above target, ongoing data 21–45	45	37	35	36	36	41	35

*also runs of 12.

practices during an expansion programme. An overhaul of the system and the introduction of safety incentives resulted in improvement so that in late 1988, an average of less than 22 was apparent. Further effort in a 'safety training and observation programme' yielded a further dramatic reduction, with an average in early 1989 of around 12 per month. At this stage a new cusum chart, with $T = 12$, was drawn up for later use.

Note the 'Manhattan diagram' superimposed on the run chart. After any change, or when the mean has been stable for some time, a horizontal line across the run chart displays the latest mean value as calculated from the cusum.

12.7 Tests for retrospective cusums

12.7.1 When 'historical' data are plotted retrospectively, interest often lies in the differences between adjacent segments rather than in discrepancies from a target – indeed, there may be no specific target value. Several methods exist for testing differences between segments, one of which is based on adapting the truncated mask to a sloping datum corresponding to the mean of one of the segments under test. Details are again available in BS 5703 (Part 2).

We present here two of the most useful methods. The first requires very little description. The values of $\bar{x}$ and s for the two segments are calculated, and used

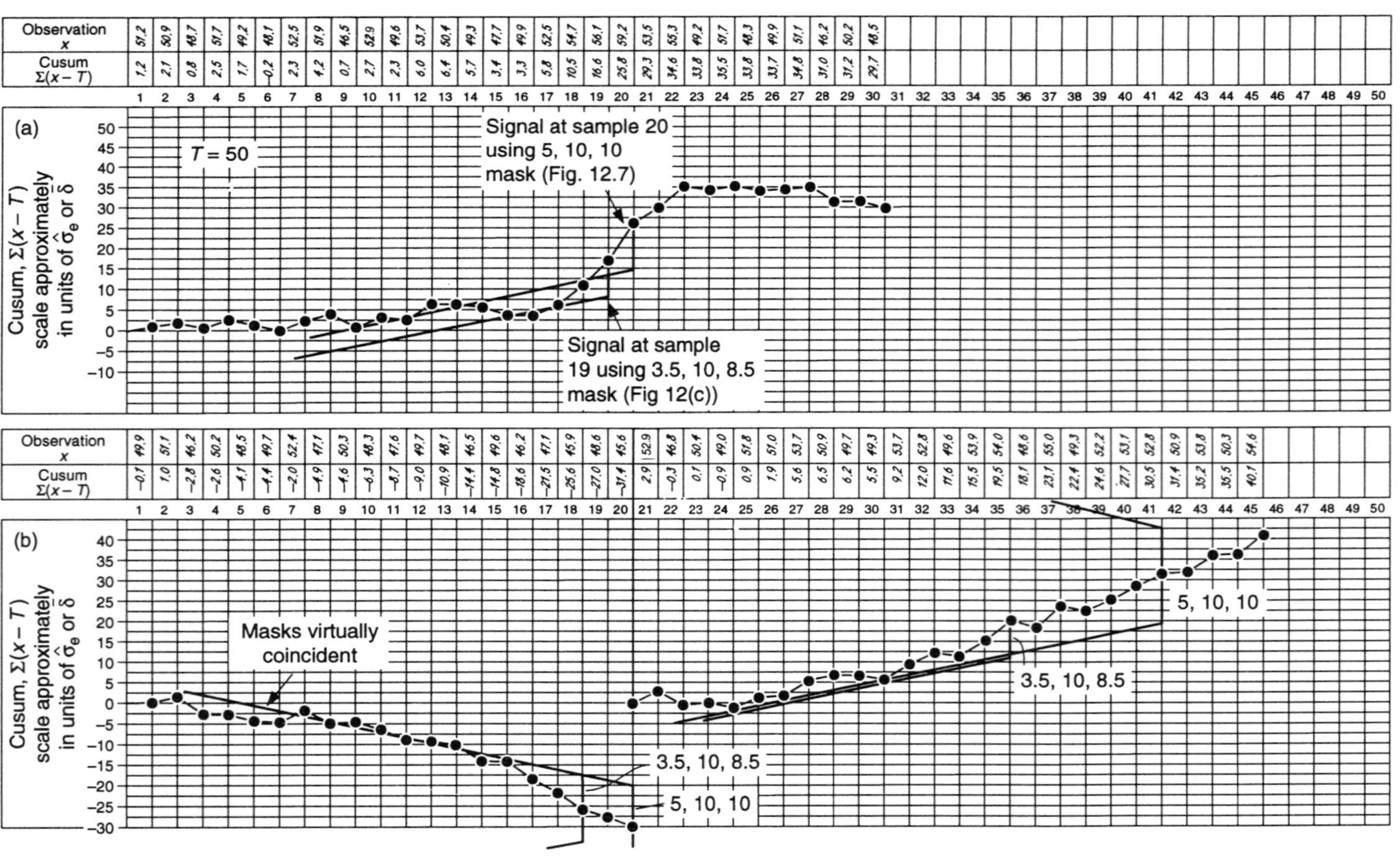

Fig. 12.13 Cusum charts for data of Figs 9.1 to 9.6. (a) Initial data of (cf. Figs 9.1–9.3); (b) on-going data (cf. Figs 9.4–9.6).

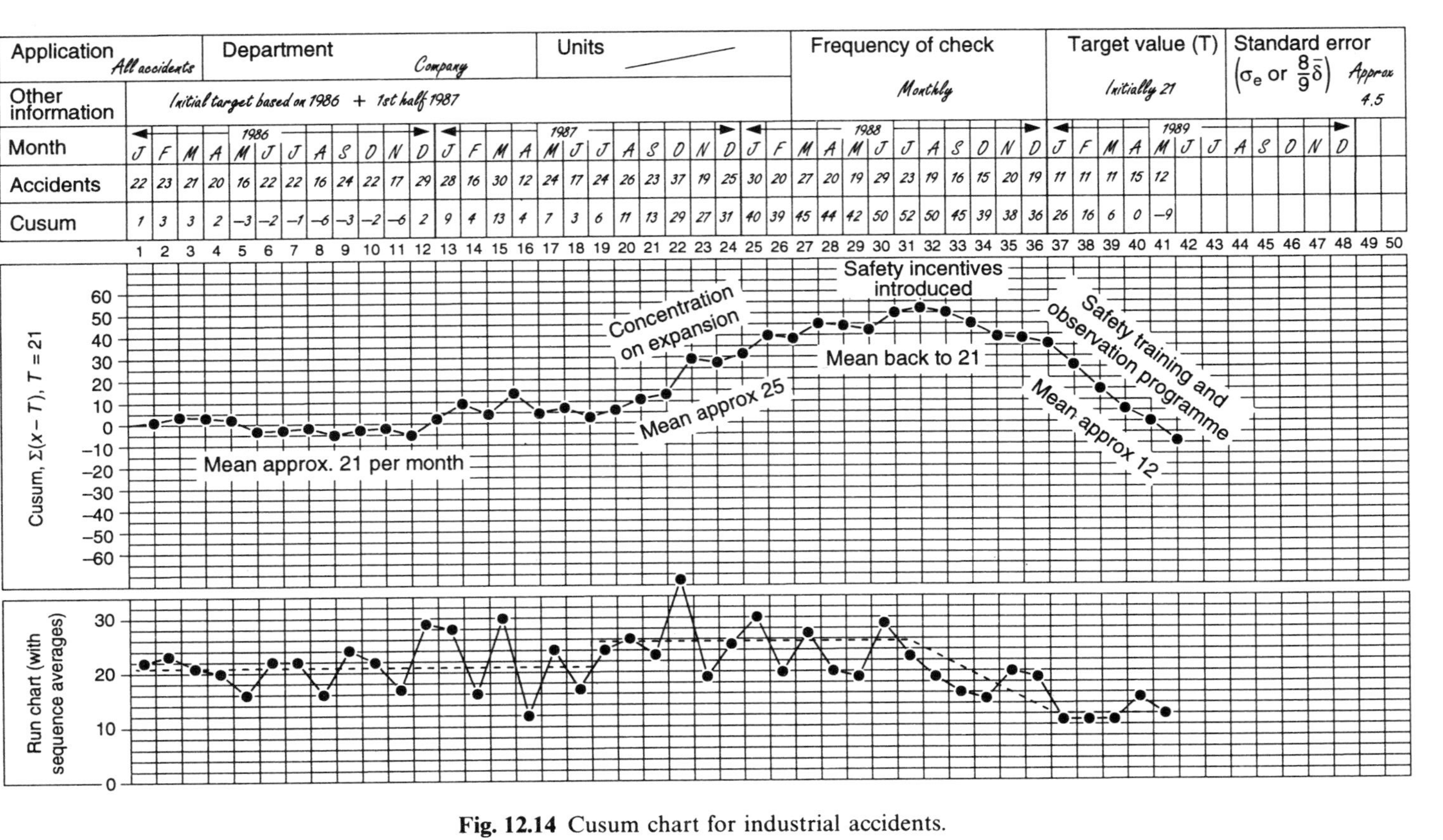

Fig. 12.14 Cusum chart for industrial accidents.

for the t or t' test of (6.6) or (6.10). Because the segments are usually chosen after inspection of the chart, it is wise to interpret the significance probability cautiously. This is especially prudent if several pairs of segments are tested within one overall sequence.

The second test was specifically developed for the retrospective cusum. Known as the **span test**, it is illustrated in Fig. 12.15. The maximum vertical distance, V_{max}, is determined from the line joining the ends of the two segments under test, and V_{max}/σ_e may then be tested for significance using the criteria in Table 12.5. V_{max} may either be estimated graphically, or from formula (12.6). Using the notation as for estimating a trend, we set i, k corresponding to the starting point and finishing point of the whole span, and j to the indicated change point between i and k. Then

$$V_{max} = C_j - C_i - \frac{j-1}{k-i}(C_k - C_i) \tag{12.6}$$

12.7.2 For retrospective plotting, the mean of the complete sequence of data is usually taken as the target for calculating the cusum. Figure 12.15 shows the data of Table 12.1 replotted using the mean, 14.831 25, to obtain the cusum. As in Figs. 12.4 and 12.5, there appear to be three possibly differing segments. The slopes in Fig. 12.13 differ slightly from the earlier cusum plots because of the change in target value, but the general pattern is similar. The apparent change points are at samples 11 and 22, but taking the first pair of segments 0–11 and

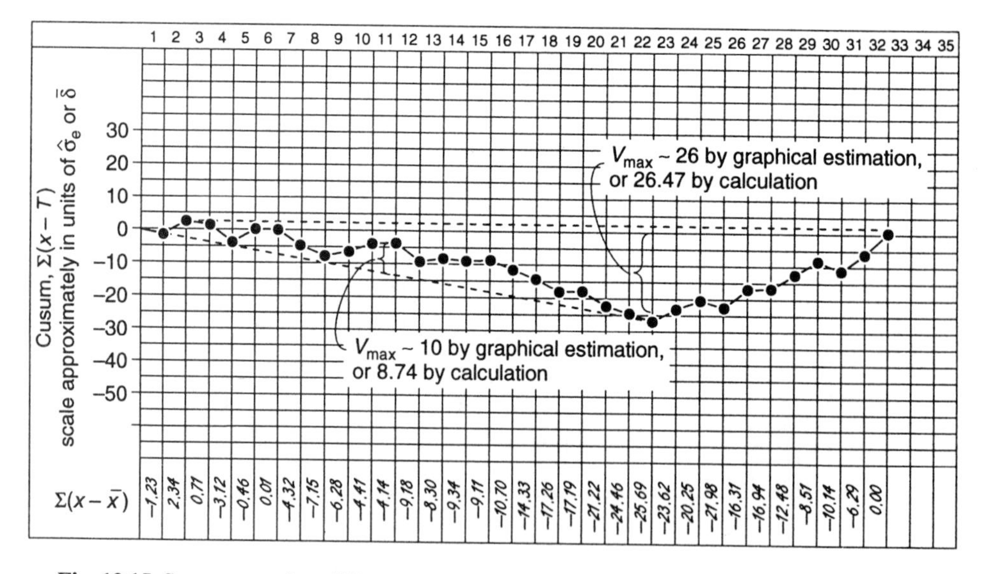

Fig. 12.15 Span tests for differences between segment averages based on data of Table 12.1. Plotted with $T = \bar{x} = 14.83$ (approx).

11–22, we find that in fact a line joining C_1 to C_{22} gives slightly greater height (or depth!) below the cusum than does C_0 to C_{22}. This height can be estimated graphically, but (12.6) gives a more reliable result. With $i=1$, $j=11$, $k=22$, and $C_1=-1.23$, $C_{11}=-4.14$, $C_{22}=-25.69$, we have:

$$V_{max} = -4.14-(-1.23)-\frac{11-1}{22-1}\{-25.69-(-1.23)\} = 8.74$$

With σ_e approximately 3.0, $V_{max}/\sigma_e = 2.91$ for a span of $k-i=22-1=21$. From Table 12.5, this does not reach the 5% significance probability, and we therefore regard the sequence from C_0 to C_{22} as one homogeneous segment. To test this segment against the last ten values, we now have $i=0$, $j=22$, $k=32$, but inspection reveals that the value of V_{max} is slightly enhanced by joining C_2 to C_{32} rather than joining C_{32} to the origin. The height at C_{22} is then about 26; Formula (12.6) this time gives 26.47, and with $V_{max}/\sigma_e = 26.47 \div 3 = 8.82$, this is significant at about the 0.5% probability level for a span of 30.

If one used t-tests, the segments would give the following values:

Observations 1 to 11; $\bar{x}=14.45$, $s=2.617$, $n=11$;
Observations 12 to 22; $\bar{x}=12.87$, $s=1.931$, $n=1$.

The pooled standard deviation is 2.300, and the test statistic is

$$t = \frac{14.45-12.87}{2.3\sqrt{\left(\frac{1}{11}+\frac{1}{11}\right)}} = 1.61$$

From Table A.1, with 20 degrees of freedom this does not reach the 5% two-tail (or even one-tail) significance level.

We therefore merge the two segments to test against the final set of observations from 23 to 32 inclusive.

Observations 1 to 22; $\bar{x}=13.66$, $s=2.386$, $n=22$;
Observations 23 to 32; $\bar{x}=17.40$, $s=2.941$, $n=10$.

Table 12.5 Critical values of V_{max}/σ_e for 'span' tests

	Significance probability (%)					
Span*	5	2.5	1	0.5	0.25	0.1
10	3.8	4.2	4.7	5.0	5.2	5.5
20	5.5	6.0	6.7	7.2	7.6	8.0
30	6.8	7.5	8.3	8.8	9.4	9.9
50	9.0	9.7	10.8	11.4	12.1	12.9
100	12.8	14.0	15.6	16.7	17.8	18.9

*Span = total observations in the two segments tested.

The pooled value of s becomes 2.565, and

$$t=\frac{13.66-17.40}{2.565\sqrt{\left(\frac{1}{22}+\frac{1}{10}\right)}}=-3.82$$

This value exceeds the 0.05% point for t with $\nu=30$, so for a two-sided test this is significant beyond the 0.1% level. Even a cautious interpretation would suggest a real difference between the segments 1–22 and 23–32, using either the span or '±'-methods.

12.8 Weighted cusums

12.8.1 The versatility of the cusum technique is further extended by its potential for dealing with weighted data, where different elements contain varying amounts of information. The most faimiliar examples to SPC users are the situations usually covered by p and u-charts for attributes and events. In order to use the counts of non-conforming items or non-conformities arising from subgroups of differing sizes, the number of occurrences is divided by the subgroup size to yield a proportion (p) or rate of occurrence (u). In doing so, the information from large subgroups receives no more attention than that from smaller ones. It is also usually recommended that subgroups should not vary by more than about 20–25% on either side of an average or nominal, unless special control limits are calculated for the very small or very large subgroups. This can be both an irksome restriction and a tedious operation if subgroup sizes vary widely. The weighted cusum provides an excellent solution.

The area of application goes far beyond that of p and u applications. Some further examples are:

(i) production monitoring where rates of use of energy, materials, manpower are estimated from batches of varying size, time units of varying length, etc.;
(ii) applications where scrap, trimmings, offcuts are weighed rather than counted, thus precluding the use of p and u charts which are restricted to counted data;
(iii) budgeting, sales or financial data based on varying time bases—months or weeks. While, for example, calendar months vary only from 28 to 31 days, creating no real problem, **working** months, allowing for weekends, public and company holidays might vary from a few days to almost a full month.

Using the weighted cusum approach, variations in subgroup or element size of even 100:1 can be readily and rigorously handled. We must first establish a suitable notation.

12.8.2 For generality we shall refer to w_i as the element size, and X_i as the element total. The rate estimate for the element is then $y_i = X_i/w_i$. In some cases, the data may comprise y_i and w_i, from which $\mathbf{X}_i$ can be synthesized as $y_i \cdot w_i$. Thus, for element number i, we have:

$$\left.\begin{array}{l}\text{Element size } = w_i\\ \text{Element total} = X_i \text{ or } y_i \cdot w_i\\ \text{Element rate } = y_i \text{ or } X_i/w_i\end{array}\right\} \quad (12.7)$$

In the situation where a weighted cusum might replace a p-chart, the element size w_i would be the number of items, n_i; the element total, X_i, would be the number of non-conforming items, np_i; and the element rate, y_i, becomes the proportion p_i.

Consider an example in weighing scrap from an industrial process. The daily process output is w_i, the weight of scrap is X_i and the scrap rate $y_i = X_i/w_i$. Again, for a monthly reporting of expenditure, the number of working days is w_i, the gross expenditure in the month X_i, and the average daily expenditure $X_i/w_i = y_i$.

For cusum purposes, we require a target value. This is best expressed as a target quantity per unit of element size. In p and u applications it will correspond to the target value for p or u (perhaps $\bar{p}$ or $\bar{u}$), in other cases to a target scrap or absence or breakdown rate per unit of production, manpower or time, respectively.

12.8.3 We are now able to formulate the weighted cusum produce. For each element of size w_i, a local target is calculated as $T \cdot w_i$, the expected element total for that element size. This is exactly analogous to the provision of differing par scores for holes of differing difficulty in golf! Extending the analogy, the cusum contribution is calculated as $X_i - T \cdot w_i$, and these deviations are then cumulated as in normal cusum practice. The only departure from the golf analogy is that the cusum $\sum(X_i - T \cdot w_i)$ is plotted against the cumulated element sizes, $\sum w_i$, instead of against the element number. In this way, the integrity of the cusum in exactly matching slope to local average is maintained. To summarize:

$$\begin{array}{l}\text{Cusum at element } i = C_i = \sum_{r=1}^{i} (X_r - T \cdot w_r)\\ \text{Sample axis coordinate} = \sum_{r=1}^{i} w_r\end{array} \quad (12.8)$$

C_i is then plotted as ordinate (cusum) against $\sum w$ as abscissa (sample axis).

In a similar notation to (12.4), the average rate over any sequence from the $(j+1)$th to kth elements is obtained as

$$\hat{y}_{j+1,k} = T + \frac{C_k - C_j}{w_k - w_j} \quad (12.6)$$

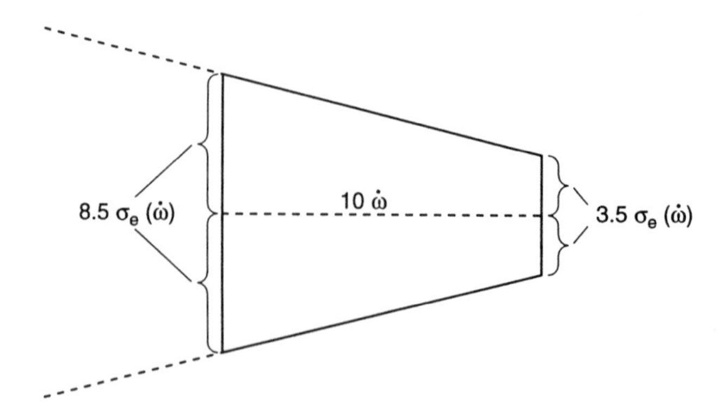

Fig. 12.16 Truncated mask for weighted cusum.

Where a decision rule is required for monitoring against a target value, the 3.5, 10, 8.5 mask of Fig. 12.9(c) is recommended. It is constructed in a similar manner to that for unweighted data by defining a nominal element size $\dot{w}$, estimating the standard error of X approximate to this element size using either (12.10) or (12.12) along with (12.11) then setting up the two vertical construction lines ($3.5\sigma_e$ and $8.5\sigma_e$) at a distance apart of $10\dot{w}$, as in Fig. 12.16.

For retrospective plotting, the varying element sizes make the application of the span test more difficult, but an example of a modified t-test is given in the example of section 12.10.4.

12.9 Estimation of variation from weighted data

12.9.1 The variations in element size cause some problems in estimating variation, but these are not insuperable. For a stable process with average rate $\hat{y}$ estimated by (12.9), the variance at the average element size $\bar{w}$ is given by

$$\hat{\sigma}^2_{X|\bar{w}} = \frac{\bar{w}}{n-1}\sum\{(X_i - \hat{y}w_i)^2/w_i\}$$

the summation extending over $i = 1$ to n.

Unfortunately the evaluation of this expression requires one 'pass' of the data to obtain $\hat{y}$ and a second pass to calculate $\hat{\sigma}^2$. A one-pass method can be developed, and requires the summations

$$\sum\left(\frac{X_i^2}{w_i}\right), \sum X_i, \sum w_i$$

Then at unit element size ($w = 1$),

$$\hat{\sigma}^2_{X|w=1} = \frac{1}{n-1}\left\{\sum\left(\frac{X_1^2}{w_i}\right) - (\sum X_i)^2/\sum w_i\right\} \tag{12.10}$$

The standard error of X at any other element size, $\dot{w}$, is then obtained as

$$\hat{\sigma}_{X|\dot{w}} = \sqrt{(\dot{w}\hat{\sigma}^2_{X|w=1})} \tag{12.11}$$

12.9.1 For an unstable process, a method based on successive differences is preferable. For $w = 1$, this can be estimated as

$$\hat{\sigma}^2_{X|w=1} = \frac{1}{n-1}\sum\left\{\left(\frac{X_i}{w_i} - \frac{X_{i+1}}{w_{i+1}}\right)^2 \Big/ \left(\frac{1}{w_i} + \frac{1}{w_{i+1}}\right)\right\} \tag{12.12}$$

The standard error for any other element size is then obtained as in (12.11). It is also worth noting that if we represent the estimate in (12.10) by σ^2_A and that in (12.12) by σ^2_D, the ratio σ^2_D/σ^2_A may be tested using the von Neumann test described in section 16.4 for the nature and extent of any instability.

12.9.3 An apparently simpler method, which may be distorted by any wide fluctuations in element sizes and which must therefore be used with some caution, is to calculate $y_i = X_i/w_i$ for each data element, and then to obtain the standard deviation $\hat{\sigma}_y$ for these ratios using either the usual *s*-value or the $\sqrt{(\frac{1}{2}MSSD)}$ of (9.2). Then for any notional element size $\dot{w}$,

$$\hat{\sigma}_{X|\dot{w}} = \dot{w}\hat{\sigma}_y \tag{12.13}$$

The examples of section 12.10 include applications of these estimation procedures.

12.9.4 Where the weighted cusum is used as an alternative to the *np, p, c* or *u*-chart, the standard error for X at the nominal element size $\dot{w}$ can be obtained from the relationships between mean and variance for the binomial or Poisson distributions.

For the *np* or *p* application, the expected number of non-conforming items, say $\dot{X}$, is estimated as $\dot{n}\hat{p}$, where $\dot{n}$ is the nominal subgroup size (and corresponds to the nominal element size $\dot{w}$ for more general applications). Then

$$\hat{\sigma}_{X|\dot{n}} = \sqrt{[\dot{n}\hat{p}(1-\hat{p})]} \tag{12.14}$$

In *c* or *u* applications, obtain $\dot{c}$ via $\dot{n}\hat{u}$, and then

$$\hat{\sigma}_{X|\dot{n}} = \sqrt{\dot{c}} \tag{12.15}$$

It is often found, however, that the actual variation in numbers of non-conforming items or non-conformities is rather greater than indicated by these theoretical expressions, due to some common-cause variation between subgroups. The presence of such common causes can be tested by comparing $\hat{\sigma}_X$ via (12.14) or (12.15) with that obtained from (12.11) in conjunction with (12.10) or (12.12), using the $\sqrt{(\chi^2/\nu)}$ test of (6.8) or (6.9). An example is given in section 12.10, and this situation is further discussed in section 16.8.

12.10 Examples of weighted cusum applications

12.10.1 We now consider some typical applications of weighted cusums. The first two illustrate the cusum replacement for a u-chart and a p-chart in cases where element sizes vary so greatly that numerous subgroups would require special control limits. In addition, some between-subgroup common cause variation is present in the 'p' example of section 12.10.3, so that calculation of mask parameters from $\hat{\sigma}$ calculated from (12.14) would yield over-control; the method of (12.12) is preferable.

The third example illustrates how the weighted cusum can be extended to deal with measured characteristics from varying element sizes; i.e. cases where neither p nor u-charts are applicable. It also deals with retrospective analysis, where differences between segments, rather than active control, are the main subject of interest.

12.10.2 Figure 12.17 illustrates an example which is the cusum counterpart to a u-chart. A group of textile machines is subject to widely varying utilization from day to day to fluctuations in the nature of orders being processed. Data collected over five weeks lists the daily machine hours and numbers of machine interruptions. These arise from needle breaks, attention to yarn faults which cannot be immediately handled by operators, and other failures. From the base period covering the first 25 observations, we have:

$$\sum X = 209; \qquad \sum w = 2685; \qquad \hat{y} = 0.07784$$
$$(\bar{w} = 107.4) \qquad (7.784 \text{ per } 100\,\text{h})$$
$$\hat{\sigma}_A(w=1) = 0.2741; \ \hat{\sigma}_D(w=1) = 0.2504.$$

Compare the value of σ for a Poisson process with $\lambda = 0.07784$, i.e. $\sqrt{0.07784} = 0.2790$.

Agreement between σ for the Poisson model and the observed $\hat{\sigma}_A$, $\hat{\sigma}_D$ indicates stability and validity of $\sqrt{\bar{c}}$ as standard error.

For $\dot{w} = 100\,\text{h}$, $\bar{c} = 7.784$; a target value of 8.0 is reasonable, and at this level the standard error $= \sqrt{8}$ or 2.83 approximately.

Figure 12.17 shows the cusum chart set up with $T = 8$ per 100 h ($\dot{y} = 8$), and a 3.5, 10, 8.5 mask constructed with $\hat{\sigma}_e = 2.83$. The first 25 points are clearly in control.

At one point in the history of this process, a faulty yarn supply (more knotted joins than permitted in the buying specification) yielded an excess of needle breakages. This sequence is illustrated by observations 26–35. The mask gives a signal at observation 28, but it was necessary to continue using the yarn already in use (subject to a compensation claim against the yarn supplier). Later, represented by observation 36–47, some process improvements (including revised maintenance practices and the use of a new needle design) were introduced, and the results appear to be promising though not yet conclusive.

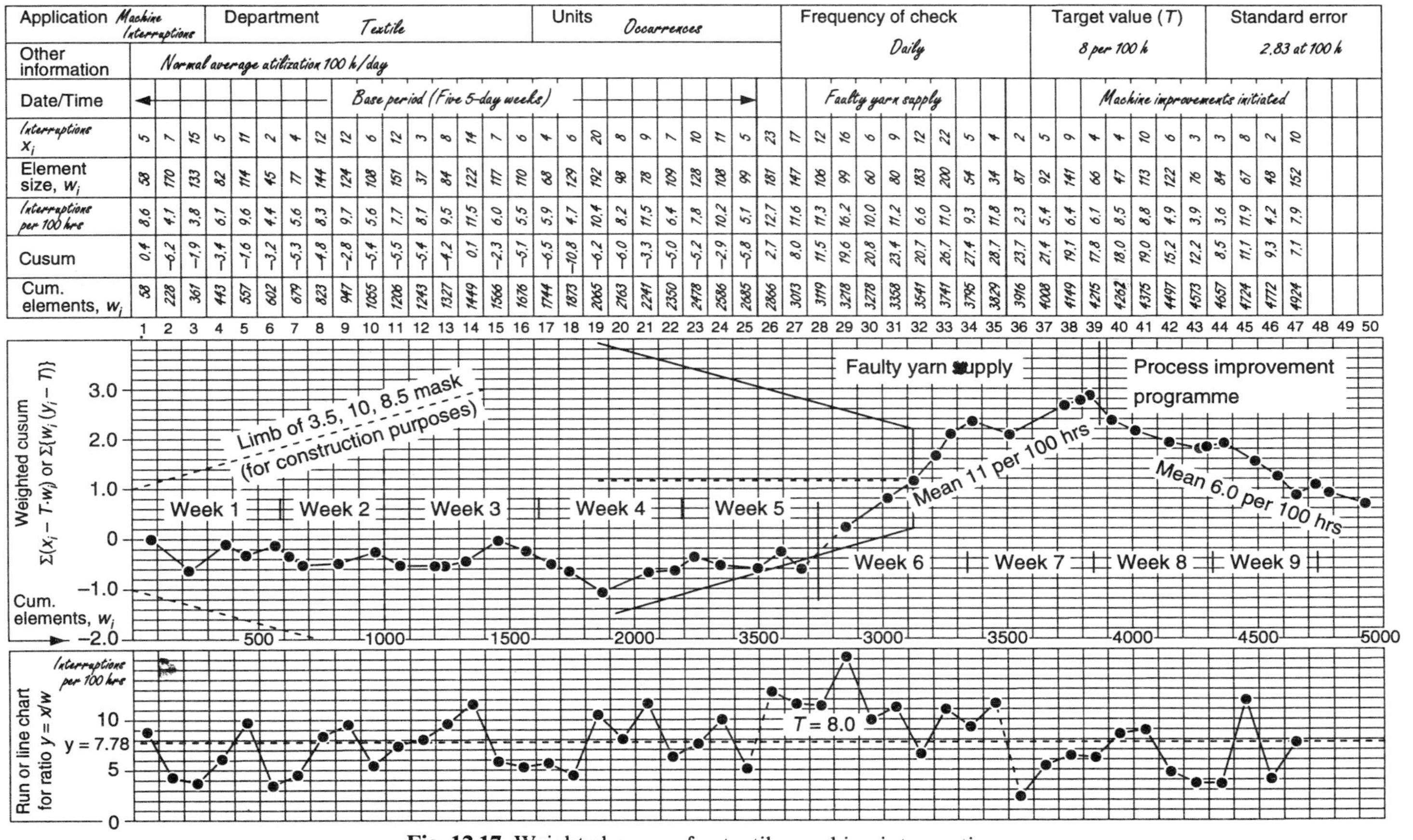

Fig. 12.17 Weighted cusum for textile machine interruptions.

For the sequences between observations 26–35 and 36–47 (inclusive), the interruption rates calculated from the cusum are:

26–35. $i = 25, j = 35; c_i = -5.8, c_j = 28.7;$

$\sum w_i = 2685, \sum w_j = 3829$

$$\hat{y} = 0.08 + \frac{28.7 - (-5.8)}{3829 - 2685} = 0.110\,16, \text{ or 11 per 100 h.}$$

36–47. $i = 35, j = 47; c_j = 7.1, c_i = 28.7;$

$\sum w_i = 3829, \sum w_j = 4924$

$$\hat{y} = 0.08 + \frac{7.1 - 28.7}{4924 - 3829} = 0.0603, \text{ or 6.0 per 100 h.}$$

Further points of interest concern the element size pattern, in this case corresponding to machine utilization. First, it is unavoidable that the points on the chart occasionally move out of phase from the observation numbers. Here, because the utilisation is in fact rather greater than 100 h per day (overall about 105 h to date) $\sum w$ moves ahead of the columns carrying the raw data. Secondly, the sparseness of information derived from small elements is highlighted by the close proximity of adjacent points – note especially days 6 and 7, 12 and 13, and 43–46. If this aspect is of particular interest, weekly (or other) markers can be noted on the horizontal axis to identify busy or slack periods as shown in Fig. 12.17. The weighted cusum thus provides useful information on *both* axes of the chart.

12.10.3 An example of a p-chart application arose in monitoring the reject rate on 100% inspection of output from a steel rolling process. The process dealt with numerous small orders each day, and it was not practicable to monitor each type of component separately. A summary of the inspection returns for ten weeks appears in Figure 12.18 along with calculation of percentage rejected and the weighted cusum with a target (based on historical data) of 1%. Note that the subgroup sizes vary enormously from 56 on day 37 to 473 on day 27. The general level is about 250–300, but there are many days with less than 150 or more 350, making a p-chart tedious to operate. A p-chart is also very noisy: there are five values beyond the UCL (after allowing for erratic subgroup sizes) and two fairly long runs below $\bar{p}$. These signals occur partly because of over-sensitivity of p-limits, based on the usual formula, to values in the upper tail, and partly because with a different mix of components from day to day, there are between-day random variations. The object here is not so much to identify bad days as to monitor longer-term stability, and the weighted Cusum satisfies this purpose.

The estimates via (12.10), (12.12) and the usual expression for $\bar{p}$, i.e. $\sum np / \sum n$, are:

$\bar{n}$ (corresponding to $\bar{w}$ for a weighted cusum) $= 267.86;$

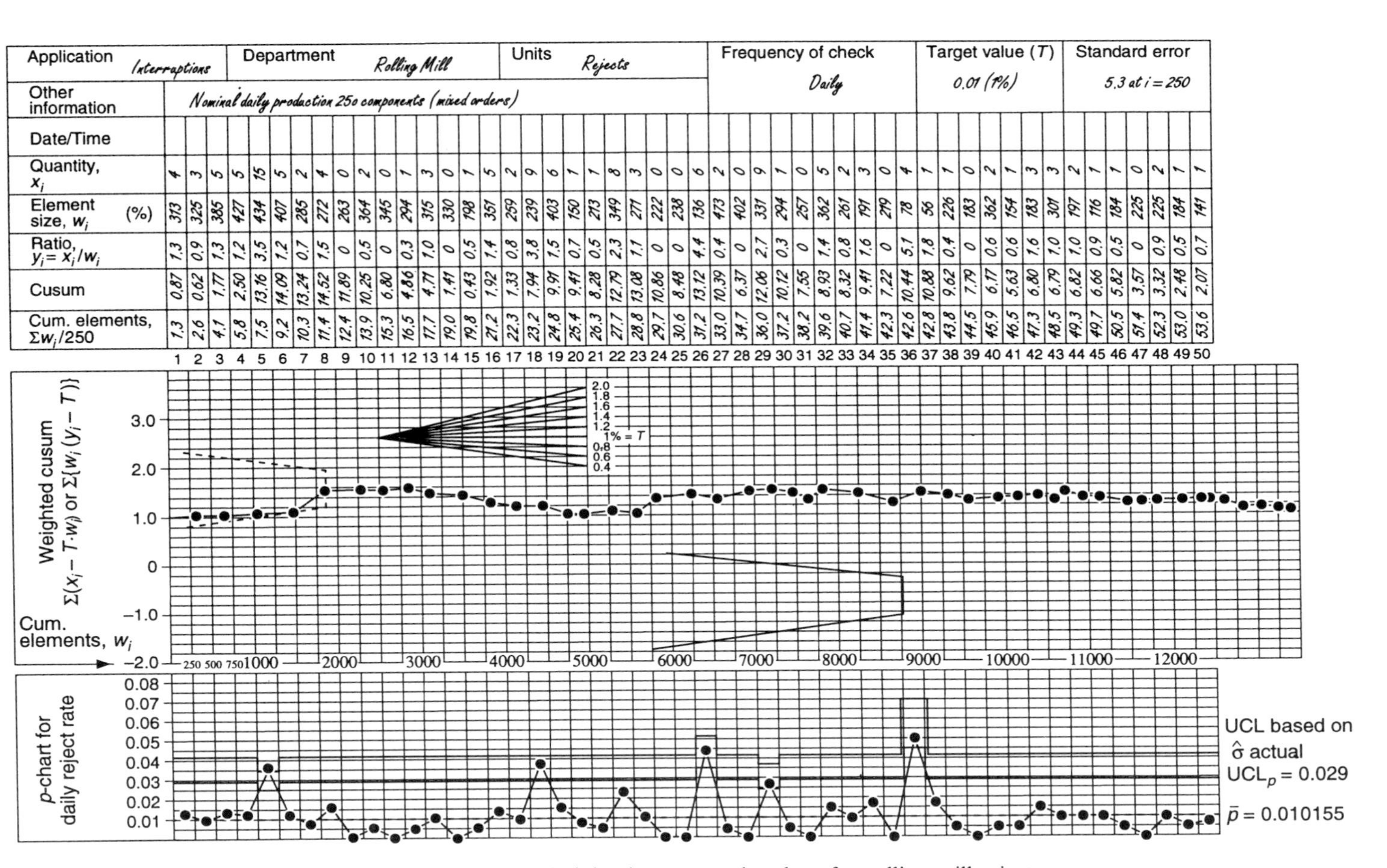

Application	Interruptions	Department	Rolling Mill	Units	Rejects	Frequency of check	Target value (T)	Standard error
Other information	Nominal daily production 250 components (mixed orders)					Daily	0.01 (1%)	5.3 at i = 250

	1	2	3	4	5	6	7	8	9	10	11	12	13	14	15	16	17
Date/Time																	
Quantity, x_i	4	3	5	5	15	5	2	4	0	2	0	1	3	0	1	5	2
Element size, w_i (%)	313	325	385	427	434	407	285	272	263	364	345	294	315	330	198	351	259
Ratio, $y_i = x_i/w_i$	1.3	0.9	1.3	1.2	3.5	1.2	0.7	1.5	0	0.5	0	0.3	1.0	0	0.5	1.4	0.8
Cusum	0.87	0.62	1.77	2.50	13.16	14.09	13.24	14.52	11.89	10.25	6.80	4.86	4.77	1.41	0.43	1.92	1.33
Cum. elements, $\Sigma w_i/250$	1.3	2.6	4.1	5.8	7.5	9.2	10.3	11.4	12.4	13.9	15.3	16.5	17.7	19.0	19.8	21.2	22.3

	18	19	20	21	22	23	24	25	26	27	28	29	30	31	32	33	34
Date/Time																	
Quantity, x_i	9	6	1	1	8	3	0	0	6	2	0	9	1	0	5	2	3
Element size, w_i (%)	239	403	150	213	349	271	222	238	136	473	402	331	294	257	362	261	191
Ratio, $y_i = x_i/w_i$	3.8	1.5	0.7	0.5	2.3	1.1	0	0	4.4	0.4	0	2.7	0.3	0	1.4	0.8	1.6
Cusum	7.94	9.91	9.41	8.28	12.79	13.08	10.86	8.48	13.12	10.39	6.37	12.06	10.12	7.55	8.93	8.32	9.41
Cum. elements, $\Sigma w_i/250$	23.2	24.8	25.4	26.3	27.7	28.8	29.7	30.6	31.2	33.0	34.7	36.0	37.2	38.2	39.6	40.7	41.4

	35	36	37	38	39	40	41	42	43	44	45	46	47	48	49	50
Date/Time																
Quantity, x_i	0	4	1	1	0	2	1	3	3	2	1	1	0	2	1	1
Element size, w_i (%)	219	78	56	226	183	362	154	183	301	197	116	184	225	225	184	141
Ratio, $y_i = x_i/w_i$	0	5.1	1.8	0.4	0	0.6	0.6	1.6	1.0	1.0	0.9	0.5	0	0.9	0.5	0.7
Cusum	7.22	10.44	10.88	9.62	7.79	6.17	5.63	6.80	6.79	6.82	6.66	5.82	3.57	3.32	2.48	2.07
Cum. elements, $\Sigma w_i/250$	42.3	42.6	42.8	43.8	44.5	45.9	46.5	47.3	48.5	49.3	49.7	50.5	51.4	52.3	53.0	53.6

Fig. 12.18 Weighted cusum and p-chart for rolling mill rejects.

$\bar{p}$ (corresponding to $\hat{y}$ for a weighted cusum) = 0.010 155, i.e. close to 1% rejections;

$$\left.\begin{array}{l}\hat{\sigma}_{\mathrm{D}} = 0.172\,17\\ \hat{\sigma}_{\mathrm{A}} = 0.163\,28\end{array}\right\} \text{ at } \dot{n} = 1$$

Now for $n = 1$, $\sigma_{np} = \sqrt{[np(1-p)]} = \sqrt{[p(1-p)]} = 0.100\,26$, much less than $\hat{\sigma}_{\mathrm{D}}$ or $\hat{\sigma}_{\mathrm{A}}$ (which are in close agreement) from the data. For medium term monitoring, a measure of variation based on the data is appropriate in this case, as the between-days component has no assignable pattern.

At a nominal subgroup size of 250, we have:

$$\left.\begin{array}{l}\hat{\sigma}_{\mathrm{D}} = 0.17217\sqrt{250} = 2.72\\ \hat{\sigma}_{\mathrm{A}} = 0.16328\sqrt{250} = 2.58\end{array}\right\} \text{ whence } 2\hat{\sigma} \sim 5.3$$

Obviously a scale factor of 5 will be suitable for the vertical axis of the cusum chart, and the resulting plot is shown in Fig. 12.18. There are no significant instabilities apart from an abnormally high reject number on day 5; the two periods around days 9–15 and 38–50 do not indicate any real improvement in performance and are presumed due to the 'luck of the draw' in the mix of orders at the time. One of the reasons for including this example is as a reminder that not all control charts or cusums give spectacular signals or new revelations about the process!

12.10.4 A firm manufacturing chassis for domestic electrical appliances had collected weekly records of trimming scrap from its panel cutting shop. The total weight of sheet metal used and the weight of trimmings sent for scrap metal sale were available for six months previous activity. The weekly throughput varies depending on the nature of items being manufactured, as well as occasional short weeks from public holidays, etc. Table 12.6 lists the 26 weeks' figures.

The appearance of the cusum in Fig. 12.19 suggests an initial segmentation as indicated on the chart. The statistics for these segments are:

A (1 to 10 inclusive) $\sum w = 136\,220$, $\bar{w} = 13\,622$, $\hat{y} = 0.063\,35$; $\hat{\sigma}_{\mathrm{D}} = 0.948\,75$, $\hat{\sigma}_{\mathrm{A}} = 0.983\,64$ @ $w = 1$.
B (11 to 14 inclusive) $\sum w = 57\,100$, $\bar{w} = 14\,275$, $\hat{y} = 0.050\,95$; $\hat{\sigma}_{\mathrm{D}} = 0.414\,17$, $\hat{\sigma}_{\mathrm{A}} = 0.594\,71$ @ $w = 1$.
C (15 to 24 inclusive) $\sum w = 126\,700$, $\bar{w} = 12\,670$, $\hat{y} = 0.071\,98$; $\hat{\sigma}_{\mathrm{D}} = 0.664\,32$, $\hat{\sigma}_{\mathrm{A}} = 0.594\,75$ @ $w = 1$.
D 25, 26 Too short for reasonable estimation.

We must now add one further expression to those of section 12.9.1. For the estimate $\hat{y}$ based on a total of W element units ($W = \sum w_i$), the standard error of $\hat{y}$ is

$$\hat{\sigma}(\hat{y}) = \hat{\sigma}_{w=1}/\sqrt{W} \tag{12.16}$$

Table 12.6 Trimming scrap from sheet metal cutting

Week	Sheet used w(tonnes)	Scrap x(kg)	Scrap (%) $y = 100x/w$	Remarks
1	15.08	1069	7.09	
2	12.17	762	6.26	
3	9.80	512	5.22 }	Easter
4	10.70	545	5.09 }	
5	16.11	988	6.13	
6	18.70	1367	7.31	
7	15.33	794	5.18	
8	6.74	438	6.50	Spring holiday
9	17.14	1220	7.12	
10	14.45	935	6.47	
11	14.98	850	5.67	
12	12.12	645	5.32	
13	13.45	613	4.56	
14	16.55	801	4.84	
15	14.79	1034	6.99	
16	11.73	810	6.91	
17	4.85	446	9.20 }	Summer holiday,
18	6.33	473	7.47 }	urgent orders only
19	17.38	1157	6.66	
20	16.22	1206	7.44	
21	9.21	719	7.81	August holiday
22	15.62	1089	6.97	
23	13.90	1002	7.21	
24	16.67	1184	7.10	
25	14.22	883	6.21	
26	12.94	772	5.97	

Average sheet used ($\bar{w}$) = 13.353 tonnes (13353 kg)
Average scrap rate = 0.064 27 (6.427%)
$\left.\begin{array}{l}\sigma_D = 0.83237\\ \sigma_A = 1.11499\end{array}\right\}$ at 1 kg $\begin{array}{l}\rightarrow 96.2\\ \rightarrow 117.9\end{array}\Big\}$ at 13.353 tonnes
$\sigma_D/\sigma_A = 0.7465$, suggests instability (see 16.4)
Cusum plotted with $T = 0.065$ (6.5% scrap) and $\dot{w} = 10.0$ tonnes

This statistic will provide the basis for t' tests between segments. First, for segments A and B,

$$\hat{\sigma}(\hat{y}_A) = 0.948\,75/\sqrt{136\,220} = 0.002\,571,$$
$$\hat{\sigma}(\hat{y}_B) = 0.414\,17/\sqrt{57\,100} = 0.001\,733.$$

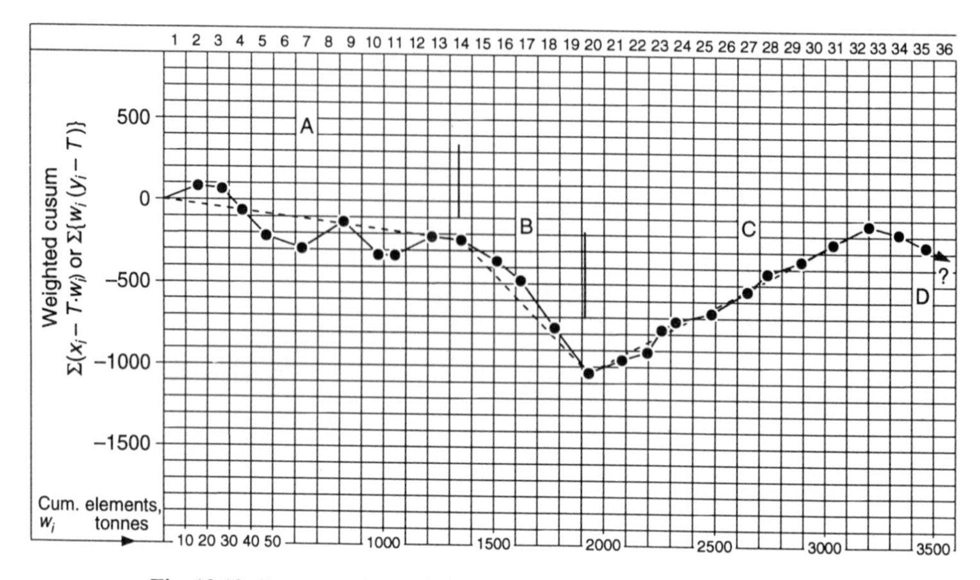

Fig. 12.19 Retrospective weighted cusum for chassis trimming scrap.

Then

$$t' = \frac{0.063\,35 - 0.050\,95}{\sqrt{(0.002\,571^2 + 0.001\,733^2)}} = 3.999$$

For around ten degrees of freedom (allowing for the discrepancy between the measures of variation in A and B) this is highly significant.

For segment C, $\hat{\sigma}(\hat{y}_C) = 0.664\,32/\sqrt{126\,700} = 0.001\,866\,3$, and for the comparison between B and C,

$$t' = \frac{0.050\,95 - 0.071\,98}{\sqrt{(0.001\,733^2 + 0.001\,866\,3^2)}} = -8.257$$

even more significant than for A–B.

Subsequent investigation suggested that the higher scrap rate during the late summer period was due to a large order for some chassis whose design inevitably produced more scrap. As this had been costed into the selling price, no further action was taken. No reason was found for the apparently lower scrap in early summer, but the cusum chart suggests that the last couple of weeks have reverted to the rate of just over 6% that prevailed from April to June. Naturally ways were being sought to reduce scrap levels, but scope is limited by the designs which involve removing material for wiring apertures, tubes and small windows.

12.11 Conclusion

Cusum methods were originally developed for quality control applications, but their efficiency, versatility and effectiveness of presentation has led to their adoption for a variety of applications outside the quality control area.

Among the applications recorded in the literature or known to the present author, the following demonstrate the wide range of problems amenable to cusum processing:

(i) monitoring rates of occurrence of customer complaints, clerical errors, accidents, etc.
(ii) checking use of resources against output, e.g. materials in industrial production, energy, fuel consumption of vehicles and many 'productivity' measures of performance.
(iii) evaluating effects of social improvements (re-housing, improved sanitation or living standards) on certain diseases linked to these conditions.
(iv) monitoring levels of insurance claims for various categories of insurance (e.g. defect levels in ocean-going vessels, effect of seat-belt laws on personal injury).
(v) controlling general levels of departmental expenditure to budgetary targets, whilst allowing for some period-to-period latitude.
(vi) assessing the effects of therapy on patients' measurable functions (blood pressure, pulse rate, haematological measurements, etc.)
(vii) investigating questions of disputed authorship, and of insertions of 'foreign' material into documents or witness's statements.

Clearly applications are not confined to industrial or technological problems, but include legal, medical, agricultural, commercial, econometric, social and even domestic applications.

13

Process capability: attributes, events, and Normally distributed data

13.1 Introduction

13.1.1 Prior to this Chapter, the capability of a process has been mentioned only in passing or (as in step 10 of the procedure for setting up control charts) it has been deferred for separate attention.

'Control' charts deal with determining whether or not a process shows **stability**, a concept which does not necessarily involve the '**capability**'. A process 'in control' is operating under the influence only of common causes of variation. The level of common cause variation may be very small, permitting adequate conformance to requirements like specifications, or it may be large, requiring fundamental improvement before such requirements can be met. '**Capability**' and '**control**' are therefore distinct facets of a process.

It is quite possible for a process that is not strictly in statistical control to meet specification requirements. However, for valid measurement of capability, a process *must* be in control. The measures of performance and capability described in this Chapter assume that this is so, and that variation is due only to common cause aspects.

13.1.2 **Capability** is defined as the measure of the extent to which a process is able to meet predetermined performance criteria. Such criteria may include:

(i) conformance to specified limits or tolerances;
(ii) satisfying maximum or minimum levels for test parameters,
(iii) the production of a high proportion of items which are acceptable to some go/no go, pass/fail assessment of quality;
(iv) the achievement of low levels of faults, defects, interruptions or other detrimental occurrences.

13.1.3 For **attributes**, performance is indicated by the proportion of non-conforming items produced when the process is in control. An associated measure of capability is then the proportion of conforming items (1-p). If preferred, both performance and capability may be expressed as percentages instead of decimal fractions.

13.1.4 For **events**, performance is assessed by the average rate of occurrence of the faults, defects, interruptions, etc. This needs to be related to some unit of assessment, such as defects per assembly, cracks per 100 metres, machine stoppages per hour, etc.

13.1.5 For **measurements**, statements of capability depend on relating some measure of common cause variation (e.g. range or standard deviation) to the appropriate performance criteria. For simplicity, these criteria will be referred to as the specification, and any maximum or minimum values as the specification limits. In some applications, there may not be a formal specification, but a target range or other requirement may sometimes have similar implications. Take care, however, that general guidelines for controlling process characteristics, like temperature, relative humidity, etc. are not interpreted in the same way as specification limits – capability indices for such measures may be meaningless, and create unnecessary confusion.

Because capability measurement is much simpler for attributes and events, these are now covered in detail before dealing with the larger subject of capability for measured data.

13.2 Capability: attributes

13.2.1 It is convenient to consider the estimation of capability for attributes data in terms of the operation of a control chart – either np or p. The control chart thus acts as both a vehicle for data collection and, provided that the process is in control, a means of readily extracting the necessary summaries for capability statements.

For each subgroup entered on the control chart, the subgroup size, n, and the number of non-conforming items, np, are required.

13.2.2 The **process performance** for attributes data is usually defined as the proportion of non-conforming items averaged over the whole period of evaluation. This is, of course, identical to the value of $\bar{p}$ that would be used for calculation of control limits, viz:

$$\bar{p} = \frac{\text{Total number of non-conforming items}}{\text{Total number of items in evaluation}} = \frac{\sum np}{\sum n} \qquad (13.1)$$

Some users prefer to express the performance as a percentage non-conforming, i.e. $100\bar{p}\%$.

13.2.3 The **capability** in this case represents achievement in terms of conforming items, and is the complement of the performance,

$$1 - \bar{p} = 1 - \frac{\sum np}{\sum n} \tag{13.2}$$

or 100(1-$\bar{p}$)% if preferred.

13.2.4 There are no generally agreed standards for attributes performance (unlike measurements, for which there is at least some agreement on satisfactory values for capability indices). For relatively minor and non-functional imperfections, a performance of one or two per cent non-conforming, hence a capability of 98–99%, may be reasonable. In other cases satisfactory performance may require non-conformance in parts per thousand or even per million, with capability consequently of 99.9% or greater. Suitable levels need to be agreed with the customer or user of the product or service.

13.3 Capability: events

13.3.1 As for attributes, the data for assessment need to be drawn from a control chart for an in-control process: in this case a c or u-chart. The number of non-conformities, c, and the number of sample units or area of opportunity, n, are noted for each subgroup.

13.3.2 The **performance** for event data represents the rate of occurrence of non-conformities per unit of observation. This corresponds to $\bar{u}$, as used for calculation of control limits for the u-chart, viz:

$$\bar{u} = \frac{\text{Total number of non-conformities}}{\text{Total units of observation}} = \frac{\sum c}{\sum n} \tag{13.3}$$

Note that 'units' may represent numbers of items or units of quantity: energy, volume, length, weight, time, etc.

As for attributes, event performance may be expressed on a 'per 100' basis, often represented by *DHU* (defects per 100 units) = 100$\bar{u}$. Note, though, that *DHU* must *not* be construed as a percentage, and that there is no direct measure of capability to be obtained either by subtracting $\bar{u}$ from 1 or *DHU* from 100. In fact, for difficult or complex operations, there may be more than one non-conformity per item ($\bar{u} > 1$) so that *DHU* exceeds 100.

13.3.3 Again as for attributes, there are no generally agreed criteria for event capability or performance. Increasingly, and especially for functional defects, performance in terms of few faults per million units is demanded.

13.3.4 As an example of the calculations, the chart of Fig. 10.3 lists extrusion faults for pieces of plastic extrusion which are 10 m long. The process is in control except for sample 16 which just violates the upper control limits. This appeared to be an unusual combination of occurrences of different faults rather than a real change in the process, so we will use the data to assess the capability. The units need to be defined, and with 50 cuts per subgroup a total of $20 \times 50 \times 10$ metres $= 10\,000$ metres is involved on the completed chart. If we choose *DHU* in terms of 100 metres, then we have 100 such units with a total of 197 defects. Hence

$$\bar{u} = \frac{197}{100} DHU = 1.97 \text{ per 100 metres}$$

Whether this is a satisfactory level of performance depends on the end-use of the material. However, as for measurements, it is important not to regard conformance to a specified defect level as the sole aim of SPC–improvement in performance by identifying causes of non-conformity and progressively eliminating them is as relevant to attributes and events as it is to measurable characteristics.

13.4 Capability: Normally distributed measurements

13.4.1 In the remaining sections of this Chapter, we consider measured characteristics for which the normal distribution provides a reasonable model. Non-Normality is discussed in Chapter 14.

The most important measures of capability for measurements are those concerned with the achievement of the **process**. In some circumstances, pre-production studies may be carried out to assess the suitability of a machine for a particular product, or to obtain a short-term measure of a process. Such studies, according to circumstances, may be termed machine capability or process potential capability. Because they may not include all elements of common cause variation, more stringent standards are often set for machine capability (C_m, C_{mk}, etc.) or process potential (P_p, P_{pk}, etc.) than for process capability (C_p, C_{pk}, etc.) as measured over an in-control period from a control chart with all sources of common cause variation present.

13.4.2 There are some small differences in calculation and interpretation of capability according to whether the specification has only an upper *or* a lower limit (one-sided specification) or a pair of limits (two-sided specification). A two sided specification, especially when linked to a characteristic expected to have an approximately Normal distribution, is often expressed as nominal value $\pm$ tolerance. In some cases, specifications such as $(\text{nominal})^{+a}_{-b}$ are encountered, $a \neq b$, and problems can then arise as to the best target value to choose for process operation, especially if the limit involving the smaller of a a and b is difficult to satisfy with the target set at the specific nominal value.

In many cases, it will be desirable to set a target or aim value at the nominal, but where non-conforming items are rectifiable if violating one limit but rejectable if violating the other, it may be preferred to aim at a process average on the less critical side of the nominal.

For a one-sided specification there may or may not be a nominal value (e.g. simply a tensile strength requirement of not less than 10 tonnes).

There are various methods of stating capability. All may, with some adaptation, be applied to one or two-sided specifications. They require a measure of the common cause variability in terms of the standard deviation of the system.

13.4.3 Direct calculation of standard deviation was defined in (3.19) as

$$s = \sqrt{\left[\frac{1}{n-1}\sum(x-\bar{x})^2\right]}$$

or equivalent formulae such as (3.20), as implemented by calculators with statistical functions or on suitable programmed computers.

This estimate of variation is appropriate for studies where the data comprises one sample of n items–a 'snapshot' as in a machine capability study. Where the data comprises k small samples each of n items, as from a control chart, other methods are preferable. For an $\bar{x}$, R or $\bar{x}$, s-chart, the standard deviation *for a process which is in-control* may be estimated by one of the three methods following. It cannot be too strongly emphasized that capability analysis is invalid for any process that is out of control, and that the following methods are not appropriate for a process that is stable but contains between-subgroup sources of common cause variation, e.g. those covered in 11.3 where the $\bar{x}$, R and δ chart can be applied. The estimate (11.2) should be used in such cases.

13.4.4 From the control chart, over an 'in-control' period covering at least 20 subgroups, calculate

Either the value of $\bar{R}$ from the $\bar{x}$, R-chart

or the value of $\bar{s}$ from and $\bar{x}$, s chart.

Reasonably, one would expect the values of $\bar{R}$ or $\bar{s}$ to depend on the actual level of common cause variation in the process, as measured by the true process standard deviation.

In order to estimate σ, mathematically derived factors are used in conjunction with $\bar{R}$ or $\bar{s}$ to yield the estimate $\hat{\sigma}$ (this symbol is used to distinguish the estimate from the explicitly calculated value of s).

For an R-chart, $\qquad \hat{\sigma} = \bar{R}/d_n \qquad (13.4)$

For an s-chart $\qquad \hat{\sigma} = \bar{s}/c_n \qquad (13.5)$

where d_n, c_n are the factors appropriate to sample size n in Table 13.1 Note that in American texts and standards, these factors are denoted by d_2 or c_2, respectively, the subscript having no connection with the sample size.

Table 13.1 Factors for estimating process standard deviation

Sample or subgroup size, n	From average range, d_n factors*	From average standard deviation C_n factors†
2	1.128	0.7979
3	1.693	0.8862
4	2.059	0.9213
5	2.326	0.9400
6	2.534	0.9515
7	2.704	0.9594
8	2.847	0.9650
9	2.970	0.9693
10	3.078	0.9727

*D_2 in US notation
†C_2 in US notation
$\hat{\sigma} = \bar{R} \div d_n$ or $\bar{s} \div C_n$

for $n > 10$, $C_n \doteqdot \sqrt{\left(\frac{2n-3}{2n-2}\right)}$.

13.4.5 A further alternative, not requiring the use of a table of factors, is to calculate the total of the squared values of sample standard deviations (the variances), divide by the number of subgroups, and finally take the square root. Symbolically, for a set of k subgroups for each of which s has been calculated, the calculation is:

$$\hat{\sigma} = \sqrt{\left(\frac{\sum s^2}{k}\right)} \tag{13.6}$$

This estimate is readily obtainable via an $\bar{x}$, s function by entering the s-values as data, then using a sequence $\sum x^2 \div n = \sqrt{}$, calling up $\sum x^2$ and n from the sums-of-squares and sample count registers.

The reader may question why $\bar{s}$ is not itself a suitable estimate. In fact, (13.6) is the definitive combined estimate: it is the square root of the average variance. If $\bar{s}$ and $\sqrt{}$(average s^2) are calculated from the same set of data, $\bar{s}$ is always the smaller, and is thus said to be biassed. The c_n factors allow for this bias and yield valid estimates of σ for the whole process.

13.5 Statements of capability

13.5.1 Having evaluated the variation of the machine or process, the next step is to calculate a concise measure of its ability to conform to the specification limits. Some of the most widely used procedures use the 6σ spread (or mean $\pm 3\sigma$) to measure what is termed the 'process (or machine) performance'. This is the region estimated to contain 99.7% of the individual values, or to exclude only 0.135% at either end. The first two measures below use the concept of the 6σ spread, and in the guise of the 99.7% central zone it also plays a role in capability statement for non-normal variables (though more or less than 6σ may be required to cover the 99.7% region).

13.5.2 The definition of 6σ as the process performance leads to the formulation of some widely used capability indices. The first indicates the ability of the process to meet a two-sided specification in terms of variation only. If the performance is such that 6σ can be accommodated *within* the specification limits, then the index

$$C = \frac{\text{USL} - \text{LSL}}{6\sigma} \tag{13.7}$$

will take a value *greater* than 1.0. If the six-sigma spread does not fit into the specification, then C will be less than 1.0. Subscripts may be added to C (or other letters used) to indicated the application of the index; thus:

- C_{m} machine capability index
- C_{p} process capability index
- P_{p} process potential index (carried out on a first production run, usually with subgroups more closely spaced than for a full process evaluation).

Now the value of σ will not be known, and must be estimated either by s (for one sample, as for a machine study) or by $\hat{\sigma}$ derived from one of (13.4) to (13.6) for a process evaluation. We thus have

$$\left.\begin{aligned} \hat{C}_{\text{m}} &= \frac{\text{USL} - \text{LSL}}{6s} \\ \text{and} \qquad \hat{C}_{\text{p}}, \hat{P}_{\text{p}} &= \frac{\text{USL} - \text{LSL}}{6\hat{\sigma}} \end{aligned}\right\} \tag{13.8}$$

in common use.

13.5.3 The C-indices tell only part of the capability story. They depend only on variation, and do not take into account the location of the process or machine mean. A process with $C_{\text{p}} = 1.0$, but with its mean displaced from the nominal towards one of the specification limits will yield more than 0.135% beyond the tolerance in one direction, but less than 0.135% in the other. To ensure overall capability, one must establish that the mean, μ, is an adequate distance from the specification limit(s). We therefore define

$$\left.\begin{aligned} \hat{C}_{\text{pU}} &= \frac{\text{USL} - \bar{x}}{3\hat{\sigma}} \\ \hat{C}_{\text{pL}} &= \frac{\bar{x} - \text{LSL}}{3\hat{\sigma}} \end{aligned}\right\} \tag{13.9}$$

Provided that each of these is at least 1.0, very few items will violate USL or LSL – less than 0.135% for a perfect Normal distribution, but only one in a few hundred in more practical terms.

For a two-sided specification it is customary to define capability in terms of whichever of $\hat{C}_{pU}$ or $\hat{C}_{pL}$ is the smaller (the 'worst-case' capability), so that

$$\hat{C}_{pk} = \text{Minimum of } (\hat{C}_{pU}, \hat{C}_{pL}) \tag{13.10}$$

In fact it can be shown that this is equivalent to defining $\hat{C}_{pk}$ in terms of $\hat{C}_p$ and the displacement of $\bar{x}$ from the midpoint of the specification, viz:

$$\hat{C}_{pk} = \hat{C}_p \left\{ 1 - \frac{|\bar{x} - \mu_0|}{d} \right\}$$

where $\mu_0 = \frac{1}{2}(\text{USL} + \text{LSL})$, usually the nominal value, and $d = \frac{1}{2}(\text{USL} - \text{LSL})$, usually the $\pm$ tolerance.

This form lends itself to theoretical analysis, but we shall not consider it further here. Yet another index of theoretical appeal which does not seem to have gained wide acceptance is

$$\hat{C}_{pm} = \frac{\text{USL} - \text{LSL}}{6\hat{\sigma}'}$$

where

$$\hat{\sigma}' = \sqrt{[\hat{\sigma}^2 + (\bar{\bar{x}} - \mu_0)^2]}$$

i.e. the displacement of $\bar{\bar{x}}$ from mid-specification is incorporated into the measure of variation in the process. A disadvantage of this index is that even when $\bar{\bar{x}}$ lies outside the specification, $\hat{C}_{pm}$ approaches zero but does not become negative (as do $\hat{C}_{pU}$, $\hat{C}_{pL}$ and hence $\hat{C}_{pk}$). This sign-reversal is a useful indicator that more than half of the process output is likely to violate one or other of the tolerances.

Other indices have been designed for multiple criteria, e.g. to express capability for hole size and location simultaneously. Time will tell whether these remain only in specialized use or will gain general acceptance.

13.5.4 *C*-indices below 1.0 correspond to a distance of less than 3σ between process mean and a specification limit. This in turn implies more than 0.135% of product outside specification, and is regarded as highly unsatisfactory, i.e. non-capable.

C-indices greater than 1.0 are regarded as demonstrating a basic level of capability, but for functional, critical or safety items much higher indices are often required, even 2.0 or more. It is impossible to offer specific criteria, as customer requirements vary considerably, but in 1993 indices of 1.33 or 1.67 represent the most widely used criteria for C_p, C_{pU}, C_{pL} and C_{pk}. For C_m and *P* indices, the criteria are generally 0.33 greater than for the 'process' indices, the additional standard deviation allowing for additional common cause variation likely to be encountered in moving from machine or pre-production to the full scale process.

The use of criteria such as 1.33 or 1.67 for these indices implies, of course, that the distance between the process mean and USL or LSL should be at least $4\sigma(1.33 \times 3)$ or $5\sigma(1.67 \times 3)$. However, the index is still defined in terms of the 3-sigma process spread as in 13.9 – it is *not* correct to insert 4σ or 5σ etc., into these formulae.

Finally, one must caution against over-interpretation of very high C-indices. For example, an index of, say, 2.0 implies that the process mean is six standard deviations from a specification. In a perfect Normal distribution, this means that one item in about a thousand million violates the specification limit, but it is ludicrous to stretch the Normal model to this extent. A more reasonable interpretation is that a shift in the process mean of 2 or even 2.5 standard deviations will not cause any real problem. Thus, very high indices represent a safety factor rather than a measure of proportions of items beyond specification limits.

The same considerations apply to the alternative ways of expressing capability described in following sections. Figure 13.1 illustrates the indices for various levels of capability.

13.5.4 An alternative and informative measure of capability, preferred by some users, is simply to use the 'z-value' which measures the distance from the process average to one or other of the specification limits,

$$z = \frac{\text{USL} - \bar{\bar{x}}}{\hat{\sigma}} \quad \text{or} \quad \frac{\bar{\bar{x}} - \text{LSL}}{\hat{\sigma}} \tag{13.11}$$

Evidently $z = 3.0$ corresponds to $C = 1.0$, $z = 4.0$ to $C = 1.33$, etc. An advantage of the z-format is that the approximate proportion of out-of-specification output can readily be estimated from a table of the Normal distribution.

13.5.5 Sometimes the capability statement is expressed in a form reciprocal to the C-index. In this case, the range covered by six (estimated) standard deviations is calculated and expressed as a fraction, or more often a percentage, of the specification.

For a just-satisfactory C-index of 1.00, this method would indicate that six standard deviations cover (or use) 100% of the specification. The general relationship is

$$\frac{6\sigma \text{ performance}}{\text{Total specification width}} = 100 \times \frac{1}{C\text{-index}}$$

Thus C-indices of 1.33, 1.67, 2.00 etc. would be rendered as 75%, 60% and 50% of specification 'used' by the process or machine. Evidently a C-index of less than 1.00 will indicate that the process uses more than 100% of the specification. Thus for $C = 0.82$, the performance uses 122% of the total tolerance.

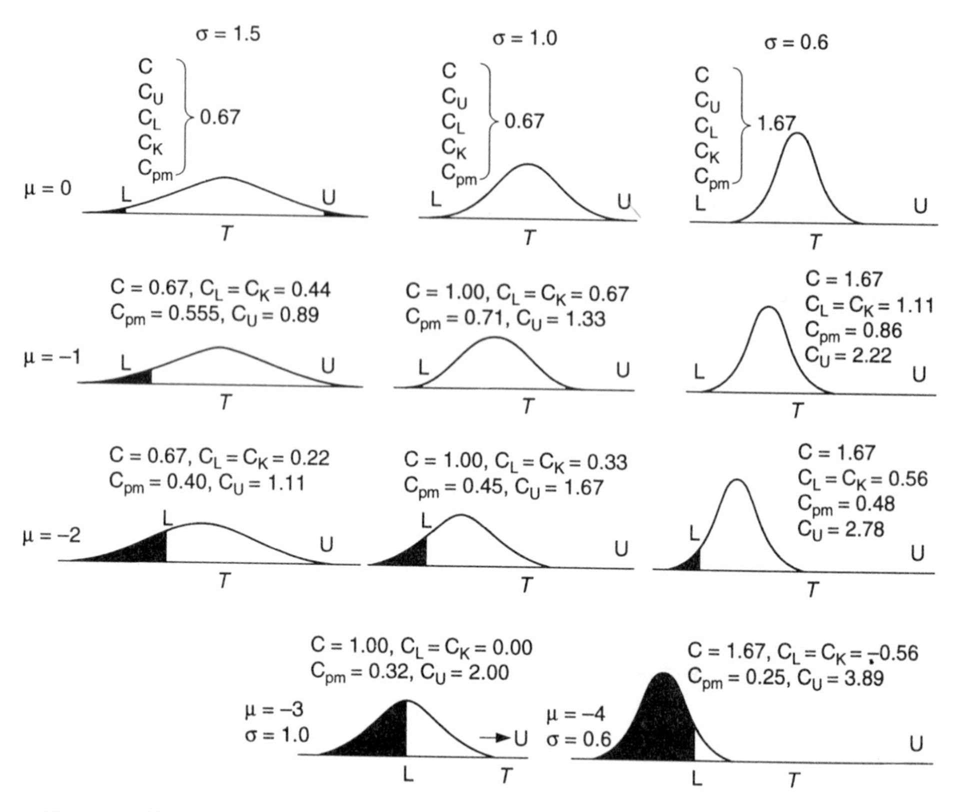

Fig. 13.1 Illustrations of various capability indices. LSL $(L) = -3$; Nominal $(T) = 0$; USL $(U) = 3$. Black areas show out-of-specification.

13.5.6 Other methods

British Standards up to the 1970's used a measure based on average range, known as the 'relative precision index' or RPI, defined as

$$\text{RPI} = \frac{\text{Total width of specification}}{\bar{R}}$$

Unfortunately this index has slightly differing interpretations depending on the sample sizes involved for $\bar{R}$ and it is much less widely used now, being superseded by C-indices. The 'standardized precision index', expressed as

$$SPI = \frac{\text{Total specification width}}{\text{Standard deviation}}$$

was designed to overcome this problem, but has not achieved widespread use. From the above expression, it is obvious that $SPI = 6 \times C_p$(or P_p or C_m).

13.5.7 It is sometimes useful to summarize the performance of a process for which there is no formal specification. A comparison of the process variation with the average or target value may serve this purpose.

(i) Where no specific target value is prescribed, the coefficient of variation (usually expressed as percentage) can be used:

$$CV = 100 \times \frac{\hat{\sigma}}{\bar{\bar{x}}}$$

(ii) Where there is a target value T, by analogy with the C_{pm} index noted in section 13.5.3 we may use

$$CV_T = 100 \times \frac{\hat{\sigma}'}{T}$$

where

$$\hat{\sigma}' = \sqrt{[\hat{\sigma}'^2 + (\bar{\bar{x}} - T)^2]}$$

There are no agreed criteria for these measures of performance, but they can be particularly useful for judging the impact of process improvements. For consistency with other indices, they are sometimes multiplied by six to express the 6σ spread (or process performance) as a proportion of the target value.

To provide an example, we may take the data from Table 8.1 as summarized in 13.6.4. With $\bar{x} = 239.53$ and $\hat{\sigma} = 1.041$, we find $CV = 0.435\%$ (or alternatively $6CV = 2.61\%$). For a target value of 240, the value of σ' would become

$$\sqrt{\{1.041^2 + (239.53 - 240)^2\}} = 1.142$$

Then $CV_T = 0.476\%$ (or $6CV_T = 2.86\%$).

13.6 Confidence limits for capability indices

13.6.1 One should note that a capability index calculated from sample or control chart data is an **estimate** of process performance, and like $\bar{x}$ and $\hat{\sigma}$ (on which it depends) it is subject to sampling variation. The reliability of C-indices therefore deserves some attention. The theory is somewhat complex, but for indices based on 50 or more individual values – as is generally the case – some excellent and reasonably simple approximations are available.

It is first necessary to determine the number of **effective** degrees of freedom on which the estimate of standard deviation is based. For a machine capability study using n items in one sample, this will be simply $n - 1$. For estimates based on k samples of n items each, the degrees of freedom are given by

$$\nu = k(n - 1) \times \text{discount factor}$$

The discount factor depends on the method used to calculate $\hat{\sigma}$ ($\bar{R}$, $\bar{s}$ or $\sqrt{\text{av}(s)^2}$), and a list of discount factors for various sample sizes appears in Table 16.1.

13.6.2 Confidence limits for C-indices can now be obtained as:

$$\hat{C}\left[1 \pm z\sqrt{\left(\frac{1}{2\gamma}\right)}\right] \tag{13.12}$$

where z is the Normal deviate corresponding to the required confidence level, e.g. 1.96 for 95% confidence. For routine use, $z = 2$ provides a convenient approximation with simple appeal.

13.6.3 For C_k indices, the expression is a little more complicated but still manageable:

$$\hat{C}_k\left[1 \pm z\sqrt{\left(\frac{1}{9knC_k^2} + \frac{1}{2v}\right)}\right] \tag{13.13}$$

The same expression applies also to C_U or C_L indices by substituting $\hat{C}_U$ or $\hat{C}_L$ for $\hat{C}_k$.

13.6.4 We consider two examples. First, we suppose that the valve liner data of section 2.2.1 represented a short-term capability study. With $n = 52$ and s calculated directly from the values in one sample, we have $v = n - 1 = 51$.

Suppose that the weight specification requires 240 ± 2.5 g. Then we have, with $\bar{x} = 239.51$ and $s = 0.9335$,

$$\hat{C}_m = \frac{242.5 - 237.5}{6 \times 0.9335} = 0.8927$$

(not a satisfactory indication of capability). We also have

$$\hat{C}_{mU} = \frac{242.5 - 239.51}{3 \times 0.9335}, \quad \hat{C}_{mL} = \frac{239.51 - 237.5}{3 \times 0.9335}$$

i.e. 1.068, 0.718, respectively, whence $\hat{C}_{mk} = 0.718$ (from bad to worse). (We may note that $\hat{C}_{pm}$ as defined in section 13.5.3 would yield an index of

$$\hat{C}_{pm} = \frac{242.5 - 237.5}{6[0.9335^2 + (240 - 239.51)^2]} = 0.79,$$

between the values of $\hat{C}_p$ and $\hat{C}_{pk}$).

How reliable are these indices? Using (13.12), the 95% confidence limits for C_m become approximately

$$0.8927[1 \pm 2\sqrt{(\tfrac{1}{102})}] = 1.069,\ 0.716$$

These limits represent $\pm 20\%$ (approximately) of the actual estimated index.

Similarly, using (13.13), the 95% limits for $\hat{C}_{mk}$ are

$$0.7177\left[1 \pm 2\sqrt{\left(\frac{1}{9 \times 50 \times 0.7177^2} + \frac{1}{102}\right)}\right]$$

which work out at 0.888 and 0.547, or $\pm 24\%$ of $\hat{C}_{mk}$ itself.

For the second example, we take the data from the $\bar{x}$, R chart of section 8.6. After discarding one subgroup on account of a wildly anomalous value, we have $k = 24$ subgroups of $n = 5$ items each, with $\bar{\bar{x}} = 239.53$ and $\bar{R} = 2.421$. We shall assume that the chart shows the process to be in control, though there are serious reservations in this case, which are further investigated in Chapter 16. From $\bar{R}$, using d_5 of Table 13.1 we estimate

$$\hat{\sigma} = 2.421 \div 2.326 = 1.041$$

Here, because $\bar{R}$ is used to estimate σ, discount factor of 0.906 applies (obtained from Table 16.1), so there are $24 \times (5 - 1) \times 0.906 = 87$ degrees of freedom. Using the same specification (240 ± 2.5) we have:

$$\hat{C}_{\mathrm{p}} = \frac{242.5 - 237.5}{6 \times 1.041} = 0.8005$$

$$\hat{C}_{\mathrm{pk}} = \mathrm{Min}\left(\frac{242.5 - 239.53}{3 \times 1.041}, \frac{239.53 - 237.5}{3 \times 1.041}\right) = 0.650$$

(Again, $\hat{C}_{\mathrm{pm}}$ lies between C_{p} and $\hat{C}_{\mathrm{pk}}$,

$$\hat{C}_{\mathrm{pm}} = \frac{242.5 - 237.5}{6\{1.041^2 + (240 - 239.53)^2\}} = 0.730)$$

From (13.12), the 95% confidence limits for C_{p} are

$$0.8005\left(1 \pm 2\sqrt{\left(\frac{1}{2 \times 87}\right)}\right) = 0.922,\ 0.679$$

and for C_{pk}, via (13.13)

$$0.650\left[1 \pm 2\sqrt{\left(\frac{1}{9 \times 24 \times 5 \times 0.650^2} + \frac{1}{2 \times 87}\right)}\right] = 0.766,\ 0.534$$

The limits are proportionately wider for C_{pk} than for C_{p} ($\pm 17.8\%$ for C_{pk}, $\pm 15.2\%$ for C_{p} in the above example) because of the sampling variation in $\bar{\bar{x}}$ as well as in $\hat{\sigma}$.

13.6.5 It is theoretically possible to monitor capability indices using control chart methods. The required standard error, which can then be used to obtain 3-sigma control limits or cusum decision rules, is:

$$C(T) \times \sqrt{\left(\frac{1}{2f}\right)} \quad \text{for } C\text{-type indices}$$

(i.e. involving only process spread) and

$$C_{\mathrm{k}}(T) \times \sqrt{\left(\frac{1}{9knC_{\mathrm{k}}^2(T)} + \frac{1}{2f}\right)} \quad \text{for } C_{\mathrm{k}} \text{ indices}$$

(involving both spread and location). In these expressions $C(T)$ and $C_k(T)$ are the target values for the index concerned.

Because any change in either mean or variation will affect C_k, chart-to-chart or month-to-month variation in the indices is often appreciably greater than the theoretical standard error (which assumes process stability). Control charts for capability indices thus tend to be 'noisy'.

As prompt corrective action is unlikely to be possible, the charts provide a medium or long term data summary and an indication of trends rather than a direct control technique.

13.7 Probability plotting and graphical presentation of capability

13.7.1 As with many other statistical methods, graphical representation can improve understanding of principles as well as providing a useful means of interpreting the data. Simple diagrams of Normal curves go some way towards achieving this, as illustrated in Fig. 13.1.

A disadvantage of such diagrams is that they do not actually display the data themselves: a histogram would be preferable for this purpose, perhaps with a corresponding Normal curve superimposed as in Fig. 13.2 which displays the same valve liner data of 2.2.1 Unfortunately, matching the Normal curve to the data requires either the use of suitable computer software or some rather tedious manual calculations.

A simple procedure for sketching (rather than precisely drawing) a normal curve where $\bar{x}$ and s have been calculated is:

(i) Using a suitable horizontal scale for x, draw a vertical line of any desired length at $\bar{x}$; if matching a histogram, the line is approximately the same height as the central column(s) of the histogram.

(ii) Mark off, at distances of s on either side of $\bar{x}$, heights of 3/5 ths (or more simply 5/8 ths) the height at the centre. Five-eighths is easily measured by halving the central height, halving the top half and again halving the second quarter from the top. The true height of N (μ, σ) at $\mu \pm \sigma$ is 0.6067 of the height at μ, i.e. between 3/5 and 5/8.

(iii) Similarly at $\bar{x} \pm 2s$, mark off one eighth of the central height (the true ratio for N (μ, σ) is 0.135). Note also that beyond $\pm 2s$, the curve slowly approaches the x-axis.

(iv) Now sketch the curve through the five points located in (i), (ii), (iii), noting that the curve is steepest (i.e. there are points of inflexion) at $\bar{x} \pm s$.

The procedure is illustrated in Fig. 13.2(b) for the valve liner data.

13.7.2 The probability plot also fulfils a useful role. It displays the data, provides an indication of the fit of a Normal distribution model, and permits an interpretation of capability against the specification limits. It is useful in many

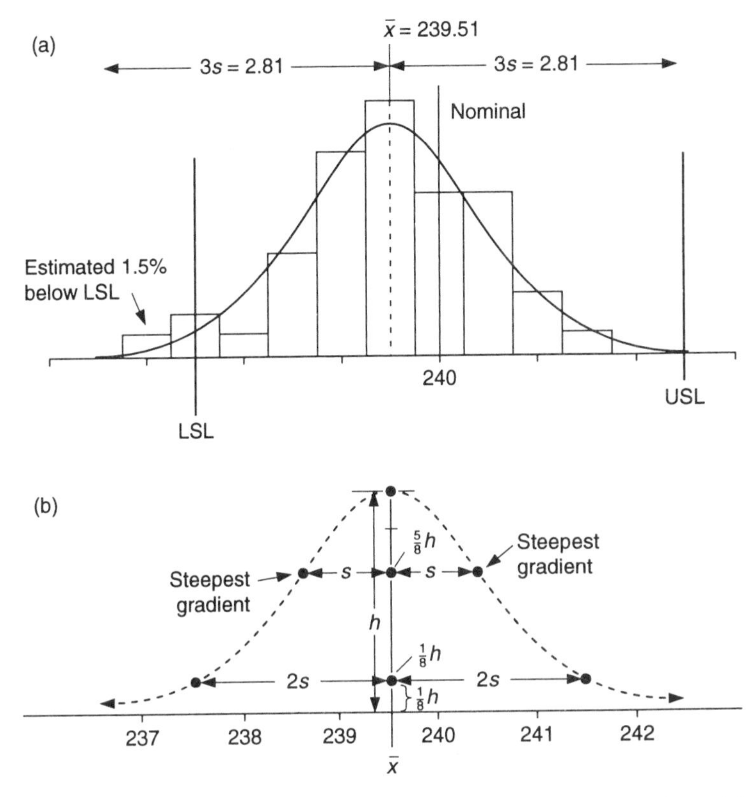

Fig. 13.2 Histogram and fitted Normal curve for valve liner data. (a) USL − LSL = 5.0; $6s = 5.62$, $C_m = 0.89$, $C_{mk} = 0.72$. (b) Approximate sketch of Normal curve with $x = 239.51$ and $s = 0.9335$; h = height at or near centre columns of histogram.

other applications besides those concerned with capability, and for many distribution models besides the Normal. It can be used either with all individual data values or with the data condensed into a frequency table containing ten or more classes, ideally 10 to 15.

The general principle underlying probability plotting is that the data are first arranged in order of magnitude. A blob diagram or stem-and-leaf is useful here, or where the data are to be grouped, a frequency table with 10–15 classes provides a set of ordered groups. Fewer than ten classes does not permit a satisfactory assessment of the fit of the distribution.

The cumulative distribution is then plotted on a special graph grid whose probability axis is scaled with the appropriate theoretical distribution function. Here we consider only the Normal distribution, but other examples are given in Chapter 14. In some cases the x-axis may have a non-linear scale, as for the log normal or Weibull distribution models.

13.7.3 The positions at which the x-values are plotted is of considerable importance. Denoting the ordered values by $x_{(i)}, i = 1$ to n, three conventions are widely used:

(i) Plot $x_{(i)}$ against $100(i/(n+1)$ on the percentage scale. This is known as the mean rank convention, and is especially recommended for Weibull plots of time-to-failure data. It suffers the complication that where n is an 'easy' number like 20, 25, 50 or 100, $n+1$ yields more awkward percentage values.

(ii) Plot $x_{(i)}$ against $100[(i-0.3)/(n+0.4)]$, the median rank convention. Although this has some minor theoretical advantages in estimating distribution parameters, these rarely outweigh the disadvantages of calculation complications when plotting manually.

(iii) A simple general-purpose method is the 'centralized distribution function' convention:

$$\text{Plot } x_{(i)} \text{ against } 100\left(\frac{i-\frac{1}{2}}{n}\right)$$

This (for easy values of n) yields convenient plotting positions, and it is only on the most extreme values in the data set that the three conventions differ perceptibly. For graphical evaluation of the fit of a model, one is usually more concerned with the values in the middle 90% of the distribution. Table 13.2 lists the plotting positions for all three conventions for samples of $n = 10$, and also the five values at the upper and lower extremes for $n = 25, 50, 100$.

Note that all three conventions give 0.5 as the plotting position for the median, when $i = \frac{1}{2}(n+1)$.

13.7.4 For simplicity, we shall therefore adopt the **centralized distribution convention (cdf)** in the examples. The final point, before describing the procedure and applying it, is to determine whether the actual data values can be plotted as they are, or whether one must make allowance for discreteness of measurement. When slightly dissimilar values are grouped into a frequency table, a class such as 239.25–239.75, with mid-value 239.5, could obviously contain values up to the class limit itself. Because the proportion plotted is cumulative, the upper class limits *must* be used for plotting. The fewer the number of classes into which the values are grouped, the more important this becomes, in order to avoid both bias and distortion of the plot.

When individual (ordered) results are plotted, one may be more flexible. In the valve liner data, for example, possible values on the scale of measurement are 236.9, 237.0, 237.1, etc. This implies that the value **recorded** as 236.9 is in fact somewhere between 236.85 and 236.95, so that the upper 'class' limit is strictly 236.95. Where the resolution of the measurements is around one tenth of the standard deviation or better, the use of actual values against 'class' limits (in

Table 13.2 Comparison of three plotting conventions

		$x_{(1)}$	$x_{(2)}$	$x_{(3)}$	$x_{(4)}$	$x_{(5)}$	$x_{(n-4)}$	$x_{(n-3)}$	$x_{(n-2)}$	$x_{(n-1)}$	$x_{(n)}$
*Mean rank	$n = 10$	9.09	18.18	27.27	36.36	45.45	54.55	63.64	72.73	81.82	90.91
†Median rank		6.73	16.35	25.96	35.58	45.19	54.81	64.42	74.04	83.65	93.27
‡Centralized *cdf*		5.00	15.00	25.00	35.00	45.00	55.00	65.00	75.00	85.00	95.00
Mean rank	$n = 25$	3.85	7.69	11.54	15.38	19.23	80.77	84.62	88.46	92.31	96.15
Median rank		2.76	6.69	10.63	14.57	18.50	81.50	86.43	89.37	93.31	97.24
Centralized *cdf*		2.00	6.00	10.00	14.00	18.00	82.00	86.00	90.00	94.00	98.00
Mean rank	$n = 50$	1.96	3.92	5.88	7.84	9.80	90.20	92.16	94.12	96.08	98.04
Median rank		1.39	3.37	5.56	7.34	9.33	90.67	92.66	94.64	96.63	98.61
Centralized *cdf*		1.00	3.00	5.00	7.00	9.00	91.00	93.00	95.00	97.00	99.00
Mean rank	$n = 100$	0.99	1.98	2.97	3.96	4.95	95.05	96.04	97.03	98.02	99.01
Median rank		0.70	1.69	2.69	3.69	4.68	95.32	96.31	97.31	98.31	99.30
Centralized *cdf*		0.50	1.50	2.50	3.50	4.50	95.50	96.50	97.50	98.50	99.50

*$100(i)/(n+1)$.
†$100(i-0.3)/(n+0.4)$.
‡$100(i-\frac{1}{2})/n$.

the sense of precision of measurement classes) makes very little difference to the appearance of the plot. Even here, however, small errors can arise if the plot is used for estimation of parameters – without adjustment the mean might appear to be 239.45 rather than 239.5, for example. The reader may make his or her own choice: in Figs 13.3 to 13.5 the true class boundaries are used.

13.7.5 We may now detail the plotting procedure, first for individual values and then for data in a frequency table.

Individual values

(i) Order the data and identify as $x_{(1)}$, $x_{(2)}$ to $x_{(n)}$.
(ii) Decide whether the data points (x_i) need upward adjustment by one half of the measurement resolution.
(iii) Plot the $x_{(i)}$ values against their respective plotting positions, obtained as

$$P_i(\%) = 100\left(\frac{i - \frac{1}{2}}{n}\right)$$

(iv) Examine the plot for linearity. If desired, draw a line of best fit by eye, concentrating especially on the 10–90% central zone of the data. Note in section 13.7.6 below that a line based on calculated $\bar{x}$ and s may also be drawn to examine the fit of the Normal distribution.

Data from a frequency table

(i) Use 10–15 classes of equal width. If the special graph grid of Fig. 13.3 is available, calculate the exact class midvalues. Thus for a class of 10.0 to 10.9 inclusive, the midpoint is $\frac{1}{2}(10.0 + 10.9) = 10.45$ (it is also the arithmetic mean of the values 10.0, 10.1, 10.2... 10.9). If this special grid, with its stepped arrows, is not available, calculate the upper limit for each class: often this is best obtained as half-way between two consecutive class midvalues.
(ii) Form the cumulative numbers of values in successive classes, starting from the smallest data values, i.e. the sum of the frequencies in any given class and all the preceding classes, $\sum f$.
(iii) Plot the upper class limits against

$$P(\%) = 100\left(\frac{(\sum f) - \frac{1}{2}}{n}\right)$$

For the special graph grid, the stepped arrows lead naturally to the correct upper class limits for plotting.
(iv) Examine for linearity, and either fit a line by eye or by calculation as for individual values.

13.7.6 Graphical estimation, if required, proceeds as follows.

(i) Follow the 50% line to the point where it reaches the sloping line fitted to the data. From this point, read off the value for x on the data axis. This is,

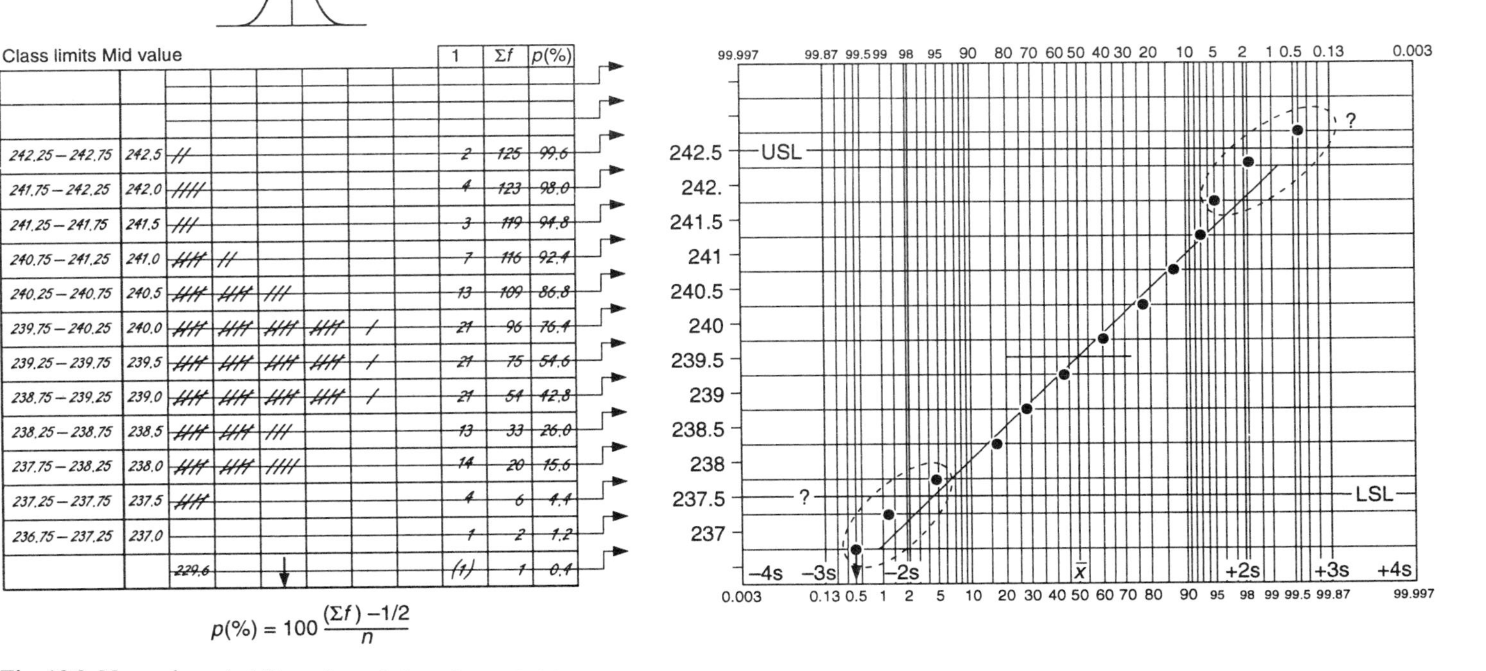

Class limits	Mid value	1	Σf	p(%)
242.25 – 242.75	242.5	2	125	99.6
241.75 – 242.25	242.0	4	123	98.0
241.25 – 241.75	241.5	3	119	94.8
240.75 – 241.25	241.0	7	116	92.4
240.25 – 240.75	240.5	13	109	86.8
239.75 – 240.25	240.0	21	96	76.4
239.25 – 239.75	239.5	21	75	54.6
238.75 – 239.25	239.0	21	54	42.8
238.25 – 238.75	238.5	13	33	26.0
237.75 – 238.25	238.0	14	20	15.6
237.25 – 237.75	237.5	4	6	4.4
236.75 – 237.25	237.0	1	2	1.2
	229.6	(1)	1	0.4

$$p(\%) = 100\,\frac{(\Sigma f) - 1/2}{n}$$

Fig. 13.3 Normal probability plot of data from Table 8.2. (Data is grouped in half gram classes). '?' indicates regions where the line deviates from some data points; see section 13.8.2.

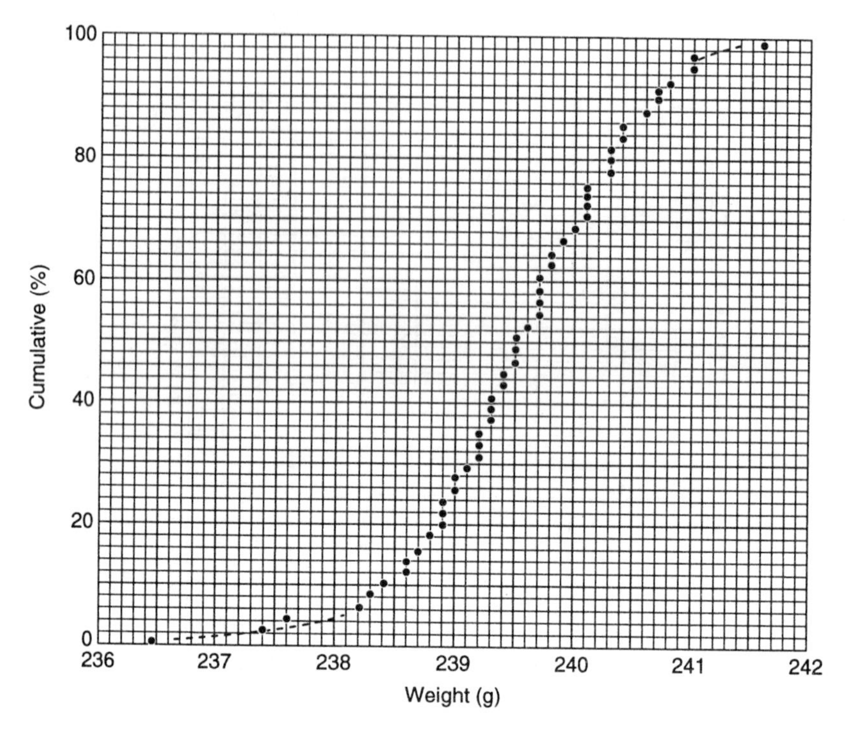

Fig. 13.4 Ogive (cumulative frequency plot) of valve liner data based on data of Table 13.3.

of course, an estimate of the **median**, but for a well-fitting line may serve as an estimate of the mean, since 'mean = median' for the Normal distribution.

(ii) Similarly, read off values corresponding to any pair of $\pm 2s$, $\pm 3s$ or $\pm 4s$ points on the percentage scale. Divide the difference between the upper and lower values of the pair by 4, 6, or 8 as appropriate. Thus if values at $\pm 3s$ are chosen, take

$$s = \tfrac{1}{6}[(\bar{x} + 3s) - (\bar{x} - 3s)]$$

This procedure cannot be recommended for calculation of capability in view of the subjective fitting of the line to represent the data, and of possible graphical inaccuracies in reading off $\bar{x}$, $\bar{x} + k.s$ and $\bar{x} - k.s$. Direct calculation from the original data values is preferable, and is rarely a problem with present-day computing or calculating resources.

The converse of graphical estimation of $\bar{x}$ and s is to use their calculated values to position the sloping line in its correct position. The value of $\bar{x}$ is plotted at the 50% probability ordinate, and those for $\bar{x} + 2s$ and $\bar{x} - 2s$ at 97.72 and 2.28%, respectively. The three points, if correctly calculated and

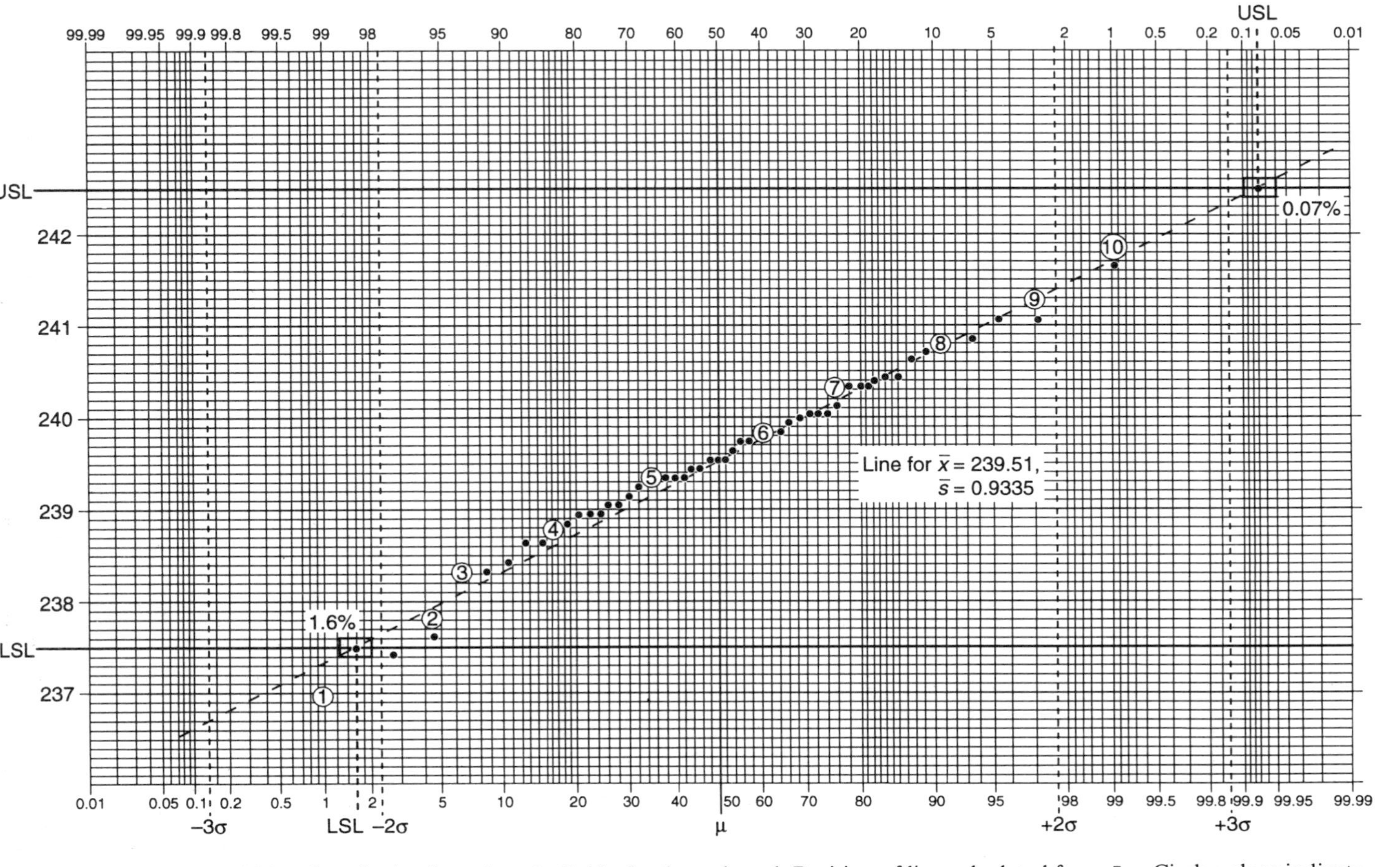

Fig. 13.5 Normal probability plot of valve liner data. Individual values plotted. Position of line calculated from $\bar{x}$, s. Circle values indicate data plotted from frequency table in Table 2.3. The line shown here is for $\bar{x} = 239.51$ and $s = 0.9335$.

plotted, will then lie on a straight line which represents the corresponding Normal distribution.

13.8 Examples of Normal probability plots

13.8.1 For the first example, we take the valve liner data and plot the original (ungrouped) values. The stem-and-leaf table of section 2.2 will provide a convenient starting point, and from this we can compile the data values with their corresponding plotting positions as in Table 13.3.

13.8.2 The individual values are shown as an **ogive** or cumulative frequency diagram in Fig. 13.4. It would evidently be difficult to assess whether this flattened *S*-shape matches what one would expect from a Normal distribution (see Fig. 4.5 for an example of the Normal cumulative curve). A Normal probability grid modifies the x-scale so that a perfect Normal distribution would yield a straight line plot. Figure 13.5 shows the individual values plotted on such a grid. There is some slight non-linearity at the three lowest values, and a line calculated from $\bar{x}$ and s lies a little below the data points in the lower quartile. However, as further examined in Chapter 14, there is no strong evidence of non-Normality.

With the specification limits drawn on the graph, it is clear that around 1.5% of the values are likely to fall below LSL if this performance is typical. Far fewer (about 0.07%) are likely to exceed USL. The situation could be improved, though not entirely remedied, by setting the mean value at the nominal value of 240. This would considerably reduce the proportion below LSL (to about 0.4%), but would increase those above USL (to 0.4%). The total (0.8%) outside both limits would thus be smaller than the present 1.6 to 1.7%.

13.8.3 It would, of course, be possible to present the valve liner data via a frequency table. This would simply reduce the number of plotting points to 10, the number of classes in Table 2.3. These points are shown by circled numbers in Fig.13.5. Most of them correspond to individual points on the graph, but 2, 7 and 10 in particular are just distinguishable from individual points because no individual values happened to be at the upper limits of their respective classes.

To provide a different example, with some diagnostic implications, we take the control chart data of Table 8.2. collected into a frequency table with half-gram class interval. This yields the table on the left side of Fig. 13.3.

Continuing with the steps listed in 13.7.5, the cumulative frequencies, $\sum f$, are converted to $P(\%)$ values, and these are plotted against the upper class limits following the stepped arrows. To avoid considerably compressing the scale, the outlying value 229.6 is not plotted but its position is indicated by the down-pointing arrow. The central part of the resulting plot is close to linear, but both tails of the distribution give rise to some doubts. These are, at least in part, due to out-of-limits sample means, but examination of the tallies suggests the

Table 13.3 Ordered valve linear data with plotting positions

Centralized cumulative distribution function $P(\%) = 100(i - \frac{1}{2})/n$
$n = 52$

i	$x_{(i)}$	$P\%$	i	$x_{(i)}$	$P\%$	i	$x_{(i)}$	$P\%$	i	$x_{(i)}$	$P\%$	i	$x_{(i)}$	$P\%$
1	236.9	0.96	12	238.9	22.12	23	239.4	43.27	34	239.8	64.42	45	240.4	85.58
2	237.4	2.88	13	238.9	24.04	24	239.4	45.19	35	239.9	66.35	46	240.6	87.50
3	237.6	4.81	14	239.0	25.96	25	239.5	47.12	36	240.0	68.27	47	240.7	89.42
4	238.2	6.73	15	239.0	27.88	26	239.5	49.04	37	240.1	70.19	48	240.7	91.35
5	238.3	8.65	16	239.1	29.81	27	239.5	50.96	38	240.1	72.12	49	240.8	93.27
6	238.4	10.58	17	239.2	31.73	28	239.6	52.88	39	240.1	74.04	50	241.0	95.19
7	238.6	12.50	18	239.2	33.65	29	239.7	54.81	40	240.1	75.96	51	241.0	97.12
8	238.6	14.42	19	239.2	35.58	30	239.7	56.73	41	240.3	77.88	52	241.6	99.04
9	238.7	16.35	20	239.3	37.50	31	239.7	58.65	42	240.3	79.81			
10	238.8	18.27	21	239.3	39.42	32	239.7	60.58	43	240.3	81.73			
11	238.9	20.19	22	239.3	41.35	33	239.8	62.50	44	240.4	83.65			

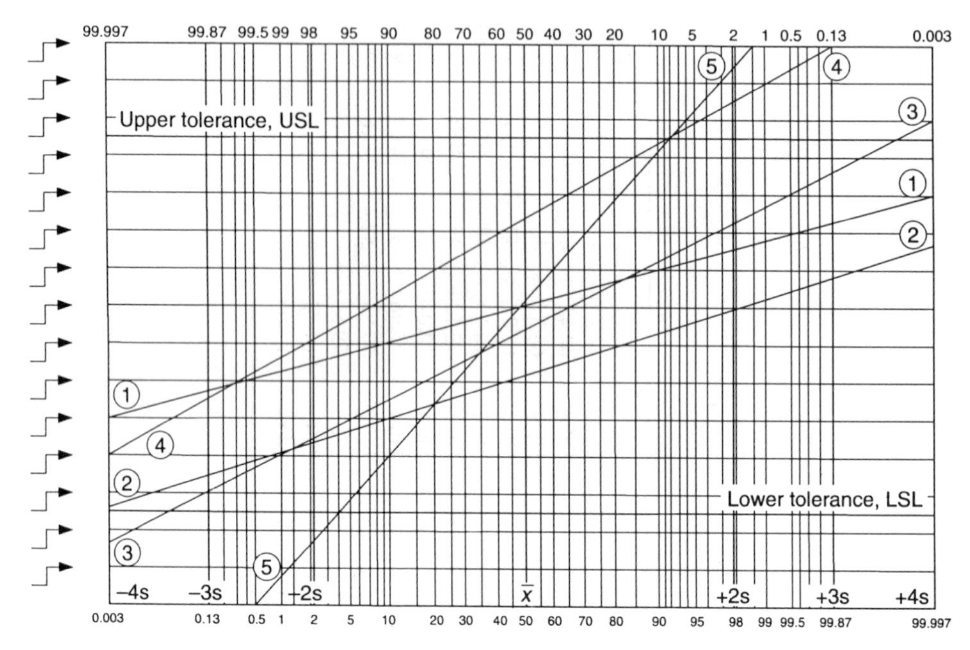

Fig. 13.6 Probability plots for various levels of capability.

upper tail of the distribution is rather longer than one would expect from a Normal curve. Some further analysis of this data will appear in Chapter 16; the probability plot is here just one of several techniques that might be brought to bear on a problem.

If, despite these indications, the specification limits were drawn on the plot, one would estimate perhaps half a percent of items would lie beyond USL and around 4% below LSL – evidently considerable room for improvement.

It would be unwise to estimate $\bar{x}$ and s from the plot, but simply to indicate the method, the intersection of the best-fit line with the 50% vertical would be used to estimate $\bar{x}$. This appears to be close to 239.5 (the stepped arrows now being ignored).

For $\bar{x} + 2s$, the line passes through the upper limit of the 241.25 to 241.75 class, so $\bar{x} + 2s = 241.75$. Similarly $\bar{x} - 2s$ is just below 237.25, say 237.2. Then,

$$\begin{aligned} \bar{x} + 2s &= 241.75 \\ \bar{x} - 2s &= 237.2 \\ \hline 4s &= 4.44, \end{aligned}$$

$$\text{whence } s = 1.1375.$$

In fact, by calculation from the frequency table data (excluding the value 229.6), $\bar{x} = 239.51$, $s = 1.144$. The Normal distribution with these as parameters (μ, σ) is practically indistinguishable from the best-fit line on the capability plot.

13.9 General interpretation of probability (capability) plots

13.9.1 Capability (absence of large variation) is indicated graphically by a flat slope of the best-fit line relative to the specification limits. Ideally (for $C_k > 1.33$) the line runs off the extreme edges, representing 0.003% or $4s$ without crossing the limits.

What is often referred to as the process or machine **setting** corresponds to the average level, and is indicated by the value at which the best-fit line crosses the 50% point (often labelled $\bar{x}$, though it is really the median) of the probability axis.

Improvement of capability results from reducing variation along with ensuring that the setting corresponds to the nominal or target value. In graphical terms, this means reducing the slope of the line ensuring that, in most cases, its 50% crossing lies midway between the specification limits.

13.9.2 As an aid to interpreting capability graphs, Fig. 13.6 illustrates some idealized plots of varying capability. The performance represented by the various lines is as follows:

(a) High capability (flat slope), well centred (crosses 50% line midway between USL and LSL). The numerical capability would be $C = C_U = C_L = 2.4$.
(b) High capability, but centred below $\frac{1}{2}(\text{USL} + \text{LSL})$. $C = 2.1$, $C_U = 2.7$, $C_L = C_K = 1.5$. The situation could be much improved by an upward adjustment of the mean level.
(c) Reasonable capability; $C = 1.3$, $C_U = 1.4$, but $C_L = C_k = 1.2$ because the mean lies nearer to LSL than to USL. An adjustment of 'setting' could yield $C_k = C = 1.3$, but there is room for further improvement by reduction of variation.
(d) Inadequate capability, $C = 1.17$, $C_L = 1.85$, $C_U = C_k = 0.68$. The discrepancy between C and C_k, as well as the fact that the 50% point is much nearer to USL than to LSL, indicate some scope for improvement by 'resetting' the average level, but capability would still not satisfy the usual requirements of the 1990's.
(e) The steepness of the line, and large proportions (around 4%) cut off by each of USL and LSL, tell their own story. The only good point is that the median is not far from the mid-specification value, but with $C = 0.57$, $C_L = 0.59$, $C_U = C_k = 0.55$, there is need for a drastic reduction in variability.

13.9.3 We close with a note of caution about the use of capability indices. Whilst they provide a useful and concise summary of performance, their validity depends on a number of factors. Among them we may particularly note the following:

(i) Is the process from which the data was obtained genuinely stable? If not, capability indices do not provide a proper measure of performance, and may even be worthless or downright misleading.
(ii) Does the data show the distribution of **individual** values to be (at least approximately) Normal? If not, the methods of Chapter 14 may be necessary, and we note therein a distinction between disturbed Normality and regular non-Normality.
(iii) Are the specification limits, generally USL and LSL, properly relevant to the situation, or merely conventions? Engineering drawings, for example, often indicate all tolerances to be ± 0.002 inch or ± 0.05 mm. For some dimensions, such accuracy may be more than adequate, for others it may be too slack a criterion. High C-indices for very wide specifications may give spurious indications, and marginal or low indices for difficult operations may fail to recognize the effort that is involved in meeting tight specifications.
(iv) The influence of measurement error (see Chapter 15) on **apparent** capability must not be overlooked. Where such errors are large relative to actual process variation, capability may be much higher than appears from the usual methods of assessing it. Thus, if repeated measurements (using the same equipment and the same operator) yield a standard error say σ_r, whilst the true process variation is σ_p, the apparent overall standard deviation will be $\sqrt{(\sigma_r^2 + \sigma_p^2)}$. For the valve liner data, suppose we have only a crude balance available with one gram indications. Then perhaps $\sigma_r = 1.0$, and with σ_p also close to 1 (actually $\hat{\sigma}$ is about 1.14 in the example of section 13.8.3) the apparent standard deviation will be around 1.4–1.5, considerably downgrading the values of C_p or C_{pk}. Improvements in processs variation may require corresponding upgrading of measurement techniques before they are fully realized.

Even given satisfactory answers to these questions, a continuous and unthinking obsession with ever higher indices may divert effort (and resources) away from more important areas of quality improvement. Increasing $C_p(C_{pk})$ from, say, 6 to 10 does not decrease violations of limits (there are virtually none anyway) though it does provide greater safety margins for 'wandering' of a process average. As with any other technological activity, capability measurement needs to combine statistical analysis with technical expertise, economics and common sense. If expensive change, re-tooling, more expensive materials, etc., are necessary for an increase from 3 to 5 in the C-index, that effort could perhaps be more usefully deployed in improving some other characteristic from a low index to a moderately good one, say from 0.9 to 1.3.

14

Capability: non-Normal distributions

14.1 Non-Normality or disturbed Normality?

14.1.1 First, we shall distinguish between applications where a truly non-Normal distribution is appropriate and those where anticipated Normality is disturbed by features in a system that can be regarded as special causes. Obvious members of this latter class are the often-cited case of two (or more) normal distributions with differing means being mixed (as when various positions of a machine have slightly differing settings) and the occurrence of outliers giving a ragged tail to an observed distribution.

Other cases include occasional shifts in mean from differing batches of materials, or from changing environmental conditions. These cannot always be regarded as special causes, and Chapter 11 has considered examples. Shifts in performance of test instruments, differences between inspectors or test operators may also yield similar influences. The approach in all such cases is not to apply non-Normal techniques unthinkingly (they may not be effective anyway) but to seek, and if possible remove or reduce, the causes. In the final event, it may prove impossible to provide valid measures of capability where uncontrollable factors continue to produce disturbed Normality.

14.1.2 It is equally erroneous to assume that all non-Normality results from anomalies or special causes. Well known cases that give rise to well-behaved but persistent non-Normal distributions include measures of distortion (eccentricity, warpage, surface finish), electrical phenomena (capacitance, insulation resistance) low levels of substances in materials (trace elements, impurities) and other physical and chemical properties (ultimate strengths, friction characteristics, time to failure under bench or real-life conditions, etc.).

These naturally non-Normal applications are the subject of the present chapter. Most non-Normal distributions differ from the Normal curve in one or both of two respects:

(i) *Skewness*: they exhibit a pattern with one tail extending further from the modal value than the other. In some cases the mode may even be at one of the extremes of the distribution. A mode at zero is a particularly common case.

When the longer tail extends to the right, in conventional graphical format, i.e. towards the larger values of the variable, skewness is said to be positive. Conversely, with the longer tail to the left (the smaller values of the variable) skewness is negative. Methods of measuring skewness will be discussed later in this Chapter.

(ii) *Kurtosis*: the balance between the peaks and the tails of the distribution. The Normal curve is said to be mesokurtic (of moderate kurtosis), whilst distributions with narrower peaks and thicker tails are leptokurtic. A broad peak with short tails gives platykurtosis. Measures of kurtosis will also be given later in the Chapter.

14.1.3 Other forms of non-Normality include bi- or multi-modality. A possible cause of extreme biomodality (the arc sine distribution) is a phenomenon like simple harmonic motion from a control device, traverse mechanism etc. This is illustrated in Fig. 14.1 in both its pure form, and when partially masked by other sources of variation. Similarly a uniform distribution (all values over a particular range being equally likely) is often masked by other variation to give a unimodal platykurtic curve.

Because the proportions of values beyond three standard deviations from the mean are markedly different for these non-Normal curves, the conventional C-index or z-value methods of stating capability are no longer valid, and no universally accepted conventions exist. We have therefore to consider several possibilities.

Among the simpler forms of non-Normality are those concerned with attri-

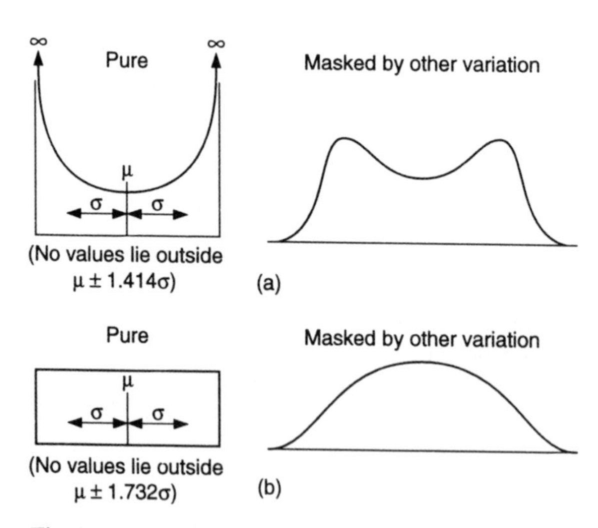

Fig. 14.1 Distribution curves: (a) arc sine; (b) uniform.

butes and events. These have already been covered in sections 13.2 and 13.3. In this Chapter, we are especially concerned with non-Normality of measurable characteristics.

14.2 Capability statements – non-Normal measured data

14.2.1 When dealing with measured data having a non-Normal distribution, there are seveal methods of stating the capability of the system to produce items within a specified range. Whilst bilateral tolerances often exist in such cases, we must also recognize that for many inherently non-Normal situations, unilateral tolerances (usually an upper limit) may apply. This occurs especially when zero represents an ideal, when negative values are impossible, and thus any non-zero measurement indicates departure from the ideal. Such cases include eccentricity, out of square, some surface roughness measurements (although not all: sometimes a degree of 'grain' is desirable or necessary, for example to provide lubricant paths), shrinkage, distortion, impurity levels, etc.

14.2.2 Among the commoner methods for stating capability, those following are the most widely used.

(i) Simply estimating the proportion of values likely to lie within or outside the tolerances. By analogy with the 3σ implied for C_p, C_{pk} in the case of Normal distributions, criteria such as 99.73% within specification (or 0.135% outside limits in either tail) or 99.994% (30 p.p.m. outside for either tail) are often used, though 99.9% (0.1% outside tolerances) is also extant.

(ii) Alternatively, one may state the percentiles of the non-Normal distribution estimated to correspond to the appropriate criterion (e.g. 99.73%, 99.9 or 99.94%), or for one-tail cases, 0.135%, 0.1% or 0.003% of violations of the limit.

(iii) A quasi-C value may be obtained. By analogy with the Normal distribution, this is defined as

$$C'_p = \frac{\text{Upper specification limit} - \text{lower specification limit}}{\text{Upper } 0.135\% \text{ point} - \text{lower } 0.135 \text{ point}}$$

The 0.135% (or occasionally the 0.1%) points need to be estimated via an appropriate (non-Normal) distribution model, not as $\pm 3s$ on either side of $\bar{x}$.

Some problems arise for C'_U and C'_L equivalents. It would at first sight appear that

$$C'_U = \frac{\text{USL} - \text{Mean}}{0.135\% \text{ point} - \text{Mean}}$$

$$C'_L = \frac{\text{Mean} - \text{LSL}}{\text{Mean} - \text{lower } 0.135\% \text{ point}}$$

with $C'_k = \text{Minimum}\,(C'_U, C'_L)$ could be used, but often for skewed distributions the median (or even the mode) may be a preferable central value; the median is in quite common use, giving

$$C'_U = \frac{\text{USL} - M}{x_{0.00135} - M}, \quad C'_L = \frac{M - \text{LSL}}{M - x_{0.99865}}$$

in widely adopted notation, where M indicates the median, $x_{0.99865}$ the value of x-exceeded by 99.865% of the distribution and $x_{0.00135}$ the value exceeded by 0.135%. Where only an upper limit is specified, only C'_U needs to be derived otherwise C'_k is the smaller of C'_U, C'_L.

(iv) The reciprocal of a C'-index is sometimes used, often with the phraseology that the process 'uses', or is capable to, $P(\%)$ of the specification. The percentage is thus

$$100 \times \frac{x_{0.00135} - x_{0.99865}}{\text{USL} - \text{LSL}}$$

for a bilateral specification. The reciprocals of C'_k, C'_U or C'_L are unlikely to convey a full summary of capability in the presence of non-Normality.

Evidently, before any of the measures (i)–(iv) can be calculated, it is necessary to identify the non-Normality, choose a suitable alternative model to the Normal distribution, and estimate the appropriate percentage points (and the median, in many cases).

14.3 Measuring non-Normality: skewness and kurtosis

4.3.1 Although there are many forms of non-Normality, for present purposes we consider only the case of distributions with a single mode, tapering off either symmetrically or asymmetrically to tails which become progressively thinner the further values are from the mode. The major features of interest are then:

(a) Is the distribution shape symmetrical or skewed. If skewed, how does one measure the skewness, and when does this asymmetry indicate significant non-Normality?

(b) Is the balance between the hump and tails of the distribution similar to that of the Normal distribution? This shape aspect is known as kurtosis; patterns with short tails and broad humps are known as platykurtic, those with long tails and narrow peaks are leptokurtic (the Normal distribution is mesokurtic).

For each of these two features, there are two measures of the shape feature concerned. In the case of skewness, one of the measures is more suited to manual analysis than to computerized methods. In the case of kurtosis, one method can be fairly easily calculated manually, the other requires greater calculation effort, preferably via a computer. We consider the manually-feasible methods first.

14.4 Skewness and kurtosis: simple measures

14.4.1 The earliest measure of skewness, devised by the eminent statistician Karl Pearson, used the fact that for smooth unimodal skewed distributions the mode, median and mean are all different. In fact they are in alphabetical order (mean, median, mode) starting from the longer tail, and the distance between mode and median is often observed to be about twice that between median and mean.

Pearson's first measure used

$$Sk = \frac{\text{Mean} - \text{Mode}}{\text{Standard deviation}}$$

taking positive values when the longer tail is to the right, negative when the left tail (i.e. towards the smaller values of the variable) is longer. However, unless very large data sets are available, location of the mode is difficult, and the relationship between the spacings of the three characteristics leads to the version now current, which is easier to determine, viz.

$$Sk = \frac{3\,(\text{Mean} - \text{Median})}{\text{Standard deviation}} = \frac{3\,(\bar{x} - M)}{\sigma_n} \tag{14.1}$$

(The measures of skewness and kurtosis in this chapter provide some rare instances of the use of σ_n, rather than s or σ_{n-1}, for sample data).

For the Normal and other distributions where the mean and median coincide, $\bar{x} - M = 0$ so the skewness coefficient is zero. Other measures of skewness where tail values play a more important role may give non-zero skewness for some cases where $\bar{x} = M$; an example follows in section 14.5.7. A further point to note when using small computers rather than manual methods is that the data require ranking to locate the median. This, and other aspects of Pearson's and Geary's statistics noted in section 14.4.3, lead to a general preference for methods based on moments, described in section 14.5.

14.4.2 A simple measure of kurtosis, or peak-to-tail balance in the distribution, is afforded by **Geary's ratio**. Often denoted by the symbol a, it uses two measures of dispersion—the mean absolute deviation and the standard deviation. The MAD is calculated as

$$MAD = \frac{\sum|x - \bar{x}|}{n}$$

where $||$ denotes absolute value (without sign). Unfortunately this cannot be calculated in one pass of the data. It is necessary first to calculate $\bar{x}$, and then to derive and add the absolute deviations. Geary's ratio is then

$$a = \frac{MAD}{\sigma_n} = \frac{\Sigma|x - \bar{x}|}{n} \Big/ \sqrt{\left(\frac{\Sigma(x - \bar{x})^2}{n}\right)} \tag{14.2}$$

For large samples from a Normal distribution the average value of a is $\sqrt{(2/\pi)}$, or 0.7979. For smaller samples, it is close to $0.7979 + 0.2/n$; thus for $n = 50$, typically a would be around 0.802. Platykurtic (short-tailed) distributions give larger values of a, leptokurtic (long-tailed) distributions yield lower values of a.

14.4.3 Both Pearson's skewness coefficient and Geary's a are influenced most by values near the centre of the distribution. This is one of their disadvantages, making them insensitive to some forms of tail-heavy skewness and kurtosis. For this reason, and for facility of one-pass calculation, measures based on moments are often preferred. We now consider the general topic of moments before returning to skewness and kurtosis.

14.5 Moments of statistical distributions

14.5.1 The general principles of moments in statistics are identical to those in mechanics or physics. Consider the system of weights illustrated in Fig. 14.2.

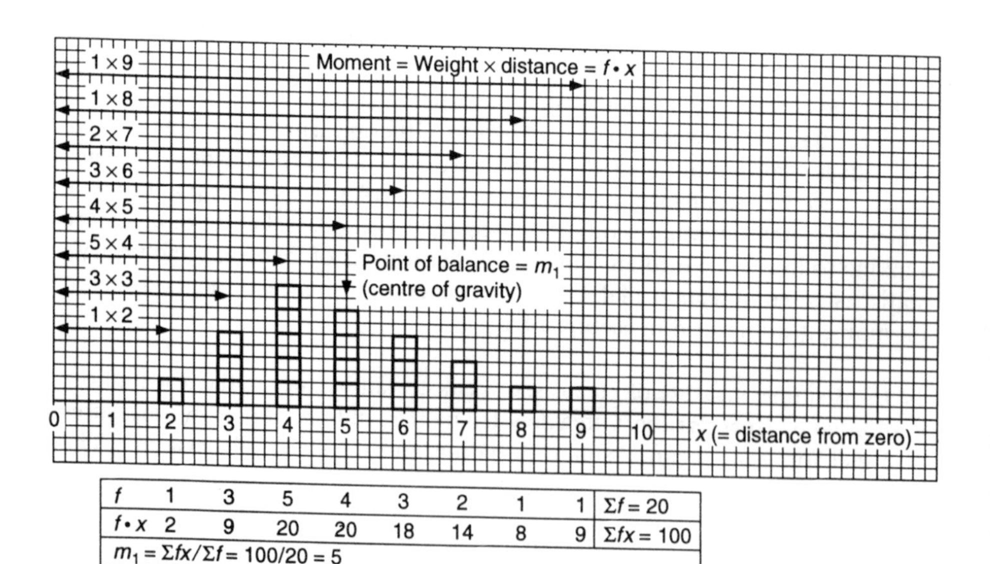

f	1	3	5	4	3	2	1	1	$\Sigma f = 20$
$f \cdot x$	2	9	20	20	18	14	8	9	$\Sigma fx = 100$
$m_1 = \Sigma fx / \Sigma f = 100/20 = 5$									

Fig. 14.2 Principle of moments.

Each column of weights acts through a point on the horizontal scale at a distance x units from an origin. The moment about zero for each column is the total weight multiplied by the distance from zero, i.e. $f \cdot x$. Thus the total moment is obtained as $\sum f \cdot x$.

If the total moment is divided by the total weight, $\sum f$, we find the point on the x-axis through which the whole system can be regarded as acting. It is, in fact, the projection of the centre of gravity of all the weights on the x-axis. It also obviously corresponds to the arithmetic mean of the system if we regard it as a histogram or frequency diagram, since $\bar{x} = \sum f \cdot x / \sum f$. Thus the mean is known as the first zero-moment of the system, and is often denoted as m_1.

14.5.2 So-called 'higher moments' are usually referred to this centre of gravity, and the second moment in statistics involves the squares of the distances of the data points from the centre, i.e. the values of $(x - \bar{x})^2$. When these values are added and averaged, we have

$$m_2 = \frac{1}{n} \sum f (x - \bar{x})^2$$

in the frequency table format of Fig. 14.2. For a set of individual values not necessarily in the form of a frequency table this becomes

$$m_2 = \frac{1}{n} \sum (x - \bar{x})^2 \tag{14.3}$$

This measure is identical to σ_n^2, where σ_n is defined as in (3.18). There is a physical/mechanical analogy again, in that m_2 is related to the moment of inertia of a revolving system and σ_n to the radius of gyration. For present purposes in measuring skewness and kurtosis, m_2 provides a scaling factor permitting these measures to be expressed as pure numbers independent of the dimensions of measurement.

14.5.3 We are now ready to define the third and fourth moments which will provide the basis of skewness and kurtosis measures. Extending the preceding concepts to third and fourth powers of deviations from the centre,

$$m_3 = \frac{1}{n} \sum (x - \bar{x})^3 \tag{14.4}$$

$$m_4 = \frac{1}{n} (\sum (x - \bar{x})^4 \tag{14.5}$$

Evidently in this form, two passes of the data are required, the first to find $\bar{x} (= m_1)$, the second to form the deviations and sum their various powers. This is the most reliable method of calculation, though equivalent one-pass methods are based on combining the 'zero-moments',

$$\left.\begin{aligned} m'_2 &= \frac{1}{n}\sum x^2, \quad m_2 = m'_2 - m_1^2 \\ m'_3 &= \frac{1}{n}\sum x^3, \quad m_3 = m'_3 - 3m'_2 m_1 + 2m_1^3 \\ m'_4 &= \frac{1}{n}\sum x^4, \quad m_4 = m'_4 - 4m'_3 m_1 + 6m'_2 m_1^2 - 3m_1^4 \end{aligned}\right\} \quad (14.6)$$

However, if the x-values contain several significant figures, their third and fourth powers may suffer rounding and yield inaccurate end results. It is prudent to check the output of any calculator or computer against the exact values (via $x - \bar{x}$ deviations) for any untried system.

14.5.4 For the data in Fig. 14.2, the moments are developed in Table 14.1. For this data, we also have:

$$\sum x = 100, \quad \sum x^2 = 562, \quad \sum x^3 = 3484, \quad \sum x^4 = 23386;$$
$$m_1 = 5.0, \quad m'_2 = 28.1, \quad m'_3 = 174.2, \quad m'_4 = 1169.3.$$

Then using (14.6),

$$m_2 = 28.1 - 5.0^2 = 3.1;$$
$$m_3 = 174.2 - 3 \times 28.1 \times 5.0 + 2 \times 5.0^3 = 2.7;$$
$$m_4 = 1169.3 - 4 \times 174.2 \times 5.0 + 6 \times 28.1 \times 5.0^2 - 3 \times 5.0^4 = 25.3.$$

(Of course, with data values having only one significant figure, no rounding problems occur here).

Table 14.1 Moments of data in Fig. 14.2

$m_1 = \bar{x} = 5.0$

x	f	$x-\bar{x}$	$f(x-\bar{x})$	$f(x-\bar{x})^2$	$f(x-\bar{x})^3$	$f(x-\bar{x})^4$
2	1	−3	−3	9	−27	81
3	3	−2	−6	12	−24	48
4	5	−1	−5	5	−5	5
5	4	0	0	0	0	0
6	3	1	3	3	3	3
7	2	2	4	8	16	32
8	1	3	3	9	27	81
9	1	4	4	16	64	256
Totals	20 $\sum f$	–	0 $\sum f(x-\bar{x})$	62 $\sum(x-\bar{x})^2$	54 $\sum(x-\bar{x})^3$	506 $\sum f(x-\bar{x})^4$

$m_2 = 62/20 = 3.1$; $m_3 = 54/20 = 2.7$; $m_4 = 506/20 = 25.3$.

14.5.5 Having described the system of moments, we are now able to define two moment ratios which are useful descriptions of skewness and kurtosis.

The third-moment skewness coefficient, usually denoted by $\sqrt{\beta_1}$ or $\sqrt{b_1}$ (or 'root-b_1') is:

$$\sqrt{b_1} = \frac{m_3}{m_2^{3/2}} = \frac{(1/n)\sum(x-x)^3}{\sigma_n^3} \tag{14.7}$$

A positive value (the positive cubed deviations outweigh the negatives) indicates longer tail towards the higher-valued observations. A negative $\sqrt{b_1}$ indicates the longer tail is towards the small-valued observations, and a coefficient near zero indicates approximate symmetry of the distribution.

The fourth-moment kurtosis coefficient, β_2 or b_2, is given by

$$b_2 = \frac{m_4}{m_2^2} = \frac{(1/n)\sum(x-x)^3}{\sigma_n^4} \tag{14.8}$$

For an exact Normal distribution, b_2 takes the value 3.0. For broad-humped, short tailed distributions b_2 is less than 3, and for $b_2 < 2$ the implication may even be of no hump at all (uniform or rectangular shape) or of modes at the extremes rather than in the centre. Conversely $b_2 > 3$ indicates a narrower peak than the Normal curve, with longer and heavier tails. Combinations of skewness and kurtosis complicate this simple classification. As indicated in Fig. 14.3, broad-humped curves are known as platykurtic, narrow peaked curves with

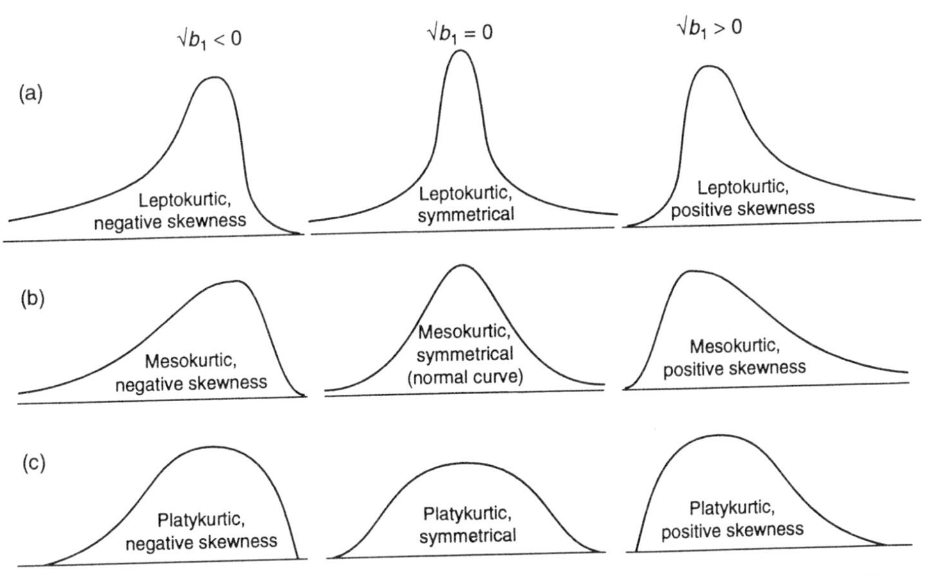

Fig. 14.3 Skewness and kurtosis. (a) $b_2 > 3, a < 0.8$; (b) $b_2 \doteqdot 3, a \doteqdot 0.8$; (c) $b_2 < 3, a > 0.8$.

long tails as leptokurtic. For $b_2 \approx 3$, the shape is mesokurtic, with the Normal distribution as the prime example.

14.5.6 The coefficients for the data of Fig. 14.2 can now be calculated. For $\sqrt{b_1}$ we have

$$\sqrt{b_1} = \frac{m_3}{m_2^{3/2}} = \frac{2.7}{3.1^{1.5}} = 0.4947$$

indicating moderate positive skewness.

For kurtosis,

$$b_2 = \frac{m_4}{m_2^2} = \frac{25.3}{3.1^2} = 2.6327$$

This would be considered mesokurtic, but the hump is slightly more dominant (especially on the lower part of the distribution) than for a Normal curve.

14.5.7 Moment coefficients do not always give the same indications as Pearson's Sk and Geary's a measures. For continuous curves, Pearson's coefficient is smaller numerically than $\sqrt{\beta_1}$, but with the same sign. For discrete data, like that of Fig. 14.2 where only integer values of x occur, this rule may fail. Thus with the median value among the four grouped at $x = 5$, we have

$$Sk = \frac{3(5-5)}{\sqrt{3.1}} = 0$$

the tail values contributing most to the $\sqrt{\beta_1}$ measure. It is interesting to note that Pearson's original measure involving the mode would give

$$Sk = \frac{5-4}{\sqrt{3.1}} = 0.568$$

Finally, Geary's measure of kurtosis for Fig. 14.1 is

$$\frac{MAD}{\sigma_n} = \frac{1.4}{1.7607} = 0.7951$$

very slightly less than the $0.7979 + (0.2/20) = 0.8079$ expected of the Normal curve, indicating slight platykurtosis (as does b_2).

14.6 Significance tests for non-Normality

14.6.1 Because of the analytical convenience of dealing with a Normal distribution of individual values, calculation of capability indices (as well as other statistical procedures) is often based on assuming Normality unless there is strong evidence to the contrary. It is therefore useful to describe some tests of the Normal hypothesis. Some are based on the skewness or kurtosis coefficients, but a simple test for more general cases is also described. There are more efficient test, but they often require special tables or computing facilities (e.g. Kolmogorov–Smirnov, Shapiro and Wilks, etc.)

14.6.2 A simple general test is based on the χ^2 'goodness-of-fit' (or preferably 'lack of fit') criterion. A particularly manageable form of this test is as follows.

(i) Calculate $\bar{x}$ and s for the n data values;
(ii) Divide the x-scale into ten zones which would have equal frequencies (each containing 10% of the x-values) for a perfect Normal curve. These zones are:
Values less than $\bar{x} - 1.282s$
$\bar{x} - 1.282s$ to $\bar{x} - 0.842s$
$\bar{x} - 0.842s$ to $\bar{x} - 0.524s$
$\bar{x} - 0.524s$ to $\bar{x} - 0.253s$
$\bar{x} - 0.253$ to $\bar{x}$
$\bar{x}$ to $\bar{x} + 0.253s$ etc.
out to values greater than $\bar{x} + 1.282s$
(iii) Count the number of values occurring in each of these ten zones. Calculate the standard deviation ($s_f = \sigma_{n-1}$) of these frequencies, and divide by $\sqrt{n}$.
(iv) Some useful percentage points of $s_f/\sqrt{n}$ are:

10%	0.365;	5%	0.395;	2.5%	0.422;
1%	0.453;	0.5%	0.475;	0.1%	0.520.

Values exceeding these criteria may indicate significant discrepancies from Normality, and the distribution of the data values should be carefully examined to diagnose the **nature** of the non-Normality.

Values of $s_f/\sqrt{n}$ of 0.266 or less indicate a good fit, and of 0.137 or less perhaps even suspiciously good! Data editing to remove outliers or other anomalous values is a possibility.

14.6.3 Significance tests can also be applied to a, $\sqrt{b_1}$ and b_2. Especially for the latter two, one or two extreme values can grossly distort these moment ratios, and with sample sizes less than about 50 they can give misleading indications. These can be particularly serious when using $\sqrt{b_1}$ and b_2 to fit other distribution forms, as mentioned in section 14.7.1.

Many SPC computer packages have facilities for testing the significance of $\sqrt{b_1}$ and either b_2 or a, and the more extensive statistical tables and some text books contain tables of percentage points. The following approximate relationships give good results for the range of sample sizes indicated and for significance levels from 5% to 1%. Thus $z \geqslant 1.64$ may be regarded as indicative, $\geqslant 1.96$ as significant and $\geqslant 2.33$ as highly significant.

(i) For $\sqrt{b_1}$, use

$$z \doteq \sqrt{b_1} \times \sqrt{\left(\frac{n+6}{6}\right)}, \quad n \geqslant 25;$$

(ii) For b_2, use

$$z \doteq (\sqrt{b_2} - \sqrt{3}) \sqrt{\left(\frac{n+25}{2}\right)}, \quad n \geqslant 50;$$

(iii) For a, use

$$z \doteq \left(a - 0.798 - \frac{0.2}{n}\right) \sqrt{[23(n+2)]}, \quad n \geqslant 25.$$

14.6.4 The example given in Fig. 14.2 is artificially constructed, and the sample size is too small to carry out a proper test of Normality. We therefore return to the valve liner data of section 2.2.1 and evaluate the fit of a Normal distribution. It is always preferable to use original data rather than a grouped frequency table for the moment calculations, and the following are based on the 52 individual observations.

$m_1(=\bar{x}) = 239.509\,615\,4$
$m_2(=\sigma_n^2) = 0.854\,715\,236$ $(\sigma_n = 0.924\,508\,1)$
$m_3 = -0.317\,059\,97$
$m_4 = 2.479\,276\,23$
$\text{MAD} = 0.717\,677\,514$
$n = 52.$

Then

$\sqrt{b_1} = -0.4012$
$b_2 = 3.3938$
$a = 0.7763$

Also with median $(\tilde{x}) = 239.5$,

Pearson's $Sk = +0.031$.

The distribution is thus slightly skewed (negatively), though Pearson's coefficient does not show this as it is the three smallest values that contribute most to m_3. It is also very slightly leptokurtic, both b_2 and a concurring on this.

However, these coefficients do not show statistically significant non-Normality. Using the z-criteria,

for $\sqrt{b_1}$, $z = -1.25$ $(p \doteq 0.1)$
for b_2, $z = 0.68$ $(p \doteq 0.25)$
and for a, $z = -0.90$ $(p \doteq 0.68)$

We may also try the general lack-of-fit test of section 14.5.2. The required zones, with the frequencies of occurrence of values in them are:

Class	Frequency
<238.31	5
238.31 – 238.72	4
238.72 – 239.02	6
239.02 – 239.27	4
239.27 – 239.51	8
239.51 – 239.75	5
239.75 – 240.00⁻	3
240.00⁻ – 240.29	5
240.29 – 240.71	8
>240.71	4
Total (n)	52

The standard deviation of these frequencies is $s_f = 1.68655$

Then $s_f/\sqrt{52} = 0.2339$

The lack-of-fit statistic does not reach any of the percentage points indicated in section 14.6.2, and in fact is slightly smaller than the 50% point of 0.2655, indicating that the Normal distribution is a satisfactory fit.

Use of the frequency table (Table 2.3), along with the class midvalues 237.0, 237.5, ..., 241.5, would yield the following moments and ratios

$$m_2 = 0.860485, \quad m_3 = -0.29646,$$
$$m_4 = 2.34626, \quad MAD = 0.70414,$$

whence $\sqrt{b_1} = -0.371$, $b_2 = 3.169$ and $a = 0.759$. The differences from the 'exact' values using individual observations, though noticeable in the second significant digit, would not in this case lead to any different conclusions.

14.7 Alternatives to the Normal distribution

14.7.1 Where significant non-Normality is indicated, it is essential to inspect the data carefully, using tallies, histograms and/or probability plots. Where irregularities are revealed ('wild' values, multiple modes, mixed distributions) further analysis is pointless, and the causes of the anomalies must be sought.

For regular curves exhibiting skewness or kurtosis, several approaches are possible, and some broad guidelines and examples can be given.

(i) It may be possible to find a data transformation which yields an approximately Normal distribution. The analysis is then carried out in terms of the transformed values, and the reverse transformation applied to interpret the final results where necessary. The example in section 14.8.1 illustrates a logarithmic transformation, via both an analysis of $\ln(x)$ and a probability plot. Other useful transformations include $\sqrt{x}$, $\sqrt[3]{x}$ (both of which, like $\ln(x)$, are useful to reduce or eliminate positive skewness), $1/x$, $1/x^2$, etc. Sometimes a physical model may suggest a possible transformation: for example a cube root transformation may be useful on weights of small beads, since weight is proportional to the cube of diameter, and the latter may well be Normally distributed. Logarithmic transformations are particularly useful for some electrical characteristics (insulation resistance, capacitance, static propensity) and for small levels of impurities or trace elements. It is interesting to note that sound levels in decibels and acidity/alkalinity measured in pH are already on logarithmic scales.

(ii) Without actually estimating parameters of other distributions, alternative probability plots may simplify analysis. Graph grids are widely available for log Normal, extreme value and Weibull distributions. It may sometimes be necessary to subtract (or less commonly, add) a constant to the data values, thus plotting $(x - c)$ where c is found by trial. The required percentage points for capability statements can be read from the grid after fitting a reasonable line to the data. The general remarks on probability plots in section 13.7 apply also to non-Normal plotting. For guidance, some platykurtic and leptokurtic distributions on a Normal grid are illustrated in Fig. 14.4. The example in section 14.8.1 illustrates the use of extreme value paper for capability analysis.

(iii) Finally, it may be possible to choose an alternative mathematical model as a suitable non-Normal distribution, and to estimate its parameters from the data. Some widely used models, in addition to the log Normal already suggested in (i) and (ii), include the extreme value distribution (extreme maximum for positive skewness, extreme minimum for negative skewness) and some families of distributions such as the gamma, Weibull, Pearson and Johnson systems. Figure 14.5 illustrates some of these forms, and Table 14.2 lists their shape properties. Selection, fitting and estimation lie beyond the scope of this book, but Hahn and Shapiro (1967) provide extensive coverage, and a useful paper by Clements (1989) gives a good account of the use of the versatile Pearson family of distribution curves.

14.7.2 Some further comment on the extreme value distribution is justified by its widespread use, especially in the automotive industry, in the form of a probability (or capability) plotting grid. The theory underlying this model is concerned with the occurrence and recording of extreme conditions such as maximum daily wave height, maximum annual wind gusting, minimum failure load of composite assemblies (e.g. chains failing at the weakest point). The model is also found to fit other situations quite well, such as eccentricity, surface roughness, unilateral distortion. Probability plotting is carried out in the usual way, and graphical extrapolation permits estimation of (for example) 0.135% points in the tails (often in only one tail for unilateral specifications). Methods based on calculation are complicated by the fact that the theoretical parameters of the extreme value distribution are its mode and a scale parameter. Tables of percentage points are available (US National Bureau of Standards 1953, 1954) but they are not easy to use. A short list of these percentage points in terms of the mean and standard deviation is given in Table 14.3.

The extreme value model cannot be regarded as an alternative for all cases of non-Normality, nor can the other families of distributions. Figure 14.6 shows

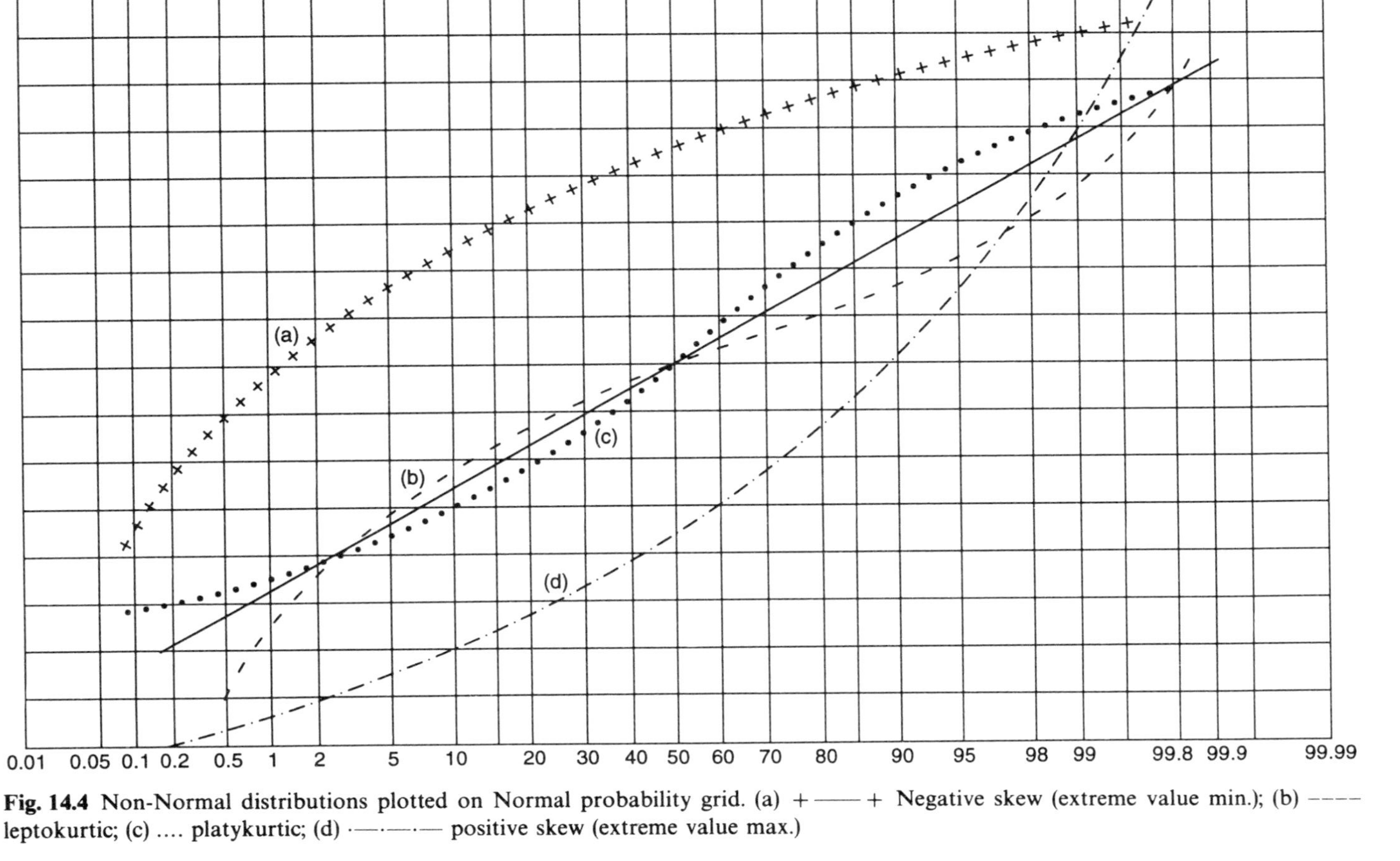

Fig. 14.4 Non-Normal distributions plotted on Normal probability grid. (a) + —— + Negative skew (extreme value min.); (b) - - - - leptokurtic; (c) .... platykurtic; (d) ·—·—·— positive skew (extreme value max.)

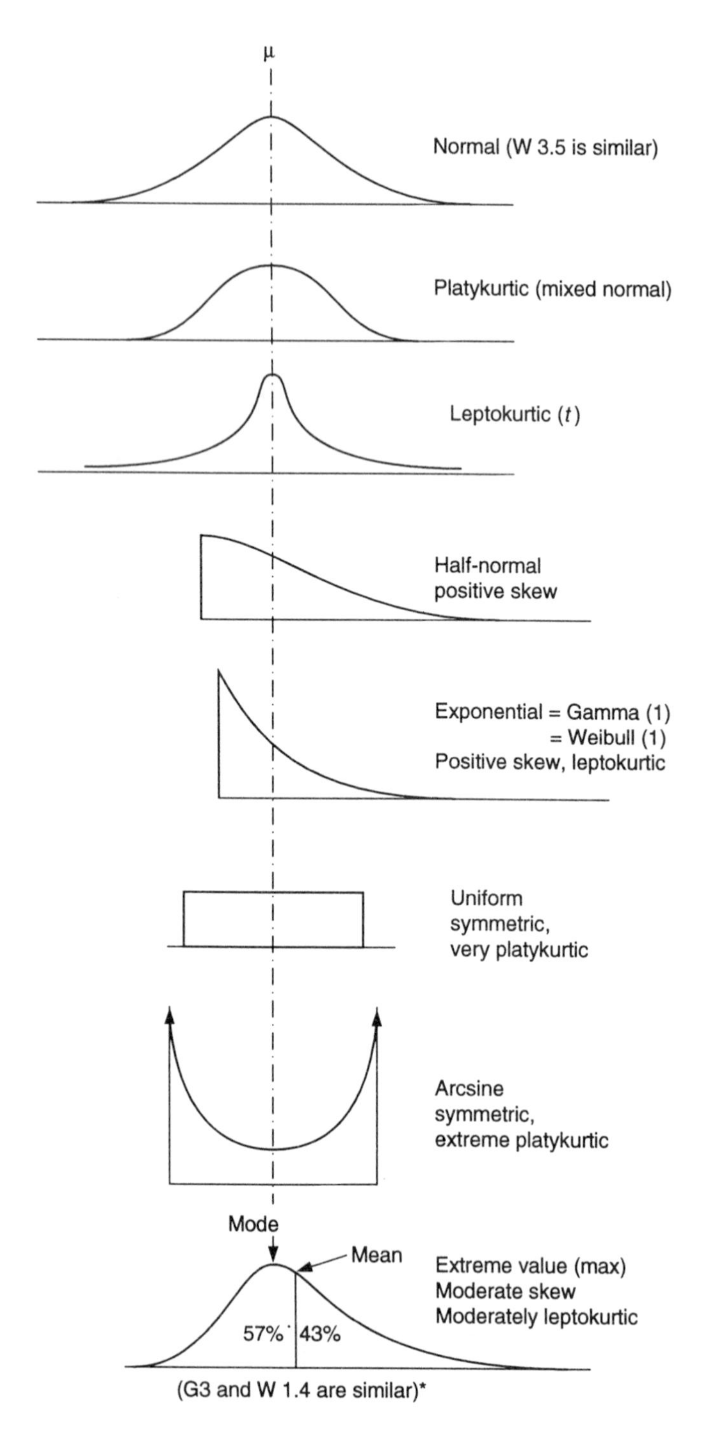

Fig. 14.5 Some non-Normal distribution shapes. Curves are scaled with (approximately) the same mean and standard deviation. (*) i.e. gamma distribution with shape parameter 3, and Weibull with shape parameter 1.4, are similar.

Table 14.2 Shape coefficients of some non-Normal distributions

μ = population mean, σ = population standard deviation, M = median

Name (and parameters)	M	Mode	$\sqrt{\beta_1}$	β_2	$\frac{MAD}{\sigma}$	$3(\mu - M)/\sigma$ (Pearson)
Normal (μ, σ)	μ	μ	0	3	0.7979	0
Uniform (0, 1)	0.5	None	0	1.8	0.866	0
Arcsine $(-1, +1)$	0	± 1	0	1.5	0.9	0
Half normal (μ, σ) (Mean $= 0.7979\sigma$, s.d. $= 0.6025\sigma$)	0.675σ	0	0.995	3.87	0.80	0.61
Extreme value (max or min)	$\mu \pm 0.24\sigma$	$\mu \pm 0.45\sigma$	± 1.14	5.4	0.77	± 0.72
Weibull, $\eta = 1$ (= Exponential, = Gamma, $k = 1$)	0.693μ	0	2.0	9.0	0.75	0.92
Weibull, $\eta = 2$	0.94μ	0.8μ	0.63	3.25	0.80	0.12
Weibull, $\eta = 3$	0.99μ	0.98μ	0.17	2.73	0.80	0.025
Weibull, $\eta = 3.5$ (close to normal)	1.001μ	1.001μ	0.025	2.71	0.80	-0.003
Weibull, $\eta = 4$	1.0066μ	1.026μ	-0.087	2.75	0.80	-0.024
Weibull, $\eta = 5$	1.012μ	1.046μ	-0.25	2.88	0.80	-0.053
Gamma, $k = 2$	0.84μ	0.5μ	1.41	6.0	0.80	0.68
Gamma, $k = 5$	0.93μ	0.8μ	0.89	4.2	0.80	0.44
Gamma, $k = 10$	0.97μ	0.9μ	0.63	3.6	0.80	0.31
Gamma, $k = 20$*	0.98μ	0.95μ	0.45	3.3	0.80	0.22

*For $k > 10$, $\sqrt{x} \sim$ Normal

Table 14.3 Percentage points of extreme value distribution in terms of mean and standard deviation (μ, σ)

%	Lower % points	Upper % points
0.1	$\mu-2.00\sigma$	$\mu+4.85\sigma$
0.135	-1.96σ	$+4.56\sigma$
0.5	-1.81	$+3.60$
1.0	-1.72	$+3.06$
2.5	-1.57	$+2.35$
5	-1.37	$+1.79$
10	-1.18	$+1.23$
25	-0.78	$+0.45$

50 (median); $\mu-0.24\sigma$; (Mode is at $\mu-0.45\sigma$).
Above data apply to *extreme* maximum. For extreme *minimum*, reverse all $\pm$ signs.
$\sqrt{\beta_1} = \pm 1.14$; $\beta_2 = 5.4$; $MAD/\sigma = 0.77$; $Sk = \pm 0.72$ (or 0.45 by Pearson's original Mean – Mode definition).
(Moderate skew, moderate lepto-kurtosis).

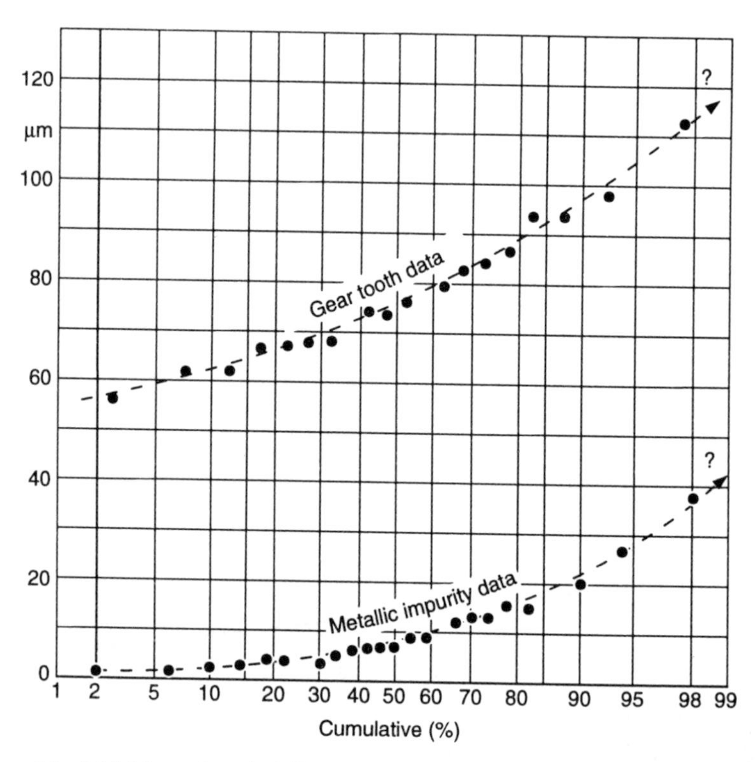

Fig. 14.6 Normal probability plots of gear tooth and impurity data.

the data of the examples plotted on an unsuitable (Normal) grid, and evidently extrapolation to the 0.135% points would be rather arbitrary.

14.8 Two examples of non-Normal situations

14.8.1 Gears are tested electronically, and a microprocessor prints out the greatest deviation of the profile from a standard pattern. The following set of results was obtained from a 20-item study:

57, 62, 62, 67, 67, 68, 68, 68, 74, 74, 76, 76, 79, 83, 84, 87, 94, 94, 98, 113

$$\bar{x} = 77.55, \quad s = 14.1625$$

Although there was no formal specification, it was required to estimate the capability (in the usual 99.73% sense) of the machine producing the gears. Treating the data as Normal would yield

$$77.55 \pm 3 \times 14.1625 = 35.06,\ 120.04,$$

i.e. 35 to 120 in actual units of measurement.

The lower limit is well below the smallest data value, but the upper limit only just exceeds the largest value. A tally, histogram or normal probability plot indicates systematic positive skewness (Fig. 14.6). For these data we have:

$\sigma_n = 13.8039, \quad m_2 = \sigma_n^2 = 190.5475$
$m_3 = 2101.28025, \quad m_4 = 112828.6892$
$MAD = 11.16$

$$a = 0.808467; \quad Sk = 3\left(\frac{\text{Mean} - \text{Median}}{\sigma_n}\right) = 0.5433$$

$\sqrt{b_1} = 0.798875; \quad \sqrt{b_2} = 3.1075$
$z(a) = 0.01; \quad z(\sqrt{b_1}) = 1.66; \quad z(b_2) = 0.146.$

For this type of data concerned with extremes, the extreme value distribution may be appropriate. The sample size is rather small to test for significance (though z for $\sqrt{b_1}$ has $p < 5\%$), but a plot on 'extreme value' paper is virtually linear (Fig. 14.7). Graphical estimation of the 0.135% points yields a 99.73% capability of 50–140 approximately. Using $\bar{x}$, s and Table 14.3 gives:

Lower limit $77.55 - 1.96s = 49.8,$
Upper limit $77.55 + 4.56s = 142.1$

To complete the example, we note that the method of Clements (1989) using Pearson curves would have yielded (taking $\sqrt{b_1}$ and b_2 into account)

Lower 0.135% point $= \bar{x} - 2.85s = 37.2,$
Upper 0.135% point $= \bar{x} + 4.68s = 143.8.$

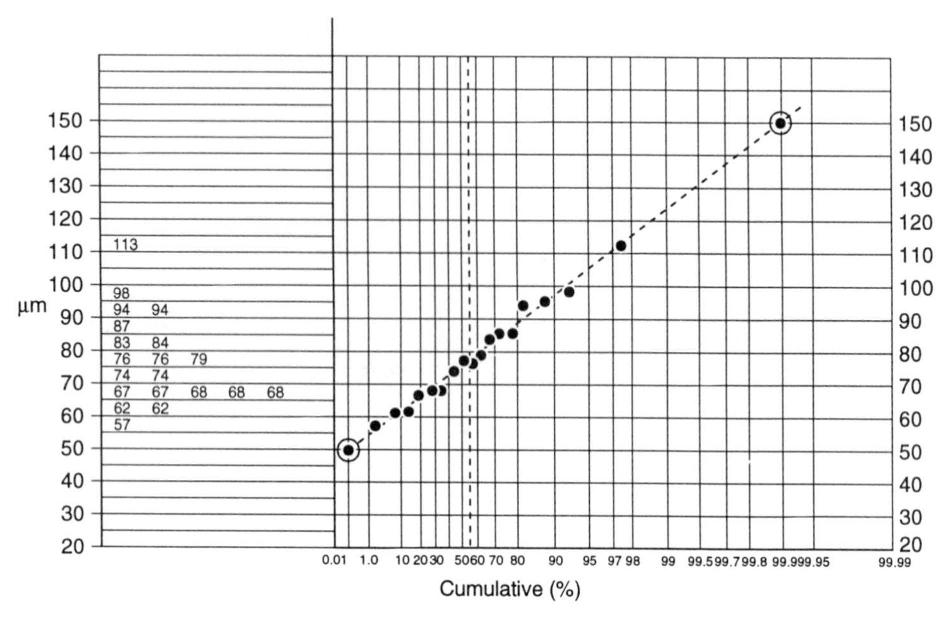

Fig. 14.7 Extreme value plot of gear tooth data.

There is quite close agreement with the extreme value distribution on the upper limit, but a moderate discrepancy on the lower limit, illustrating (as in the next example) the problems of handling non-Normality by rule-of-thumb methods.

14.8.2 A material for use in surgical appliances may not contain more than 40 p.p.m. of metallic impurities. A sample of 25 items gave the following impurity levels in parts per million (p.p.m.):

2, 2, 3, 3, 4, 4, 4, 4, 5, 6, 7, 7, 7, 9, 9, 10, 12, 13, 13, 15, 15, 19, 20, 27, 38

$$\bar{x} = 10.32, \quad s = 8.57.$$

The increasing gaps between the upper values indicate strong positive skewness, as does the probability plot (Fig. 14.6). Limits of $\bar{x} \pm 3s$, giving -15 to $+36$ are obviously absurd. The measures of skewness and kurtosis are:

$\sigma_n = 8.3987, \quad m_2 = 70.5376, \quad MAD = 6.3296$
$a = 0.75364$
$\sqrt{b_1} = 1.6452$
Also

$$3\left(\frac{\text{Mean} - \text{Median}}{\sigma_n}\right) = 3\left(\frac{10.32 - 7}{8.4}\right) = 1.19$$

$b_2 = 5.6532$

Here $z(a) = -1.305$; $z(\sqrt{b_1}) = 3.74$; $z(b_2) = 3.228$.

Although sample size is again too small for the z-approximations to be accurate, there is strong evidence of non-Normality in both skewness and kurtosis. Note that b_2 shows the leptokursis more strongly than a.

Using Pearson curves for these values of $\sqrt{b_1}$, b_2 gives a lower limit of $\bar{x} - 1.69s = -4.2$ and an upper limit of $\bar{x} + 5.447s = 57.0$.

While the upper limit looks more plausible, the lower is still unacceptably negative. However, a log Normal plot yields a satisfactory line, and provides a good basis for estimating capability (Fig. 14.8).

Any lower limit will now become a fraction (of a part per million) rather than negative, and an upper 0.135% point of about 82 p.p.m. is indicated by the plot. More precisely, taking natural logarithms of the data gives a mean (log) of 2.04 and a standard deviation of 0.79. An upper (and lower, if required) 0.135% point may now be obtained via

$$2.04 \pm 3 \times 0.79 = -0.33, 4.41$$

and transforming back via e^x gives 0.72, 82.3 as the 99.73% span. This upper limit considerably exceeds the specified maximum, and the graph suggests nearly

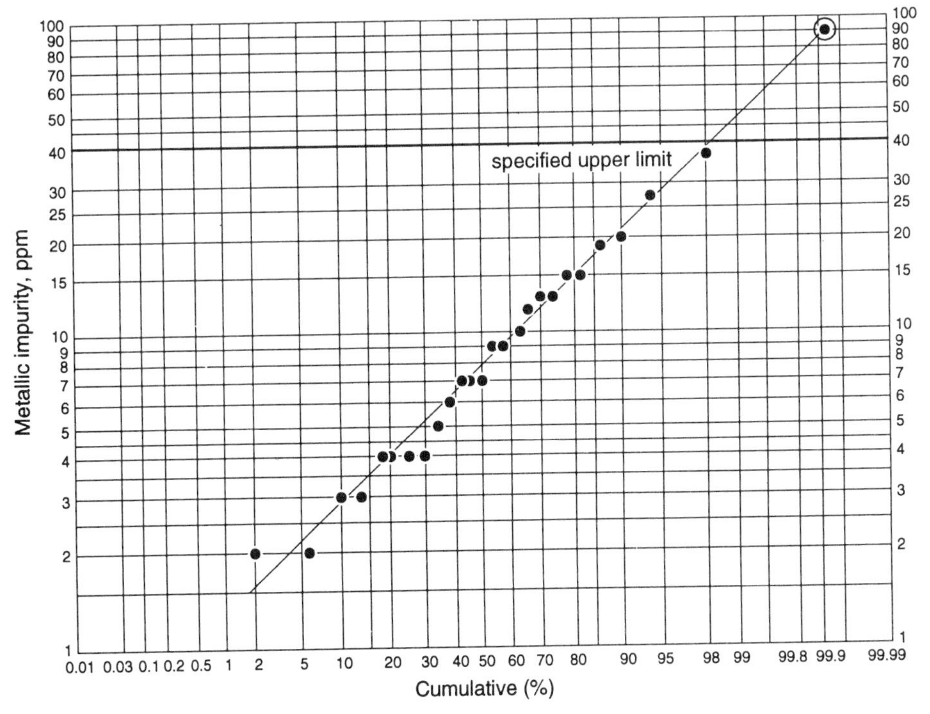

Fig. 14.8 Log Normal probability plot of impurity data.

2% out of specification. By calculation,

$$z = \frac{\ln(40) - 2.04}{0.79} = 2.09, \quad \text{with } P_{(z)} = 1.83\%$$

14.9 Confidence limits for non-Normal indices

14.9.1 A method for obtaining approximate confidence limits for C-indices was given in section 13.6. The question of corresponding limits for the C' indices described in section 14.2.2 (iii) leads to severe complications. In those cases where a Normalizing transformation can be found, one may calculate C-indices in terms of the transformed variable by similarly transforming the specification limit(s). The methods of section 13.6 are then valid.

Taking the example of 14.7.2, with USL = 40 p.p.m., for $y = \ln(x)$ we had $\bar{y} = 2.04$, $s_y = 0.79$. Then,

$$C_k = \frac{\ln(40) - 2.04}{3 \times 0.79} = 0.696$$

and (13.13) may be used to obtain confidence limits. We note that for one sample of 25 items, $k = 1, n = 25$ and $v = 24$; using $z = 2$, we have 95% limits of

$$0.696\left[1 \pm 2\sqrt{\left(\frac{1}{9 \times 25 \times 0.696^2} + \frac{1}{2 \times 24}\right)}\right]$$

$$= 0.455,\ 0.937 \quad (= C_k \pm 35\% \text{ approximately}).$$

14.9.2 Where transformation to Normality is impracticable and some non-Normal distribution is fitted using $\bar{x}$, s, $\sqrt{b_1}$ and b_2 (for example, using Pearson curves as mentioned in section 14.6.1(iii)), an approximate adjustment can be applied to (13.12) or (13.13). This involves substituting $B \cdot z$ for z in these formulae, where

$$B = \sqrt{[\tfrac{1}{2}(b_2 - 1)]} \tag{12.9}$$

(When the distribution is normal with $b_2 = 3$, $B = 1$ and no adjustment is required).

If the example of section 14.8.2 had been tackled via Pearson curve-fitting, the upper 0.135% point is 57.0 and the estimated median is 8.6. Then

$$C'_k = C'_U = \frac{40 - 8.6}{57 - 8.6} = 0.649$$

(slightly lower than the log Normal approach above).

Now with $b_2 = 5.6532$, we modify (13.14) using

$$B = \sqrt{\tfrac{1}{2}(5.6532 - 1)} = 1.5253 \text{ to give, using } z = 2,$$

$$0.649\left[1 \pm 1.5253 \times 2\sqrt{\left(\frac{1}{9 \times 25 \times 0.649^2} + \frac{1}{2 \times 24}\right)}\right]$$

$$= 0.298,\ 1.000 \quad (= C_k \pm 54\%)$$

As well as a lower value for the index than that for the log Normal approach, the confidence limits are wider because of the leptokurtic shape of the original distribution ($b_2 > 3$). For platykurtosis, the confidence limits would be narrower than for the Normal distribution.

14.9.3 No rigorous method or recommended approximation can be offered where other methods are used for non-Normal capability estimation (e.g. graphical extrapolation, curve fitting that does not involve $\bar{x}$ and s, etc.). The use of (13.12), (13.13) modified by (14.9), which necessitates calculating b_2, may provide a rough guide to the precision of a C' index, but cautious interpretation is advised.

4.10 Conclusion

As is evident from the preceding sections, the subject of dealing with non-Normality takes one into quite deep statistical analysis. No one approach can be recommended for all cases. Sometimes the source or cause can be eliminated, at other times it is an inherent aspect of variation. Of the various measures of non-Normality (and the corresponding significance tests), some may give different or stronger indications than others on particular sets of data. Methods based on transformation, non-Normal curve-fitting and empirical methods all have their respective applications.

It is important to recognise non-Normality to avoid gross errors, not only in capability measurement but also for correct prescription of control chart techniques and other aspects of statistical analysis. As in most other areas of statistics, a thorough understanding of the process or system that generates the data is invaluable.

15

Evaluating the precision of a measurement system (gauge capability)

15.1 Introduction

15.1.1 Any statistical conclusion or decision is conditioned by the quality of the data on which it is based. In some areas of application, the precision of measurement is vital, as in clinical or pathological tests, where diagnosis and treatment will depend on the test results. In other cases, such as weights and measures, sampling for conformance to contract or for pollution in effluents, there may be legal issues at stake. In the SPC field, decisions on control adjustments, the sentencing of batches of material, or rejection of individual items for violation of tolerances, may be involved.

Prudence therefore dictates that all possible sources of bias, both in sampling and measurement, should be considered and eliminated if possible. Common sense suggests that the remaining measurement error should be assessed so that both the capability and the limitations of the measurement system are known.

15.1.2 Measurement errors fall into two broad classifications, generally known as accuracy and precision (though possibly inaccuracy and imprecision would be more meaningful terms). **Inaccuracy** is taken to indicate a discrepancy between the average value obtained for any specific item and the 'true' value. Sometimes the true value can be defined, or at least closely approximated, by a reference source of measurement, e.g. master gauge or certified standard. Working instruments may give different (average) values from the reference source, and are then biassed. A micrometer with a worn screw thread or vernier caliper with strained jaws would tend to give rather higher-than-true readings. A weighing device with a zero error might give distorted values.

Steps to avoid such inaccuracy include regular calibration, training in the proper use of equipment, checking zero settings and observing standard conditions for carrying out tests. In many other cases, there may be no true value, or it may be impossible to measure it. What is the true hardness of a piece of metal or plastic? A test in any particular spot cannot be repeated, so any repeat evaluation involves a different part of the material surface. Inaccuracy is then much more difficult to measure, though reference samples may often give a near approach.

Imprecision is defined as the extent of variation among repeated measurements or tests on the same specimen or item. Again this is not always exactly determinable if a procedure is destructive: there may be an unavoidable element of material variation included in the apparent test variation. For many situations, however, tests or measures can be repeated, especially for physical dimensions, weights or other non-destructive characteristics. Imprecision can be further analysed into various contributions, of which the two most important are the inherent variation in the test or measurement equipment itself – repeatability – and the variations produced by different users, or in different locations – the reproducibility.

15.1.3 The two aspects of accuracy and precision can be represented diagrammatically in various ways. For many measurements involving only one 'dimension', Fig. 15.1(a) provides a simple illustration. For two-dimensional measurements (e.g. coordinate measuring instruments), errors in two directions yield an overall pattern of radial error; accuracy and precision can then be represented as in Fig. 15.1(b). The three coordinate case can be visualized as a cloud of particles in a three dimensional space.

Some other features of measurement error include the following, which are also presented in Fig. 15.2.

Resolution: although many measurements (length, weight, etc.) have a theoretically continuous scale, the equipment used generally has some minimum observable increment. Thus a vernier may have a resolution of 0.01 mm, meaning that values such as 17.33 and 17.34 can be observed, but not intermediate values. Evidently any instrument used to measure a quality characteristic should be able to resolve the specification width, or the process variation, into an adequate number of increments.

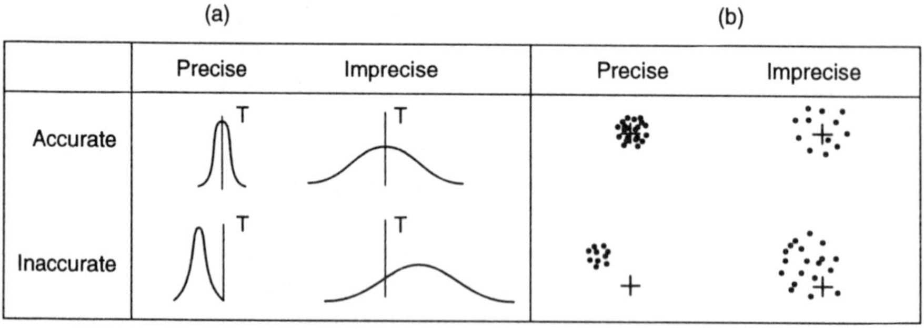

	(a) Precise	(a) Imprecise	(b) Precise	(b) Imprecise
Accurate	T	T		+
Inaccurate	T	T	+	+

Fig. 15.1 Accuracy and precision of measurement. (a) Unidimensional with true values (T); (b) two-dimensional with true values (+).

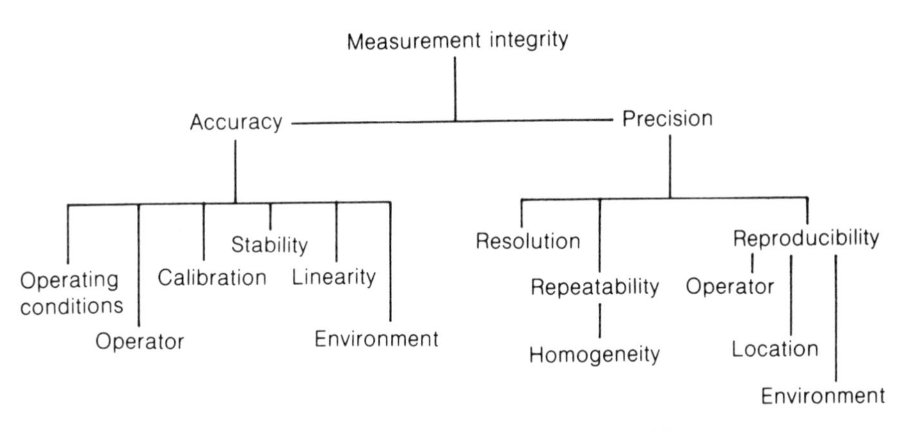

Fig. 15.2 Components of accuracy and precision.

Stability: for any given specimen, a gauge or instrument might give different readings at different times. Gradual drift in calibration, cyclic effects due to ambient temperature or other external conditions, may produce instability.
Linearity: the extent of any bias may change over the operating range of the equipment. Thus a weighing device might have a negative bias at the lower end of its operating range but a positive bias at the top end. This would constitute nonlinearity; for linearity, any bias would remain constant over the whole range.
Homogeneity: a homogeneous system preserves the same precision over the whole operating range. If errors of measurement tend to be greater in some parts of the range than in others, the system is non-homogeneous. This would apply, for example, to a dual or multirange device, where the standard deviation of the measurement errors is proportionately greater on the higher ranges than on the lower. Homogeneity is thus an aspect of precision, whereas linearity is a feature of accuracy.

This Chapter will not probe in greater detail into accuracy, stability, linearity and homogeneity, but various national, international and industry standards provide further information (e.g. British Standard 5497, ISO 5725, and the *Measurement Systems Analysis Reference Manual* (1990) published jointly by Ford, Chrysler, General Motors and the American Society for Quality Control).

15.2 Repeatability and reproducibility

15.2.1 The determination of repeatability and reproducibility requires a deliberately planned series of observations under controlled conditions. This in fact constitutes a designed experiment of a specific kind.

Such experiments can involve interlaboratory or between-site comparisons, the evaluation of accuracy, stability, linearity and homogeneity, and other possible factors (such as the ability of the equipment to deal with different materials, or determination of its response to temperature or humidity changes). In this Chapter we confine our attention to 'in-house' rather than multisite experiments, and to the evaluation of the repeatability and reproducibility only.

The simplest experiments provide a composite measure of both reproducibility and repeatability. A little more organizational effort permits separation of these two aspects of performance.

15.2.2 The statistical basis of the procedures (albeit often simplified) is the estimation of two components of measurement variation:

σ_r the **repeatability element**, sometimes called the equipment variation or EV.

σ_L the **reproducibility element**, sometimes called the appraiser variation or AV (but it may contain other aspects, e.g. time, environment as well as operator-to-operator differences).

The short-term element of repeatability indicates the basic precision of the system under controlled, homogeneous conditions. For destructive tests it inevitably includes an element of material variation, but this should be reduced to a minimum, for example by taking specimens from one sheet of material, one length of tube, sub-samples from a carefully blended master, etc.

Similarly, whilst the reproducibility element usually measures the effect of different operators or users at the same time under the same conditions, a realistic appraisal may necessitate use of a piece of equipment at different workstations (those at which the operators normally carry out the measurements), possibly even on different shifts and therefore with some (uncontrollable) changes in ambient conditions.

The combined effect of repeatability and reproducibility is measured by

$$\sigma_R = \sqrt{(\sigma_L^2 + \sigma_r^2)}$$

This is sometimes called the (system) reproducibility on the grounds that it measures the overall effect of both components in the system. Other users call it the 'repeatability and reproducibility', abbreviated to *R* and *R*.

15.2.3 These standard errors, σ_r, σ_L and σ_R, provide the basic yardstick for measurement precision. In some industries, they are quoted in this form, i.e. as standard errors. The combined measure, σ_R, indicates the overall reliability of the procedure, but σ_r and σ_L can be useful in diagnosing problems and improving the system. For example, if σ_r is the dominant component, it is the equipment or procedure itself that needs improvement; if σ_L is the greater, it suggests that operators or users may need training, supervision or more carefully drawn up instructions to improve consistency.

In other cases, the standard error may be multiplied by a factor to provide

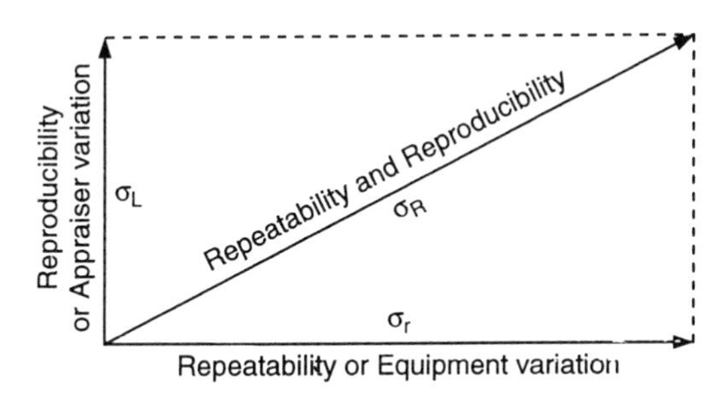

Fig. 15.3 Graphical representation of repeatability and reproducibility.

some indication of confidence in the system. For example, BS 5497 uses the estimate of the discrepancy between a pair of repeated determinations that would be exceeded by chance in only 5% of cases. Assuming a Normal distribution of individual measurements, this becomes $(1.96 \times \sqrt{2})\sigma = 2.77\sigma$ (generally using σ_R in this expression).

The automotive industry, on the other hand, uses the 99% spread of individual (repeated) measurements. This, in the absence of bias, becomes

$$\text{True value} \pm 2.576\sigma$$

so that the total width of the 99% band is $2 \times 2.576\sigma$, or 5.15σ approximately.

Because of these differences in mode of expression, this Chapter will concentrate on obtaining estimates of the standard errors themselves; they can then be multiplied by any appropriate factor. One should note, however, that some simplified automotive industry methods have the 5.15 factor 'built in' the calculations, so that σ_r, σ_L and σ_R are not obtained directly, but only in the form of $5.15\sigma_r$, $5.15\sigma_L$ and $5.15\sigma_R$. This can inhibit their use in assessing the effects of measurement variation on the apparent overall variation of a process or system. This point will be taken up later in section 16.6.

15.2.4 Where two sources of variation such as σ_r and σ_L contribute independently to the overall variation in a system, the components can conveniently be represented by line segments at right angles to each other. The hypotenuse (in this case σ_R) then represents the overall variation from the combined sources, as in Fig. 15.3. The idea can be extended to three or even more dimensions.

15.3 Experimental designs for precision evaluation

15.3.1 Although a wide range of experimental designs may be involved in the more complex aspects of studying linearity, stability, etc., we consider here two simple designs concerned only with evaluation of precision: namely repeatability and reproducibility.

The first design measures only the overall precision representing the combined effect of repeatability and reproducibility. The second, necessarily somewhat more elaborate though still simple to analyse, permits separation of the two components and their subsequent assembly into overall '*R* and *R*'.

15.3.2 For the smaller experimental design, a number, n, of items or specimens (typical of those routinely measured) is chosen. Each of several, say k, operators measures each item once only. There are thus $n \times k$ measurements in all. As detailed in section 15.4, there are two principal methods for analysing the results, using either ranges or standard deviations, though the analysis of variance (ANOVA) can be used especially if computer software is available.

15.3.3 The more useful experiment requires that for each of the n items, each of the k operators makes several, say m, measurements. It is important that m should be constant for all the $n \times k$ operator/item combinations.

15.3.4 Some considerations apply to both formats, and embody fundamental principles of good experimental design.

(i) There should be no collusion between operators, as this is likely to yield edited results that will fail to reveal the true variations of either the instrument itself or of independent users.
(ii) The operators should not be aware of the true values (if any) for the items or specimens. This can be achieved by presenting them to each operator in a different random order, using items that cannot be identified by characteristics like colour, shape, considerable differences in size, or obvious part identity numbers.
(iii) For the larger experiment, it is also necessary to ensure that the operator does not see or remember previous measurements when carrying out later checks on the same items. Randomization along with an identification system known only by the organiser will achieve this.
(iv) To ensure that the true precision of the system is evaluated, the effect of other variables should be controlled. For example, a mixture of trained and untrained operators should not be introduced; the trial should be completed within a short time, not spread over several shifts or days; for chemical analyses or destructive physical/mechanical tests, specimens or materials need close control. If equipment or instruments have to be moved to the operators (rather than vice versa), zero checks, levelling, etc. must be rigorously executed.
(v) It is obvious, but must be stated, that recording of results must be meticulous. All readings should be legible, contain the same number of significant digits, and checked for validity at the time of recording. It creates problems later if a value of, say, 173.2 looks implausible – should it have been 137.2? If a value is recorded as 168, whilst others contain one or two

decimal digits, is the observation actually 168.00 or have the decimal digits simply not been noted?

15.4 Analysis of precision data

15.4.1 Readers familiar with analysis of variance (ANOVA) will recognise that either type of experiment can be readily analysed by these techniques. Most statistical software packages include ANOVA and simply require logical entry of data and the identification of the factors for analysis.

For the smaller experiment, the classification factors are operators (1 to k) and items (1 to n). The analysis of variance will yield mean squares for operators, items and residual. In measurement precision evaluation, the residual is often ignored, partly on the grounds that it contains an operators × items interaction, which is supposed not to exist. It can, nevertheless, provide a check on the validity of the experiment and possibly a crude measure of repeatability.

The mean square for operators is used as the measure of combined R and R_1 and its square root gives the estimate of σ_R.

For the larger experiment, in ANOVA terminology we also have repeat measurements nested within operators × items. The residual mean square provides an independent estimate of $\hat{\sigma}_r^2$. Components can also be estimated for the variation between operators, between items, and a possible operators × items interaction. If the latter occurs it requires investigation or explanation before a proper interpretation of the operator effect can be made; it means that differences between operators are not consistent over the various items, giving an additional source of uncertainty in the results obtained in routine use of the system.

15.4.2 Under the assumption that no operator × item interaction exists, two simplified methods of analysis are available. The most widely used is based on average ranges over subsets of the experiment. The other is somewhat more efficient in its use of the data, and in the absence of operator × item interaction gives identical results to ANOVA. It uses standard deviations in lieu of ranges. We now describe both methods, for each of the two types of experiment.

15.4.3 Short experiment, standard deviation method
(n items k operators)

(i) For each item, calculate the variance, s^2 of the k results for the various operators (on most statistical calculators, this can be obtained by finding s and pressing either x^2 or $\times =$).
(ii) Average these s^2 values and take the square root of this average. This gives $\hat{\sigma}_R$ directly as $\sqrt{\text{Av}(s^2)}$.

Table 15.1 d^* factors for estimation of σ from $\bar{R}$ (Also for estimating $\sqrt{V}$ from a single range)

Sample	Number of ranges contributing to $\bar{R}$										
size	1	2	3	4	5	6	8	10	15	20	>20
2	1.414	1.279	1.231	1.206	1.191	1.18	1.17	1.16	1.15	1.14	1.128
3	1.912	1.805	1.769	1.750	1.739	1.73	1.73	1.72	1.71	1.70	1.693
4	2.239	2.151	2.121	2.105	2.096	2.09	2.09	2.08	2.07	2.06	2.059
5	2.481	2.405	2.379	2.366	2.358	2.35	2.35	2.34	2.34	2.33	2.326
6	2.672	2.604	2.581	2.570	2.563	2.56	2.55	2.55	2.54	2.54	2.534
8	2.963	2.906	2.886	2.877	2.871	2.87	2.86	2.86	2.86	2.85	2.847
10	3.179	3.129	3.112	3.103	3.098	3.09	3.09	3.09	3.09	3.08	3.078

Divide $\bar{R}$ (or R) by d^* to estimate $\hat{\sigma}$ (or $\sqrt{V}$).
Values for d^* for columns 6–20 are not available to greater accuracy.

15.4.4 Short experiment, range method

(i) For each item, calculate the range of results of the k operators.
(ii) Find the average of these ranges, and divide by the d^* factor for the appropriate combination of sample size k and number of samples n. This gives the overall reproducibility estimate $\hat{\sigma}_R$. The d^* factors are listed in Table 15.1.

15.4.5 Full experiment, standard deviation method

(m measurements for each of k operators on each of n items, $m \times k \times n$ measurements in all).

(i) For each item and operator combination, calculate the variance, s^2, of the m measurements. There will be $k \times n$ of these variances.
(ii) Average the variances and take the square root. This provides the estimate $\hat{\sigma}_r$.
(iii) For each operator, average all results (including m measurements on each of n items. This will give k 'operator averages'.
(iv) Calculate the variance of the operator averages, say V. Then the operator component is given by

$$\hat{\sigma}_L = \sqrt{\left(V - \frac{\hat{\sigma}_r^2}{m \times n}\right)} \tag{15.1}$$

(If V is less than $\hat{\sigma}_r^2/(m \times n)$, set $\hat{\sigma}_L = 0$).
(v) The overall reproducibility estimate is then

$$\hat{\sigma}_R = \sqrt{(\hat{\sigma}_L^2 + \hat{\sigma}_r^2)} \tag{15.2}$$

15.4.6 *Full experiment, range method*

(n items $\times$ k operators, m measurements each)

(i) For each item and operator combination, calculate the range of the m measurements. There will be $k \times n$ of these ranges.
(ii) Find the average of these ranges, and divide by the d^* factor for sample size m and number of samples $k \times n$. This estimates $\hat{\sigma}_r$.
(iii) For each operator, average all the $m \times n$ measurements, giving k 'operator averages'.
(iv) Calculate the ranges of these operator averages, and divide by the d^* factor for sample size k, number of ranges = 1. This estimates $\sqrt{V}$; square the result to obtain V. Then obtain $\hat{\sigma}_L$ using formula (15.1).
(v) The reproducibility is then obtained, as above, using (15.2).

15.4.7 The methods in sections 15.4.4 and 15.4.6 based on ranges require special range-to-standard deviation conversion factors. The values of $d_n(d_2)$ in Table 13.1 assume that the mean range, $\bar{R}$, has been obtained from a fairly large number (more than twenty) individual ranges. For smaller numbers of ranges contributing to $\bar{R}$, slightly larger factors apply. In measurement precision studies, the average range from which $\hat{\sigma}_r$ is estimated may contain an adequate number of groups to permit the use of the usual factors, but for the 'short' procedure of section 15.4.4 there may only be 5–10 ranges. In estimating $\sqrt{V}$ in the procedure of section 15.4.6, there is only one range of perhaps 3 to 5 operators. Table 15.1 lists the necessary factors.

The methods based on ANOVA or on the use of standard deviations provide the definitive procedure, though the speed and simplicity of the range methods makes them more attractive in many cases for manual calculation. It is worth noting that modern calculators permit rapid calculation of $\sqrt{\mathrm{Av}(s^2)}$, or average variance itself. If values of s (preferably with four or five significant digits) are entered via the $\bar{x}, s$ procedure, the following sequence of keystrokes applies:

Recall $\sum x^2$ from its storage location,
then $\div\, n =$ gives average variance,
and $\sqrt{\ }$ as an additional keystroke will yield $\hat{\sigma}_R$ for the smaller experiment (section 15.4.3) or $\hat{\sigma}_r$ for the larger experiment (section 15.4.5).

15.5 Worked example: coating thickness measurements

15.5.1 A process involves coating metal parts with a layer of plastic material. Occasional (destructive) tests are necessary to ensure that non-destructive thickness measurement instruments are correctly calibrated and operated. The test requires cutting a section of the metal + plastic, viewing a magnified ($\times$ 500) image on a screen and reading off the thickness against a calibrated scale. The metal-plastic boundary is not always crisply defined, and is subject to slight depressions and bulges at the microscopic level, so that operator judgement has some influence on the measurements. The scale is calibrated in 0.01 mm divisions, and with a typical thickness of half a millimetre the 50 scale divisions occupy about 25 cm on the screen. As each division occupies 5 mm, test operators interpolate to 1/10th of a scale division, so that results are expressed in microns (0.001 mm), e.g. 485 for 0.485 mm etc.

To measure the precision of the system, ten typical test pieces were cut. Each of five operators made three independent thickness determinations, with the

Table 15.2 Measurement Precision—Coating Thickness (microns). $n = 10$, $k = 5$, $m = 3$

Item	Operator (R, s^2 in brackets) 1	2	3	4	5	Item mean
1	485 485 484 (1, $0.\dot{3}$)	482 484 484 (2, $1.\dot{3}$)	486 485 484 (2, 1.0)	485 487 486 (2, 1.0)	485 490 489 (5, 7.0)	485.40
2	495 493 493 (2, $1.\dot{3}$)	490 490 490 (0, 0.0)	492 492 494 (2, $1.\dot{3}$)	495 492 492 (3, 3.0)	494 496 497 (3, $2.\dot{3}$)	493.00
3	503 503 502 (1, $0.\dot{3}$)	499 501 500 (2, 1.0)	502 500 500 (2, $1.\dot{3}$)	505 502 502 (3, 3.0)	506 503 503 (3, 3.0)	$502.0\dot{6}$
4	478 479 479 (1, $0.\dot{3}$)	475 474 473 (2, 1.0)	476 474 477 (3, $2.\dot{3}$)	479 478 478 (1, $0.\dot{3}$)	478 479 480 (2, 1.0)	$577.1\dot{3}$
5	487 485 485 (2, $1.\dot{3}$)	483 482 484 (2, 1.0)	482 481 484 (3, $2.\dot{3}$)	486 487 485 (2, 1.0)	488 488 489 (1, $0.\dot{3}$)	$485.0\dot{6}$
6	493 494 494 (1, $0.\dot{3}$)	491 498 491 (3, 3.0)	489 492 491 (3, $2.\dot{3}$)	491 494 494 (3, 3.0)	497 494 496 (3, $2.\dot{3}$)	492.60
7	498 499 498 (1, $0.\dot{3}$)	497 497 496 (1, $0.\dot{3}$)	496 498 497 (2, 1.0)	499 498 500 (2, 1.0)	501 501 500 (1, $0.\dot{3}$)	$498.\dot{3}$
8	456 457 461 (5, 7.0)	457 453 457 (4, $5.\dot{3}$)	458 458 458 (0, 0.0)	458 459 459 (1, $0.\dot{3}$)	462 462 461 (1, $0.\dot{3}$)	458.40
9	464 466 467 (3, $2.\dot{3}$)	462 464 463 (2, 1.0)	463 464 463 (1, $0.\dot{3}$)	466 467 466 (1, $0.\dot{3}$)	469 468 470 (2, 1.0)	$465.4\dot{6}$
10	499 500 499 (1, $0.\dot{3}$)	495 495 497 (2, $1.\dot{3}$)	496 494 494 (2, $1.\dot{3}$)	497 497 496 (1, $0.\dot{3}$)	500 499 499 (1, $0.\dot{3}$)	$497.1\dot{3}$
Operator $\bar{x}$ ($\bar{R}$, Av(s^2))	$486.0\dot{3}$ (1.8, 1.4)	$483.1\dot{3}$ (2.0, $1.5\dot{3}$)	484.00 (2.0, $1.\dot{3}$)	486.00 (1.9, $1.\dot{3}$)	$488.1\dot{3}$ (2.2, 1.8)	485.46 (1.98, 1.48)

results listed in Table 15.2. A trial of this magnitude (150 determinations) meant that the operators were aware that it was a special study, and the well-known 'Hawthorne effect' is likely to result in rather better quality of measurement than would apply in routine testing. However, care was taken, in presenting the specimens and organizing the trial, to avoid collusion and to conceal the identity of the items on repeat measurements. With each measurement requiring about two minutes to set up the test piece and estimate the number of scale divisions, the whole study occupied one working day. It was considered essential to evaluate the reliability of the measurement system to assess its effect on interpretation of routine determinations on the process.

15.5.2 Although many readers may not be familiar with the analysis of variance, we present the results that would be obtained by calling up a computer ANOVA procedure with the factors specified as:

ITEMS; OPERATORS; ITEMS * OPERATORS; ERROR

The precise details of instructions will vary between systems: for some, it will not be necessary to specify 'ERROR', for example. The 'ITEMS * OPERATORS' factor represents the interaction, or non-independence, between the ITEM and OPERATOR factors. In this example, it may be that for items with fuzzy margins between coating and metal, some operators tend to give the benefit of the doubt to the coating, while others give it to the metal base. In other tests involving using a micrometer on a slightly soft material, operators who tend to tighten the micrometer more will obtain smaller readings especially on softer specimens. An advantage of the ANOVA procedure is that such interactions can readily be included in the MODEL or FACTORS instruction, whereas they are not usually measured in the simpler range or standard deviation methods. Any interaction that may be present is thus overlooked, though it may represent a source of variation that should be taken into account.

Table 15.3 presents the ANOVA in a widely used format. The F-ratios can be used to indicate whether factors are statistically significant, but in this case it is the mean squares that are of primary interest. They provide the means of estimating σ_r, σ_L and σ_R as well as the item-to-item variation (if required) σ_I

Table 15.3 ANOVA table for full experiment

Factor	Sum of squares	Degrees of freedom	Mean Square	F	Significance level
Items	28 306.06	9	$3\,145.11\dot{7}$	2 125.079 6	$P = < 0.001$
Operators	459.36	4	114.84	77.594 6	$P = < 0.001$
$I*O$	65.84	36	$1.82\dot{8}$	1.235 7	$P = 0.205$
Error	148.0	100	1.48		
Total	28 979.26	149			

and the interaction σ_{IO}. The estimates are obtained as follows.

$\hat{\sigma}_r^2$ (repeatability) = MS (ERROR);
MS (I*O) corresponds to $\hat{\sigma}_r^2 + m\hat{\sigma}_{IO}^2$, so that
$\hat{\sigma}_{IO}^2 = \{\text{MS(IO)} - \text{MS (ERROR)}\} \div m$;
$\text{MS(OP)} = \hat{\sigma}_r^2 + m\hat{\sigma}_{IO}^2 + mn\hat{\sigma}_L^2$, so that
$\hat{\sigma}_L^2 = \{\text{MS(OP)} - \text{MS(I*O)}\} \div mn$.

In some cases, an estimate of the item-to-item variation is also required. Since

MS (ITEMS) $= \hat{\sigma}_r^2 + m\hat{\sigma}_{IO}^2 + mk\hat{\sigma}_I^2$, we have
$\hat{\sigma}_I^2 = \{\text{MS (ITEMS)} - \text{MS(I*O)}\} \div mk$.

If the IO term is small, i.e. non-significant, the latter two components can be reasonably measured with MS (ERROR) substituted for MS (I*O) in the formulae. The absence of an interaction is implicitly assumed when using the range or standard deviation methods.

In the case of the smaller experiment (one test per operator per item, as in section 15.3.2) the interaction and error cannot be separated, so that the residual variation is a compound of $\sigma_r^2 + \sigma_{IO}^2$. When using range or standard deviation methods, this is also combined with σ_L^2 and only the combined repeatability and reproducibility, σ_R^2, is obtained from the analysis. Table 15.5 shows the ANOVA that would result from the reduced experiment involving only the first measurement by each of the first two operators on each of items 1, 3, 5, 7, 9, as in Table 15.4.

In fact the interaction is non-significant, so the range and standard deviation methods will give very similar results to the ANOVA. We find the components are:

$\hat{\sigma}_r^2 = 1.48$, so $\hat{\sigma}_r = 1.2166$;

$$\hat{\sigma}_{IO}^2 = \frac{1.82\dot{8} - 1.48}{3} = 0.1163, \text{ giving } \hat{\sigma}_{IO} = 0.3410;$$

Table 15.4 Example of smaller precision evaluation

Operators 1, 2; items 1, 3, 5, 7, 9; one measurement by each operator on each item
$n = 5$, $k = 2$, $m = 1$

Item	Operator 1	Operator 2	Range R	Standard deviation S	Variance S^2
1	485	482	3	2.12132	4.5
3	503	499	4	2.82843	8.0
5	487	483	4	2.82843	8.0
7	498	497	1	0.70711	0.5
9	464	462	2	1.41421	2.0
Mean	487.4	484.6	2.8	–	4.6

Table 15.5 ANOVA for reduced experiment

(See Table 15.5)

Factor	Sum of squares	Degrees of freedom	Mean square	F	Significance level
Items	1 787.00	4	446.75	525.588	$P < 0.001$
Operators	19.6	1	19.6	23.059	$P = 0.008\,64$
Error	3.4	4	0.85		
Total		9			

$\hat{\sigma}_r^2 = 0.85$, $\hat{\sigma}_r = 0.9220$.
$\hat{\sigma}_L^2 = (19.6 - 0.85)/2 = 9.375$; $\hat{\sigma}_L = 3.061\,9$.
$\hat{\sigma}_I^2 = (446.75 - 0.85)/5 = 89.18$; $\hat{\sigma}_I = 9.443\,5$.
$\hat{\sigma}_R^2 = 3.061\,9^2 + 0.922\,0^2 = 10.225\,3$, $\hat{\sigma}_R = \sqrt{10.225\,3} = 3.197\,7$.

$$\hat{\sigma}_L^2 = \frac{114.84 - 1.82\dot{8}}{3 \times 10} = 3.7670,\ \hat{\sigma}_L = 1.9409;$$

$$\hat{\sigma}_I^2 = \frac{3145.11\dot{7} - 1.82\dot{8}}{3 \times 5} = 209.5526,\ \hat{\sigma}_I = 14.4759.$$

$$\hat{\sigma}_R = \sqrt{(\hat{\sigma}_L^2 + \hat{\sigma}_r^2)} = \sqrt{(1.9409^2 + 1.2166^2)} = 2.2906.$$

Note that if the interaction term in the analysis is ignored, the $\hat{\sigma}_L^2$ component is estimated as

$$(\text{MS(OPERATORS)} - \text{MS(ERROR)})/(m \times n) = \frac{114.84 - 1.48}{30} = 3.778\dot{6},$$

and $\hat{\sigma}_L$ increases slightly to 1.9439. This also has an effect on $\hat{\sigma}_R$, which becomes 2.2932. In this case the effect is small, but can be larger when a significant interaction is present. This 'condensed' analysis, which discards the I∗O effect and its degrees of freedom, is identical to that obtained from the s-method of section 15.5.3 (and close to that for the range procedure of section 15.5.4).

15.5.3 For range and standard deviation methods, $\hat{\sigma}_L$ and $\hat{\sigma}_r$ are not separated in the smaller experiment, and the effect is as if the OPERATORS and ERROR sums of squares were combined (along with their degrees of freedom) to give

$$\text{MS}(R + R) = (19.6 + 3.4) \div (1 + 4) = 4.6$$

This is used as the estimate of $\hat{\sigma}_R^2$, and is identical to the average variance at the foot of Table 15.4. Evidently with much less data, the short procedure gives less reliable results, and in this case the R and R is over-estimated (compared with the full experiment) while σ_I is smaller than for the full experiment mainly because the thinnest specimen (item 8) happens not to have been included.

15.5.4 We now re-work the example using the *s*-method, following the steps of section 15.4.3 for the short experiment of Table 15.4 and section 15.4.5 for the full experiment, Table 15.2.

For the short experiment, the combined *R* and *R* is simply the square root of the average of the variances across the two operators for the five items, viz. $\sqrt{4.6} = 2.1448$.

For the full experiment, the average within-cell variance of 1.48 provides $\hat{\sigma}_r^2$, so that $\hat{\sigma}_r = 1.2166$ (as for the ANOVA method). To obtain $\hat{\sigma}_L$, we first require the variance of the five operator means, which is 3.827 975. This is inserted as the value of V in formula (15.1) giving

$$\hat{\sigma}_L = \sqrt{\left(3.827\,975 - \frac{1.48}{10 \times 3}\right)} = 1.943\,9$$

Finally, using (15.2),

$$\hat{\sigma}_R = \sqrt{(1.9439^2 + 1.48)} = 2.2932$$

as would be obtained from ANOVA if the interaction is ignored.

15.5.5 The range method proceeds along similar lines. We note in passing that whilst several sets of three measurements give the same range, their standard deviations or variances may differ. Thus both 486, 485, 484 (item 1, operator 3) and 492, 492, 494 (item 2, operator 3) give $R = 2$, but their standard deviations are 1.0 and 1.1547, respectively. The standard deviation (and therefore the variance) makes explicit use of the third result, thus using rather more of the sample information content.

For the short experiment, the combined *R* and *R* is obtained via the average of the five operator-to-operator ranges, 2.8. As there are only five ranges of two items each, the d^* factor is 1.191, so that

$$\hat{\sigma}_R = 2.8 \div 1.191 = 2.351$$

(cf. 2.1448 via the *s*-method and 2.29 from ANOVA)

The full experiment provides fifty within-cell ranges of three items each; the d^* factor then becomes the usual 1.693 for samples of three. With $\bar{R} = 1.98$, we then have

$$\hat{\sigma}_r = 1.98 \div 1.693 = 1.1695$$

(cf. 1.2166 for the *s* and ANOVA methods).

The range of the five operator means is $488.1\dot{3} - 483.1\dot{3} = 5.0$. With only one range of five items $d^* = 2.481$, and this yields $\sqrt{V} = 5.0 \div 2.481 = 2.0153$, whence $V = 4.0615$ (cf. 3.828 approximately for the *s*-method). Then, using (15.1),

$$\hat{\sigma}_L = \sqrt{\left(4.0615 - \frac{1.1695^2}{10 \times 3}\right)} = 2.0040$$

a little larger than the 1.9439 via the *s*-method or 1.9409 from ANOVA.

Table 15.6 Summary of measurement precision estimates

	R method	s and ANOVA methods
Short experiment	$\hat{\sigma}_R = 2.35$	$\hat{\sigma}_R = 2.14$
Full experiment	$\hat{\sigma}_r = 1.17$ $\hat{\sigma}_L = 2.00$ $\hat{\sigma}_R = 2.32$	$\hat{\sigma}_r = 1.21$ $\hat{\sigma}_L = 1.94$ $\hat{\sigma}_R = 2.29$

The R and R standard error is finally derived via (15.2) as

$$\hat{\sigma}_R = \sqrt{(1.1695^2 + 2.0040^2)} = 2.32$$

no greater accuracy in the result being justified by the approximate nature of the method. The estimate is similar to that from the s-method (2.2932) and ANOVA (2.2906).

15.5.6 It is useful to collect the results from these alternative analyses together. If no interaction between operators and items is assumed, ANOVA and the s-method give identical values, so they are merged in Table 15.6.

The breakdown of $\hat{\sigma}_R$ into two components for the full experiment is very useful. In this case it suggests that differences between operators is the larger source of imprecision. This might be construed as indicating that more precise standards for identifying metal/coating boundary, or for locating an 'average' region in cases where the thickness appears to vary, is a first priority. A halving of the variation from this source would reduce the overall σ_R to around 1.5, a considerable improvement. Figure 15.4 illustrates the components as obtained from the full experiment using the s or ANOVA methods.

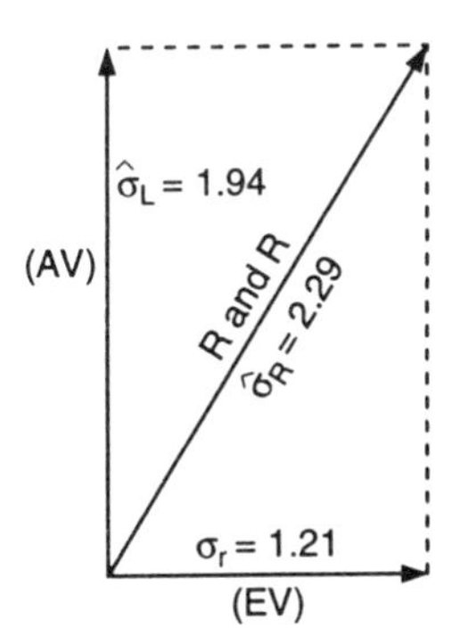

Fig. 15.4 Graphical representation of R and R with data from Table 15.6. (Full experiment: s-method).

15.6 Automotive industry procedures

15.6.1 Major automotive companies have recognized the importance of measurement precision, generally known in the industry as 'gauge capability'. The *Measurement Systems Analysis Reference Manual* (1992) represents the latest collective thinking on the subject. It adopts the usual overall approach to evaluation via simple experimental design, and the ANOVA procedure as the most informative means of analysis, but there are some differences in presentation that should be noted.

While not excluding other designs, two are especially suggested:

(i) for a rapid and approximate indication of combined R and R, the short experiment with $n = 5$, $k = 2$ and $m = 1$, i.e. five items measured once each by two operators;
(ii) The full experiment to separate repeatability (EV, or equipment variation) from AV (appraiser variation). The optimum conditions are regarded as $n = 10$, $k = 3$, and $m = 2$ or 3.

15.6.2 The results of the short experiment are typically analysed by the range method described in section 15.4.4. However, because the motor industry has adopted the 5.15σ measure of precision (99% band width), the d^* factors of Table 15.1 are combined with 5.15σ to give a factor $5.15 \div 1.191 = 4.33$ approximately. Thus for the example of section 15.5.3 (Table 15.4) the overall R and R would be obtained as $2.8 \times 4.33 = 12.124$. This is approximately 5.15 times the σ_R obtained in section 15.5.5 using the d^* factor directly. The 99% band-width value of 12.124 would be quoted as the GR & R (gauge repeatability and reproducibility).

15.6.3 For the larger experiment, the equivalent of section 15.4.6 is described as the 'average and range' method. Again the factor 5.15 is merged with d^*, so that for $n = 10$, $k = 3$, $m = 2$ (ten items, three operators, two repeat measurements) we have $K_1 = 4.56$ from $5.15 \div 1.128$, and $K_2 = 2.70$ from $5.15 \div 1.912$ (the d^* for one range over three operators). If three repeats per item are used, K_1 becomes 3.05, and if only two operators are involved, K_2 is 3.65.

The results for the full experiment are not susceptible to analysis using these constants, as *five* operators were involved. In fact, a K_2 factor of $5.15 \div 2.481$ would be required, i.e. 2.08 approximately.

The observant reader may note some minor discrepancies arising from rounding in these expressions. In fact a more accurate value for the 99% spread is $2 \times 2.5758 = 5.1516$, and the exact range-to-standard deviation conversion factor for many ranges of two items each is $\sqrt{(4/\pi)}$. Dividing 5.1516 by $\sqrt{(4/\pi)}$ gives 4.5655, nearer to 4.57 than 4.56. Similarly, for *one* range of two items, $d^* = \sqrt{2}$, and $5.1516 \div \sqrt{2} = 3.6427$, nearer to 3.64 than to 3.65. These minor discrepancies are of no practical importance, in that values will never be **exactly**

Normally distributed, and in any case the precision estimates are sample values, not exact 'population' standard errors.

15.6.4 The larger experiment may also be analysed via the ANOVA method, and the reference manual suggests that the operator × items interaction should be evaluated. When using the ANOVA, the variance components themselves, σ_r^2, σ_L^2 etc., are obtained, and their square roots are multiplied by 5.15 to obtain EV, AV and GR & R, the latter as $5.15\sqrt{(\sigma_r^2 + \sigma_L^2)}$. Some other points covered in the manual include confidence intervals for the components, and the estimation of item-to-item variation. This chapter would need considerable extension to deal with these aspects, and the interested reader is referred to the manual.

15.6.5 Finally, the automotive industry relates $\hat{\sigma}_R$ to the spread of item-to-item variation. This spread corresponds either to the specification width USL – LSL, or to the 'process performance' of $6\hat{\sigma}_p$, where $\hat{\sigma}_p$ is the actual item or process standard deviation. The **smaller** of USL – LSL or $6\hat{\sigma}_p$ is adopted as the yardstick. The GR & R, as measured by $5.15\hat{\sigma}_R$, is expressed as a percentage of specification width or process performance, % GR & R.

It is suggested that a GR & R of less than 10% represents a good measurement system, 10 to 30% may be acceptable depending on the application, costs, etc., but a value greater than 30% is unacceptable.

An alternative measure of effectiveness is provided by an assessment of the number of distinct categories into which the items can be separated by the measurement system. This is given by $[1.41 \times (\sigma_p/\sigma_R)]$, where the square brackets indicate 'integer part only', i.e. the result is rounded down to the next integer.

Let us assess the example of 15.5 assuming that the item-to-item variation, $\hat{\sigma}_I = 14.4759$, is a good estimate of the variation in the process. Then with $\hat{\sigma}_R = 2.2906$, we have the following results:

$$\%\text{GR \& R} = \frac{5.15 \times 2.2906}{6 \times 14.4759} \times 100 = 13.58\%$$

This is reasonably good, but not ideal: it exceeds 10%, although it is considerably better than 30%.

The estimate of 'distinct categories' is

$$\left[1.41 \times \frac{14.4759}{2.2906}\right] = [8.91] = 8$$

so the measurement system (including both equipment and operators) can resolve the process spread into at least eight categories. Ten or more categories is regarded as good, three or fewer as unacceptable.

15.7 Weights and measures applications

15.7.1 Another area of application in which specific requirements for measurement precision are laid down is that of quantity control in weights and measures. Here the motivation comes not only from general SPC considerations but also from the need to conform to legal requirements. In carrying out official tests, for example, trading standards, officers are required to ensure that the **overall** measurement error (which can include resolution and the effect of container weight variation as well as *R* and *R*) does not exceed one fifth of the 'tolerable negative error'. This is usually taken to imply that the statistical error must be such that σ_R does not exceed one tenth of the tolerance.

When checkweighing equipment is used as part of a packing process, e.g. to reject or segregate underweight packages, there are again legal requirements for the performance of the instruments.

The *Code of Practical Guidance for Packers and Importers* (HMSO, 1979) gives details of procedures and checks, and the *Manual of Practical Guidence for Inspectors* (HOSO, 1979) provides further statistical background. Here we consider two of the checks on measuring equipment that may have wider applications.

15.7.2 It should be noted that modern weighing devices involve little or no operator (appraiser) involvement. When using a vernier or micrometer, different operators may apply different pressures to the item being measured, and read the instrument scale in different ways. When carrying out a chemical or physical test involving sample preparation, mixing, titration, measuring quantities of reagents and sometimes interpreting an 'end point', different operators may show variations. Weighing items on an automatic, usually electronic balance, with digital display leaves little scope for operator variation. The main purpose of a precision check is thus to determine the instrument error (or EV), measured fundamentally by $\hat{\sigma}_r$. Somewhat simpler methods can therefore be used.

15.7.3 The first method is to take one or more items which are typical of those likely to be measured in routine operations, and make a moderate number of repeated determinations, say $n = 25$ or 50 on each item. The mean and standard deviation (for each item) are then calculated. If a reference method is available, the discrepancy between $\bar{x}$ and the reference value provides a measure of bias. The repeatability, σ_r, is estimated by s, if necessary averaged over the several items as $\sqrt{(\text{Av}(s^2)}$.

In weights and measures applications, the 99.7% spread of the Normal distribution, i.e. six standard errors of measurement in this case, is known as the 'zone of uncertainty' of the equipment.

While mostly applicable to weighing devices, the method is also useful with other measurements where digital display or printed output is provided by the instrument, so that no operator effect is likely.

Table 15.7 Data for 'attribute' measurement study

Set point 500 g

Test weight	Accept	Reject	Total passes	*Plotting (%) (a)	(b)
497.5	$a=0$	$r=50$	$n=50$	Ignored	Ignored
498	0	50	50	Ignored	1
498.5	1	49	50	2	3
499	5	45	50	10	11
499.5	9	41	50	18	19
500	19	31	50	38	39
500.5	35	15	50	70	69
501	42	8	50	84	83
501.5	48	2	50	96	95
502	49	1	50	98	97
502.5	50	0	50	Ignored	99
503	50	0	50	Ignored	Ignored

*See text (section 15.7.4) for plotting percentage details
(a) 'Weights and Measures' convention.
(b) Automotive convention.

15.7.4 The second method is used where a simple accept/reject indication is obtained from the instrument. Thus a checkweigher can be arranged so that its 'set point' coincides with a tolerance or other important point on the scale. Items which appear (to the 'mind' of the checkweigher) to be on one side of the set point are separated from those on the other side. Typically, underweight items are moved away from the acceptable product stream.

A similar system arises in engineering industries with 'attribute' gauges of the go/no go type, especially electronic devices which exclude operator effects.

Several test items, preferably around ten, are measured by a reference method. Their 'true' values should lie in equal steps in a narrow band centred on the set point. Each item is passed over the checkweigher or through the attribute gauging system at least ten times, preferably 25 times or more. The decision of the instrument, accept or reject is noted for each pass of each item. A set of results like that in Table 15.7 will be obtained.

For a two-sided specification, or if a checkweigher segregates into three or more streams, a similar procedure is followed for each set point.

The results are plotted on Normal probability paper, plotting percentage accepted against the true value of the test item. Figure 15.5 illustrates the plot for the data of Table 15.7.

There is some difference in treatment of the data in converting to percentage for plotting. The two 'Weights and Measures' codes cited in section 15.7.1, generally using $n=100$, use the simple percentage accepted. The automotive

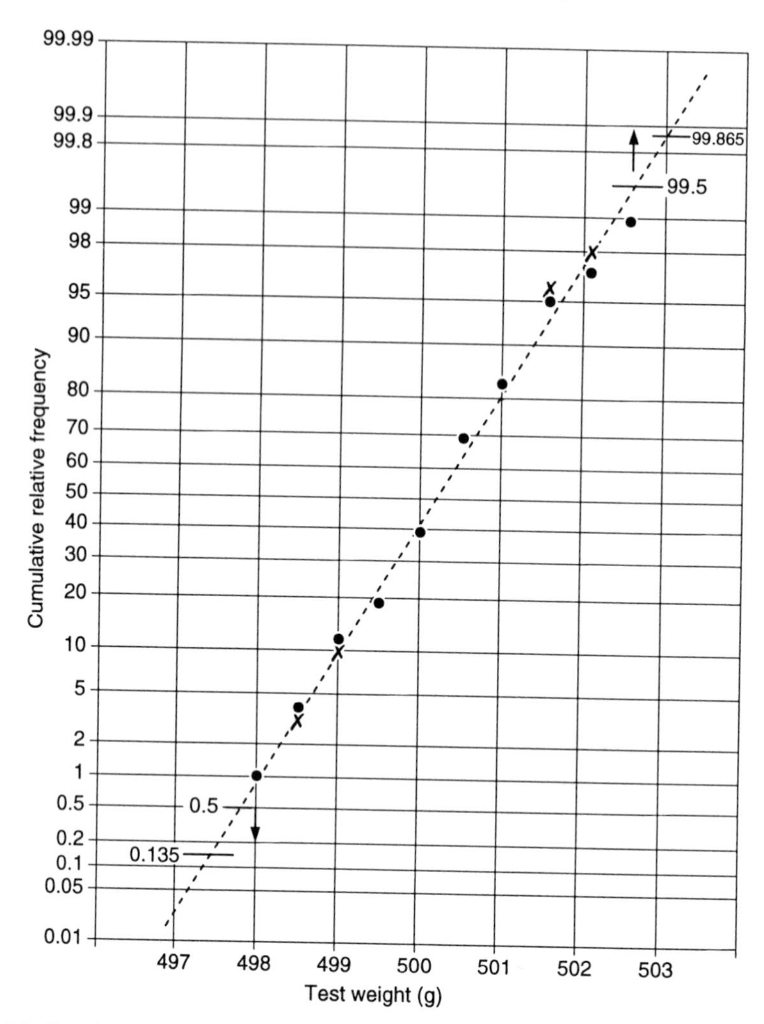

Fig. 15.5 Graphical representation of precision (attribute gauges or checkweighers) ↑ ↓ × = weights and measures % values; ● = automotive % values. Between 15 and 85% the two methods are almost indistinguishable. See section 15.7.4 for details.

reference manual, generally using rather smaller numbers of passes, uses

$$\left.\begin{array}{ll} (i+\frac{1}{2})/n & i<0.5n \\ i/n & i=0.5n \\ (i-\frac{1}{2})/n & i>0.5n \end{array}\right\} \qquad (15.1)$$

These latter are shown under (b) in Table 15.7. For the 'automotive' method, only the least extreme of the 'all pass' or 'all fail' data is used, so that in Table 15.7 the values for 497.5 and 503 are not plotted, but those for 498 and 502.5 are used after adjustment by (15.1). As there are no 0% or 100% points on the Normal probability scale, the 498 and 502.5 data cannot be used under the 'weights and measures' procedure.

In a manner similar to that in graphical capability analysis, the precision of measurement can be evaluated from the plot. For repeatability using the automotive procedure, this is the distance between the 0.5% and 99.5% points, and provides a graphical estimate of $5.15\sigma_r$. The weights and measures 'zone of indecision' covers the $\pm 3\sigma_r$ range from 0.135 to 99.865%. For the data of Table 15.7 and Figure 15.3, we have:

0.135% point	497.4
0.5%	497.8
50%	500.2
99.5%	502.5
99.865%	502.9

Repeatability (99% or $5.15\hat{\sigma}_r$) = 502.5 − 497.8 = 4.7 g
Zone of indecision (99.7% or $6\hat{\sigma}_r$) = 502.9 − 497.4 = 5.5 g
Bias = 50% point − set point = 500.2 − 500 = +0.2 g.

15.8 Outliers and the use of control charts

15.8.1 As in any experiment or data collection activity, occasional untypical values may be observed. Because differing items and operators are involved, the one-sample tests for outliers of 6.7 are not appropriate. Where a measurement study uses the ranges or standard deviations of numerous 'subgroups', i.e. the items × operators cells of the study, the control chart can usefully be adopted as a means of assessing the homogeneity of the level of repeatability.

The usual control limits, using the factors of Table 8.1 Part 1, are set up based on overall average range, $\bar{R}$. Where the pooled standard deviation, rather than $\bar{s}$, is used to determine $\hat{\sigma}_p$, the factors listed in Table 15.8 (rather than those in Table 8.1 Parts 3/4) are required.

Table 15.8 Control limits for s (based on $\sqrt{\text{Av}(s^2)}$ as overall estimate)

Lower and upper 0.1% and 2.5% points

Sample size, n	LAL	LWL	UWL	UAL
2	O	0.031	2.241	3.291
3	0.032	0.159	1.921	2.628
4	0.090	0.268	1.765	2.329
5	0.151	0.348	1.669	2.149

As well as checking for values outside control limits, one should look out for sequences of high or low R (or s) values associated with particular operators or items. The reasons for such patterns should be investigated, as the overall measure of repeatability, and even reproducibility, may be invalidated.

15.8.2 The data of the measurement study of Table 15.2 are used to illustrate the applications of both the R and s-charts. Because a substantial number of cells are involved, the risk of false indications of outliers is rather high when using the US control limits, and the UK action and warning procedure is used in this example.

From Table 15.2, we have for the range analysis:

$\bar{R} = 1.98$, so that with $m = 3$ values per cell,
LAL $= 1.98 \times 0.04 = 0.08$,
LWL $= 1.98 \times 0.18 = 0.36$,
UWL $= 1.98 \times 2.17 = 4.30$,
UAL $= 1.98 \times 2.98 = 5.90$.

As the data are only expressed to the nearest whole micron, LAL and LWL are not appropriate. Any range of 5 or more violates UWL, and any of 6 or more exceeds UAL.

Similarly for the s-chart,
$s_{\text{pooled}} = 1.2166$, so that
LAL $= 1.2166 \times 0.032 = 0.039$,
LWL $= 1.2166 \times 0.159 = 0.193$,
UWL $= 1.2166 \times 1.921 = 2.337$,
UAL $= 1.2166 \times 2.628 = 3.197$.

Again LAL and LWL are inappropriate, though as for R, if one operator tends to produce several zero ranges or standard deviations, one would suspect some irregularity in the procedure.

15.8.3 Inspection of the control charts in Fig. 15.6 shows no apparent anomalies. No ranges or standard deviations exceed UAL, but there are two warning values for each of R and s – not surprising in a sequence of 50 points. No obvious differences are seen between the operators. The fairly coarse resolution results in several strings of three identical R or s values, and indeed even one string of five values for $R = 2$ (here the s values vary between 1.0 and 1.15). As is also typical of R and s-chart, more values lie below the centre-line than above (though for R, 18 of the 50 are almost exactly *on* the centre-line).

15.9 Conclusion

The analysis of measurement precision is a very diverse topic, and we can only indicate some of the more widely used procedures. It will be noted that the

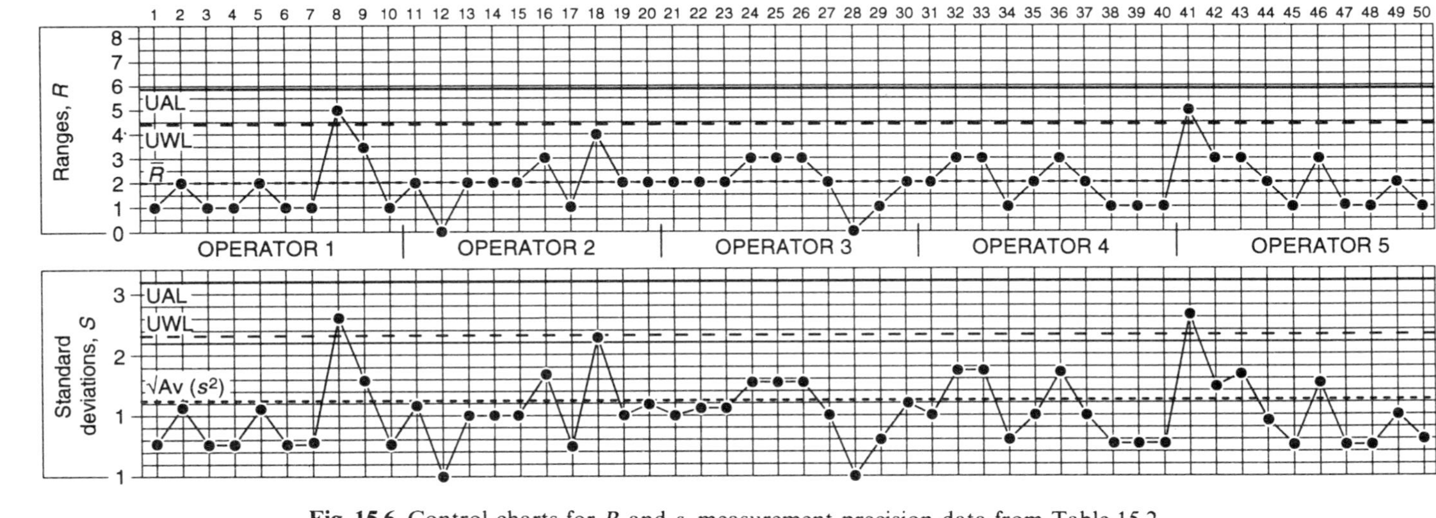

Fig. 15.6 Control charts for R and s, measurement precision data from Table 15.2.

analysis of variance is a particularly useful and versatile approach, providing the key to several different models that may be adopted to represent the measurement system. While the range, standard deviation and probability plotting methods can simplify some of the calculations, they do not offer such a powerful means of investigating interactions, isolating specific effects or combining components as does the ANOVA.

In the following Chapter, we shall present various methods for analysing control chart data which are closely related to those for measurement precision. Again, an understanding of ANOVA techniques, especially in conjunction with computer procedures, can permit more thorough and revealing analysis of this kind of data.

16

Getting more from control chart data

16.1 Introduction

16.1.1 The control chart, in various forms, has been introduced in Chapters 7–12, and Chapters 13–15 have also had connections with the control chart or data obtained from it. This Chapter is especially connected with charts of the $\bar{x}$, R; $\bar{x}$, s type or their cusum counterparts, though some of the methods are applicable to charts for individual values or moving averages, and others to charts for attributes such as the np, p, c and u-types.

The main purpose of the control chart is, of course, to detect changes (whether adverse or advantageous) as soon as possible 'on-line', so that appropriate action can be taken especially when correction or investigation is required. When control charts are completed, apart from calculating capability indices they are often simply filed away, although one of the important SPC principles is that such data provides valuable information for correction, improvement, re-design, experimentation and generally developing an understanding of the process.

In this Chapter we present some simple, rapid and readily programmable methods for reviewing control charts and extracting useful summaries of their information content.

16.1.2 Some of the questions one may wish to ask are listed below. The techniques of section 16.2 onwards are useful for answering them via a completed control chart. They are thus largely retrospective, but can often detect features and process instabilities that the conventional rules for chart interpretation may not detect in day-to-day operations.

(i) Is the process completely stable, i.e. strictly in-control?
(ii) If not, can features be identified that cause or contribute to the instability?
(iii) Are these features inherent in the process (system) or are they special causes?
(iv) If they are inherent, can their contributions to overall variation be measured or monitored?
(v) What is the overall variation in the system, and how can it be used to measure the true process capability?

(vi) Is the sample size/sampling frequency regime the best for the system, or could a change of format provide greater sensitivity, improved efficiency or lower costs?

16.1.3 The first requirement is a measure of the within-sample element of variation. Subject to considerations in earlier chapters (e.g. the R or s-chart in-control, no within-sample special causes arising from structured samples, etc.), the methods of Chapters 8 and (especially) 13 are useful. We have three possibilities:

(i) from a range chart using

$$s_0 = \bar{R} \div d_n \text{ (or } d_2 \text{ in USA notation)} \qquad (16.1)$$

This is the usual method for calculating $\hat{\sigma}$ for capability estimation, but here we use s_0 to indicate the lowest level of variation in the system (but see section 16.1.4).

(ii) from a standard deviation chart using

$$s_0 = \bar{s} \div c_n \text{ (or } c_2 \text{ in USA notation)} \qquad (16.2)$$

The c_n factors, like those for d_n, were introduced in Table 13.1, but are repeated in Table 16.1 along with further useful constants.

Table 16.1 Conversion and discount factors for s_0

	$\bar{R}$ method			$\bar{s}$ method		
Sample size, n	Conversion factor, d_n	Discount factor	C_R $(=\sigma_R/\mu_R)$	Conversion factor, d_n	Discount factor	C_s $(=\sigma_s/\mu_s)$
2	1.128	0.876	0.756	0.7979	0.876	0.756
3	1.693	0.907	0.525	0.8862	0.915	0.523
4	2.059	0.913	0.427	0.9213	0.936	0.422
5	2.326	0.906	0.372	0.9400	0.949	0.363
6	2.534	0.893	0.335	0.9515	0.957	0.323
8	2.847	0.862	0.288	0.9650	0.968	0.272
10	3.078	0.828	0.259	0.9727	0.975	0.239
12	3.258	0.796	0.239	0.9776	0.979	0.215
15	Not recommended			0.9823	0.983	0.191
20				0.9869	0.987	0.163
25				0.9896	0.990	0.145

For k samples of n items each, $k \geqslant 20$, $s_0 = \bar{R} \div d_n$ or $\bar{s} \div c_n$.
Effective degrees of freedom for $s_0 = k(n-1) \times$ discount factor.

Check for stability: test s_R or s_s against $\bar{R} \times \dfrac{\sigma_R}{\mu_R}$ or $\bar{s} \times \dfrac{\sigma_s}{\mu_s}$.

Methods (i) and (ii) assume that at least 20 values of R or s are available, and also that individual values are, at least approximately, Normally distributed. The third method, as introduced in section 13.4.5, does not depend on these assumptions, needs no table of factors, is fully efficient and can be used whenever all of the individual values or the sample standard deviations are available. This is, for k samples,

(iii) $$s_0 = \sqrt{\left(\frac{\sum s^2}{k}\right)} \tag{16.3}$$

Examples of some of these calculations were provided in section 13.6.4, using data from section 8.6 (Fig. 8.1, Table 8.2). To recapitulate, for $n = 5$

$$s_0 = \bar{R} \div d_n = 2.421 \div 2.326 = 1.041$$

or

$$\bar{s} \div c_n = 0.979 \div 0.940 = 1.041,$$

or

$$\sqrt{\left(\frac{\sum s^2}{k}\right)} = \sqrt{\left(\frac{26.7195}{24}\right)} = 1.055$$

In each case, an out-of-control range or standard deviation (from sample 23 containing a grossly anomalous value) has been omitted.

Before proceeding further, for example to obtain a confidence interval for σ_0 based on s_0, or to test whether it has changed from an earlier value, we shall require the new material of section 16.2. We need to note one further point before proceeding.

16.1.4 It is indicated in section 16.1.3 that s_0 is the lowest level of variation in the **system** or **process**. It may not necessarily be free of measurement error, and in deducing what is the true process variation one may wish (or need) to remove the contaminating effect of repeatability and/or reproducibility. Suppose that in the above example, a measurement study has provided an estimate $\hat{\sigma}_R = 0.53$, and one measurement is made on any one item in routine operations, then the true within-sample between-item variation could be estimated as

$$\sqrt{(s_0^2 - \hat{\sigma}_R^2)} = \sqrt{(1.055^2 - 0.53^2)} = 0.912$$

The actual item-to-item variation is thus over 10% smaller than the apparent variation due to the influence of measurement error.

Although it may not be valid to state the smaller value for contractual or capability purposes, it should be remembered that measurement error, rounding and other factors may inflate the apparent base-level variation in the system.

16.2 Efficiency and degrees of freedom

16.2.1 If any further analysis is to be undertaken, e.g. estimation of confidence intervals for mean or variation, tests of stability, etc., it will be necessary to know the number of degrees of freedom on which s_0 is based. The various methods of estimation use the data in different ways, and some extract more of the available information than others, i.e. they are more efficient. Thus for example, when using ranges of subgroups rather than standard deviations, the $n-2$ values lying between x_{max} and x_{min} are not explicitly used. Of course, *some* of their information is used in the sense that none of these $n-2$ items lies above x_{max} nor below x_{min}, but ranges are rather less efficient than standard deviations.

We have looked (in (16.2) and (16.3)) at two ways of using standard deviations. Even here there are differences in efficiency, with the 'average variance' method of (16.3) using slightly more of the information content than the $\bar{s}$ method of (16.2).

If one is absolutely sure that a process is completely stable (in control), the most efficient method of all is to pool the $k \times n$ items (k subgroups of n items each) into one set and to calculate the overall standard deviation. There will then be $kn-1$ associated degrees of freedom. Unfortunately, this cannot be done until *after* the stability tests described later have established stability, and we must therefore consider how many effective degrees of freedom are available from the $k \times n$ sample structure, bearing in mind the method of calculation used.

16.2.2 When method (16.3) is used to obtain s_0, there are $n-1$ degrees of freedom contributed by each of the k subgroups, so that $v = k(n-1)$ for this 'average variance' method.

The simplest way of evaluating n when (16.1) or (16.2) is used to calculate s_0 is to discount $k(n-1)$ using the factors of Table 16.1. It will be noted that $\bar{s}$ provides more degrees of freedom than $\bar{R}$ except for $n = 2$, when the two methods are equivalent.

Returning to the data from Chapter 8 used in section 16.1.3, with $k = 24$, $n = 5$ the degrees of freedom are as follows:

$\bar{R}$ method: $v = 24 \times 4 \times 0.906 \doteqdot 87$;
$\bar{s}$ method: $v = 24 \times 4 \times 0.949 \doteqdot 91$;
s^2 method: $v = 24 \times 4 = 96$.

16.2.3 A simple check for the stability (in-control state) of the R or s-chart may also be carried out. The standard deviation of the R or s values is divided by their mean ($\bar{R}$ or $\bar{s}$) and this coefficient of variation is compared with the theoretical value of C_R or C_s in Table 16.1. Table A.2 gives % points for $\sqrt{(\chi^2/v)}$, which may be used to test observed/theoretical C_R or C_S for significance, using the entry for $v = k-1$ degrees of freedom.

Thus from Table 8.2, in addition to $\bar{R}$ or $\bar{s}$ we may also calculate $s_R = 1.0875$ or $s_s = 0.4020$. Then,

$$\frac{s_R}{\bar{R}} \div C_R = \frac{1.0875}{2.421} \div 0.372 = 1.208$$

or

$$\frac{s_s}{\bar{s}} \div C_s = \frac{0.402}{0.979} \div 0.363 = 1.131$$

With $k = 24$ (omitting the anomalous sample 23) we have $\nu = 23$. There is no entry for this number of degrees of freedom in Table A.2, but by comparison with the entries for $\nu = 24$ and 19, the observed ratio for $s_R/\bar{R}$ lies between the upper 10 and 5% points, and the ratio for $s_s/\bar{s}$ lies inside the upper 10% point. Whilst there is some possibility that the variation is slightly unstable, the indication is not statistically significant.

16.2.4 Provided that the R (or s) chart is in-control, confidence intervals may now be estimated for the process within-sample standard deviation, σ_0.

A simple method is to specify a dummy overall sample size equal to $\nu + 1$, and then to use the appropriate multipliers from Table 5.7. Continuing the example of the preceding section:

$\bar{R}$ *method*: $s_0 = 1.041$; $\nu = 87$ so dummy sample size $= 88$.
Interpolating in Table 5.7 for 95% confidence, the multipliers are approximately 0.875, 1.17, giving

$$0.91 \leqslant \sigma_0 \leqslant 1.22$$

$\bar{s}$ *method*: $s_0 = 1.041$; $\nu = 91$, and for a dummy sample size of 92 the multipliers and confidence limits will be as for the $\bar{R}$ method.
s^2 *method*: Here with $\nu = 96$ and a dummy sample size of 97, the factors for $n = 100$ in Table 5.7 are valid, giving 1.055×0.88 and 1.055×1.16 as the confidence limits, so that $0.93 \leqslant \sigma_0 \leqslant 1.22$ with 95% confidence, a slightly narrow interval than for $\bar{R}$ and $\bar{s}$ because of the more efficient use of degrees of freedom.

16.3 Between-sample variation

16.3.1 For an in-control process the $\bar{x}$-values plotted on a control chart represent the means of random samples drawn from a stable population. As such, the variation in these x-values can be described by their standard error (see section 5.2 and Table 5.4). In a stable system the standard error will be $\sigma_0/\sqrt{n}$ where σ_0 is the true within-sample variation of which s_0 is an estimate.

Thus if we calculate $s_{\bar{x}}$, the actual observed variation of the $\bar{x}$-values, and compare it with $s_0/\sqrt{n}$, we have a simple test for stability that is at the same time more powerful than the combined control chart rules. Of course, it can only be used retrospectively, but it can reveal subtle instabilities including 'random' between-sample variation amounting to longer-term common causes.

To calculate $s_{\bar{x}}$, simply enter the $\bar{x}$-values through the standard deviation routine of a calculator or computer. The test involves the $\sqrt{F}$-ratio (derived from the analysis of variance briefly noted in section 15.5), here simplified for the specific application of control chart review. The ratio is

$$\sqrt{F} = \frac{s_{\bar{x}}}{(s_0 \div \sqrt{n})}$$

which may be written more simply as

$$\sqrt{F} = \frac{s_{\bar{x}}\sqrt{n}}{s_0} \tag{16.4}$$

In order to assess whether $\sqrt{F}$ differs significantly from its presumed value of 1.0, we require the degrees of freedom appropriate to both $s_{\bar{x}}$ and s_0. Those for $s_{\bar{x}}$, which we denoted by ν_1, are simply $k-1$, and those for s_0 (denoted by ν_2) are, as in section 16.2, obtained as $k(n-1) \times$ discount factor.

The level of significance for $\sqrt{F}$ may now be evaluated from Table A.3, using the block appropriate to ν_1 and the line within that block for ν_2.

16.3.2 For the data of Table 8.2, again omitting sample 23 containing a rogue value, we have $s_{\bar{x}} = 0.646$ with $k = 24$.

Then depending on whether $\bar{R}$, $\bar{s}$ or s^2 is used to obtain s_0, we have:

$\bar{R}$ *method*: $\sqrt{F} = \dfrac{0.646\sqrt{5}}{1.041} = 1.388,\ \nu_1 = 23,\ \nu_2 = 87;$

$\bar{s}$ *method*: $\sqrt{F} = \dfrac{0.646\sqrt{5}}{1.041} = 1.388,\ \nu_1 = 23,\ \nu_2 = 91;$

s^2 *method*: $\sqrt{F} = \dfrac{0.646\sqrt{5}}{1.055} = 1.369,\ \nu_1 = 23,\ \nu_2 = 96.$

The nearest entry in Table A.3 is for $\nu_1 = 24$, $\nu_2 = 99$, and for these degrees of freedom the significance probability for $\sqrt{F}$ is between 2.5% and 1%. There is a fairly strong indication of instability, but inspection of the control charts in Fig. 8.1 shows one $\bar{x}$ value above UAL. Perhaps this is the sole cause of the significant value of $\sqrt{F}$? If sample 14 is also omitted, the value of $s_{\bar{x}}$ becomes 0.5304, and those for $\sqrt{F}$ fall in the range 1.12–1.14 depending on the method used. With a significance probability between 25 and 10%, there is only slight evidence of further instability, and for this example one might not wish to proceed further with the analysis. We therefore introduce a new example via the data of Table 16.2 and Fig. 16.1.

16.3.3 The control chart for R in Fig. 16.1 shows good stability, but that for $\bar{x}$ is somewhat 'wobbly'. The only clear signal is a run of seven consecutive values above the target value for samples 24–30, but the last two are very close to 50 and no obvious action is called for. Nevertheless a pattern of alternating sequences above and below 50 looks possible.

Table 16.2 Control chart data

$k = 30$, $n = 5$

Subgroup	$\bar{x}$	R
1	49.74	1.7
2	50.20	0.9
3	50.60	0.8
4	50.02	2.5
5	50.32	1.9
6	50.62	2.8
7	50.40	1.1
8	49.88	1.2
9	50.20	1.4
10	50.10	2.2
11	49.76	3.0
12	49.92	1.8
13	49.20	1.5
14	49.56	2.0
15	50.24	0.7
16	49.32	1.5
17	49.24	1.1
18	49.20	0.6
19	49.98	2.0
20	50.34	2.4
21	49.80	1.0
22	50.76	2.1
23	49.90	1.3
24	50.04	1.4
25	50.70	1.6
26	50.28	2.0
27	50.68	1.2
28	50.38	0.8
29	50.02	3.3
30	50.04	1.4

$\bar{\bar{x}} = 50.048$; $\bar{R} = 1.64$; $s_0 = 1.64/2.326 = 0.705$
$s_{\bar{x}} = 0.44208$; $\bar{\delta}_{\bar{x}} = 0.42414$; $\bar{\delta}_{\bar{x}}/1.128 = 0.37588$
$\sqrt{(\frac{1}{2}MSSD)} = 0.35056$.

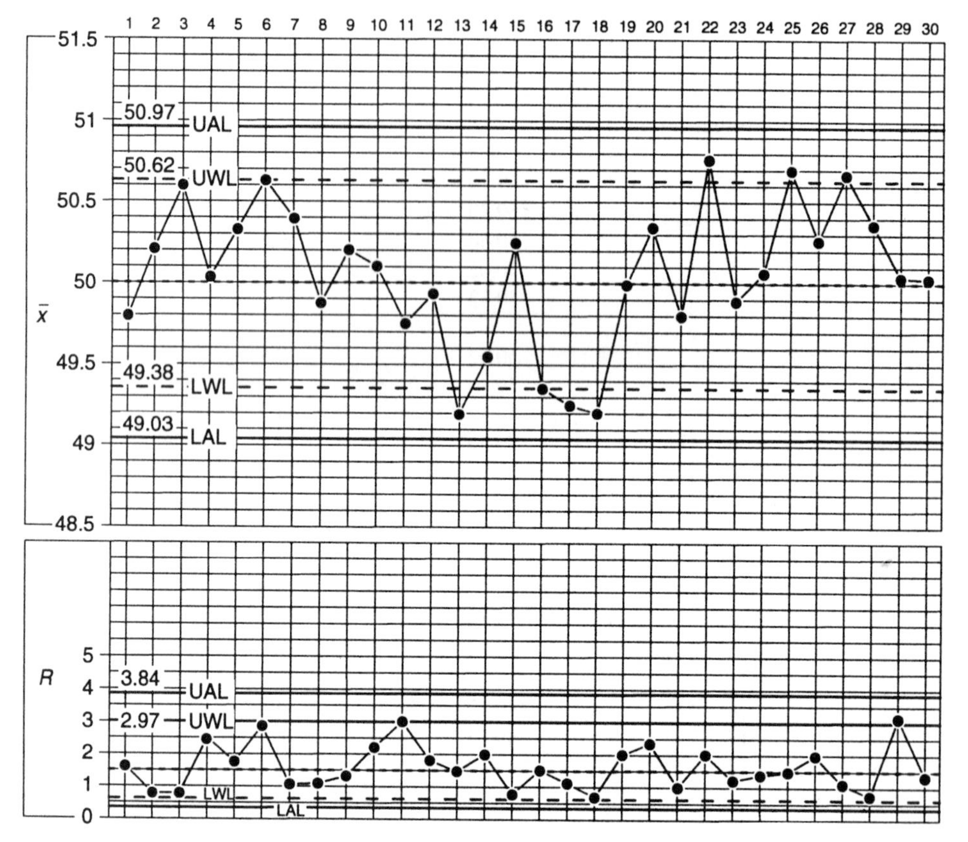

Fig. 16.1 Control chart for data of Table 16.2. ($T = 50.0$; UK control limits). For R: LAL $= 0.10$, LWL $= 0.61$.

From the data at the foot of Table 16.2, we have

$$s_0 = 0.705, \text{ with } \nu_2 = 30 \times 4 \times 0.906$$

(the latter being the discount factor for the $\bar{R}$ method with $n = 5$). Then with $s_{\bar{x}} = 0.442\,08$, $\nu_1 = 29$,

$$\sqrt{F} = \frac{0.44208\sqrt{5}}{0.705} = 1.402, \; \nu_1 = 29, \; \nu_2 = 109$$

From Table A.3, the significance probability lies between 1% and 0.5%, a strong indication of instability which is not due simply to the influence of one or two unusual values. In section 16.4 we shall look at some further diagnostic methods that may yield further clues to the nature of the instability.

16.3.4 It may occasionally be found that $\sqrt{F}$ is appreciably *less* than 1.0. In this case the reciprocal ratio is evaluated in Table A.3, with the degrees of freedom (v_1, v_2) interchanged, viz:

$$\sqrt{F'} = \frac{s_0}{s_{\bar{x}}\sqrt{n}}, \; v_1 = k(n-1) \times \text{discount}, \; v_2 = k-1$$

Where $\sqrt{F'}$ is significantly large, some possible explanations may include:

(i) there is a within-sample special cause operating. This topic is covered in section 11.4, and is the most frequently encountered cause of a significantly small $\sqrt{F}$ (or large $\sqrt{F'}$) ratio.
(ii) some control device or system lag exists which inhibits some of the random variation in the $\bar{x}$ values. Careful investigation is then needed to identify the cause, and conventional control methods may not be suitable for such applications.
(iii) The data may have been edited, or the $\bar{x}$ values rounded always towards the central or preferred value. Obviously gross falsification can also produce this effect, when values in the outer thirds of the chart are deliberately adjusted to give a spurious impression of good control.

All three phenomena may yield an excess of plotted points within the middle third of the control chart, and a corresponding shortage of values in the outer third. On UK control charts, there may be an absence of warning values, whereas for a stable process occasional excursions into the warning region should occur.

16.4 Tests and diagnostics for non-randomness: v^2 and v'

16.4.1 Once the stability of the process has been examined, and especially if instability is indicated, it may be useful to establish whether sample-to-sample variation is of a random nature or whether some non-random pattern can be diagnosed. If there are such patterns, it may be possible to eliminate them or reduce their effects. If purely random variations between samples are greater than would be expected from the within-sample variability, it would appear that there are common causes in the system that are not measured by s_0.

For random between-sample variation it will be necessary to measure the excess (over and above s_0) and allow for it when setting control limits and when evaluating process capability. Section 11.3 covered this situation, with the $\bar{x}$, R and δ chart providing one means of dealing with both control and capability measurement.

In addition to control charts, techniques like cusum analysis will often help to identify between-sample patterns. They will be illustrated later, but firstly we consider a simple numerical method that provides a ‘portmanteau’ test with useful diagnostic properties.

16.4.2 The basis of the test, originally developed in a slightly different form by von Neumann, and thus often referred to as von Neumann's ratio, is to compare the conventionally estimated standard deviation of a series of values with the estimate based on squared successive differences. These estimators have already been introduced in Chapter 9 (section 9.1) and (section 9.2), and also in conjunction with $\bar{x}$, R and δ in section 11.3.

The test for randomness can be applied to individual values in a series, and can therefore be useful in reviewing charts for individual values. Here we are concerned with its use with the $\bar{x}$-values from an $\bar{x}$, R or $\bar{x}$, s-chart, and the estimates we require are:

(i) the actual standard error of the $\bar{x}$-values, $s_{\bar{x}}$, obtained as in section 16.3.1 by applying the s-function of a calculator to the series of sample means.

(ii) Preferably, the estimate corresponding to $\sqrt{(\frac{1}{2}MSSD)}$ of (9.2), but using $\bar{x}$-values in place of individuals. For convenience, we shall identify this as

$$\sqrt{(\tfrac{1}{2}MSSD)_{\bar{x}}} = \sqrt{\left[\frac{1}{(2(k-1)}\sum_{j=1}^{k-1}(\bar{x}_j - \bar{x}_{j+1})^2\right]} \qquad (16.5)$$

(iii) an alternative (though less efficient) to (ii) is the $\bar{x}$ analogue of (9.1) using absolute successive differences (moving ranges of two) between sample means. We identify this as

$$\frac{\bar{\delta}_{\bar{x}}}{1.128} = \frac{1}{1.128(k-1)}\sum_{j=1}^{k-1}|\bar{x}_j - \bar{x}_{j+1}| \qquad (16.6)$$

or the useful approximate alternative $8\bar{\delta}_{\bar{x}}/9$

One would not normally calculate both (16.5) and (16.6), and (16.5) has the advantage of both 10% greater efficiency in its use of data and applicability to counted (attribute and event) data as well as to measurements.

The test for non-randomness compares either MSSD/2 with $s_{\bar{x}}^2$ (note the use of variances in this case) or $\bar{\delta}_{\bar{x}}/1.128$ with $s_{\bar{x}}$. We therefore define the ratios

$$v^2 = \tfrac{1}{2}MSSD \div s_{\bar{x}}^2 \qquad (16.7)$$

and

$$v' = \bar{\delta}_{\bar{x}} \div 1.128 s_{\bar{x}}$$

or

$$\tfrac{8}{9}\bar{\delta}_{\bar{x}} \div s_{\bar{x}} \qquad (16.8)$$

16.4.3 For a random series of observations, both (16.7) and (16.8) will lie close to 1.0. Under repeated sampling from the same system of (approximately) normally distributed values, v^2 and v' are themselves approximately normally distributed with simple means and standard errors:

$$v^2\text{: Mean} = 1.0,\ \sigma(v^2) \doteqdot \sqrt{\left(\frac{1}{k+2}\right)}, \qquad k \geqslant 10$$

$$v': \text{Mean} \doteqdot \frac{3k+1}{3k}\left(=1+\frac{1}{3k}\right)$$

$$\sigma(v') \doteqdot \sqrt{\left(\frac{1}{3(k+1)}\right)} \qquad k \geqslant 10$$

The alternative expression for the mean of v' emphasizes how rapidly it tends towards 1.0; even for $k = 20$, the mean is approximately $1.01\dot{6}$.

These simple properties of v^2 and v' mean that an observed value of either ratio can be assessed for statistical significance via the usual standard Normal variable. Alternatively a pair of two-standard error limits will provide a guide to the typical range of the ratios and highlight, by exception, any indications of non-randomness. Table 16.3 summarizes these points.

16.4.4 As well as signalling any significant non-randomness, v^2 or v' can also provide clues as to its nature:

(i) A value of v^2 and v' significantly greater than 1.0 suggests an excessively regular (as opposed to random) alternation of values above and below the overall average. Over-control, instrumental hunting, or a cycle whose half-duration approximately coincides with the sampling interval, are possible causes. A phenomenon known as negative autocorrelation is

Table 16.3 Summary of v^2 and v' tests for non-randomness

	Statistic used	
	v^2 (preferred)	v'
Definition	$\frac{\frac{1}{2}MSSD}{s_{\bar{x}}^2}$	$\frac{\frac{8}{9}\bar{\delta}_{\bar{x}}}{s_{\bar{x}}}$ or $\frac{\bar{\delta}_{\bar{x}}}{1.128s_{\bar{x}}}$
Mean value	1.0	$1.0+\frac{1}{3k}$
Standard error	$\sqrt{\left(\frac{1}{k+2}\right)}$	$\sqrt{\left(\frac{1}{3(k+1)}\right)}$
z-test*	$(v^2-1)\sqrt{(k+2)}$	$\left(v'-\frac{3k+1}{3k}\right)\sqrt{[3(k+1)]}$
95% zone	$1 \pm 2\sqrt{\left(\frac{1}{k+2}\right)}$	$\frac{3k+1}{3k} \pm 2\sqrt{\left(\frac{1}{3(k+1)}\right)}$

*For the z-test, find P_z from a table of the Normal distribution to obtain the significance probability for the departure of v^2 or v' from its expected value.

another possibility. As an example, if usage of a material is estimated by difference between a current and a previous stock estimate (perhaps obtained by reading a gauge or dipstick), then any overestimate on one occasion is likely to be followed by an underestimate on the next.

(ii) A value of v^2 or v' significantly smaller than 1.0 suggests features like a trend in the series of data points, a slow cyclic pattern where several samples are taken in each cycle, or a process whose mean shifts periodically – perhaps when different shifts take over a process, or successive batches or deliveries of raw material may differ slightly from each other. Environmental conditions, process interruptions and many of the between-sample common causes mentioned in 11.3 may give these effects.

Figure 16.2 illustrates some of the features for which v^2 or v' may give useful indications. Unfortunately, mixtures of causes can be largely self-cancelling, and over-control combined, for instance, with occasional shifts in average, may give ratios close to 1.0. Careful examination of the plotted points is the only way in which such disturbances can be detected.

16.4.5 We again consider the two examples introduced in section 16.3.2. For the control chart data of Fig. 8.1 and Table 8.2, the following values are obtained:

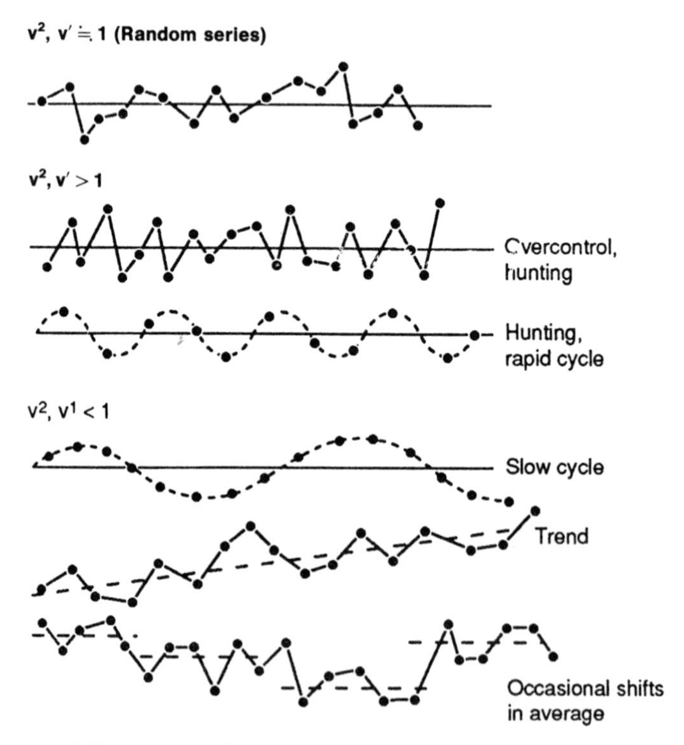

Fig. 16.2 Patterns of non-randomness and von Neumann ratios.

(a) All 25 $\bar{x}$-values included; $s_{\bar{x}} = 0.812\,73$.

$\sqrt{(\frac{1}{2}MSSD)} = 0.909\,42$, giving $v^2 = 1.252\,1$ ($u = 1.31$),

$\bar{\delta}/1.128 = 0.871\,75$, giving $v' = 1.072\,6$ ($u = 0.52$).

v^2 gives some indication of oscillation, but is not significant at the 5% level. v' gives little indication of non-randomness.

(b) Sample 23 omitted; $s_{\bar{x}} = 0.646\,00$.

$\sqrt{(\frac{1}{2}MSSD)} = 0.7392$, giving $v^2 = 1.309\,4$ ($u = 1.58$),

$\bar{\delta}/1.128 = 0.707\,74$, giving $v' = 1.095\,6$ ($u = 0.71$).

Possibly significant ($P = 0.05$) indication of oscillation from v^2, but little indication from v'.

(c) Omitting sample 23 and 14; $s_{\bar{x}} = 0.530\,40$.

$\sqrt{(\frac{1}{2}MSSD)} = 0.557\,68$, with $v^2 = 1.105\,5$ ($u = 0.53$),

$\bar{\delta}/1.128 = 0.590\,50$, with $v' = 1.113\,3$ ($u = 0.84$).

There is now little indication of non-randomness from either v^2 or v' (though interestingly v' is now marginally larger than previously). It thus appears that the slight appearance of over-control or hunting was almost entirely due to the two extreme $\bar{x}$-values.

Note that one would not generally calculate both v^2 and v' for the same data, and v^2 offers a better diagnostic resulting from its greater efficiency. It is also arguable that a test involving successive differences should not be carried out after the series has been broken by deletion of extreme observations, and this example should be regarded as purely illustrative.

Turning now to the data of Table 16.2, the values of $s_{\bar{x}}$, $\sqrt{(\frac{1}{2}MSSD)}$ and $\bar{\delta}/1.128$ listed at the foot of the table yield:

$$v^2 = 0.6288 \quad (u = -2.10),$$

$$v' = 0.8503 \quad (u = -1.55).$$

Again v^2 gives a clearer signal than v', and in this case v^2 differs from 1.0 at the 5% (two-tail) significance level. With the ratio less than 1.0, the diagnostic points to trend, a cyclic pattern or changes in average. There is no evidence of a trend in Fig. 16.1, but either a cycle or possibly three different averages are possible: say about 50.2 until sample 10, then about 49.7 or 49.8 from 11 to 20, and back to 50.3 from 21 onwards. Further examination, as in Fig. 16.3 using a five-point moving average or a cusum, indicates a cycle as more likely, and this was confirmed by further observation.

16.5 Components of variation

16.5.1 The notion of separate contributions to the overall variation in a system has already been introduced at several points in this book. Chapter 11.3 considered between-sample and within-sample sources, and (11.2) indicated how

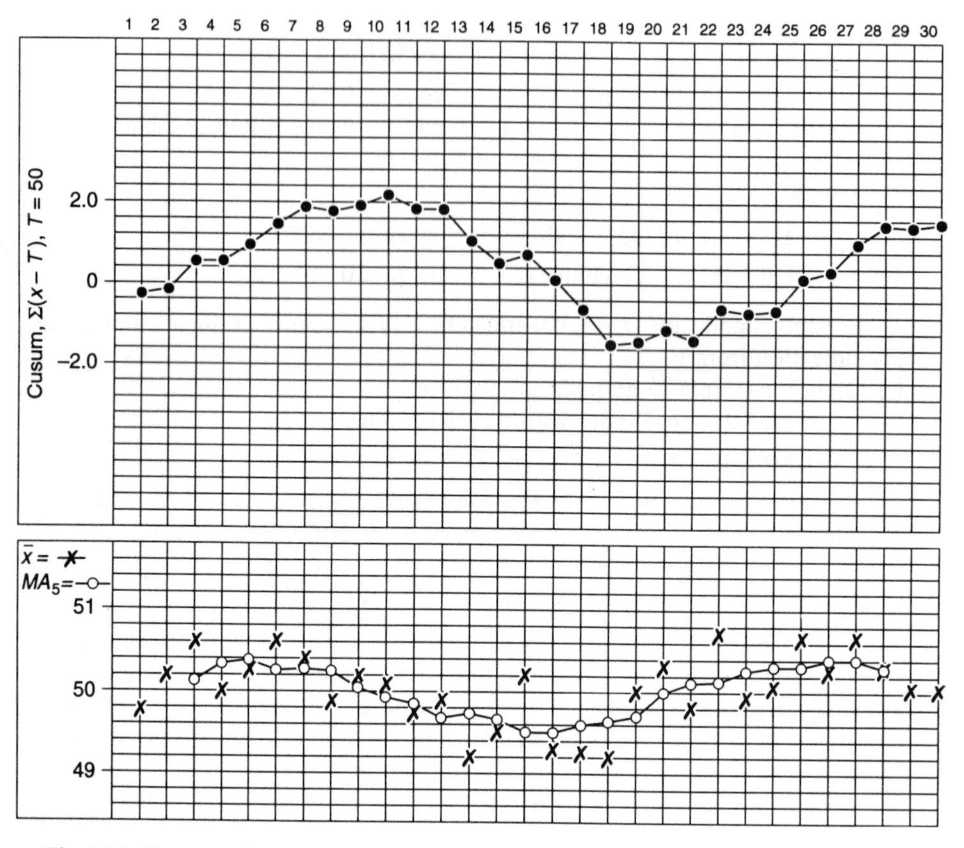

Fig. 16.3 Cusum and moving average (MA_5) charts of control chart data from Table 16.2.

both contributions could be combined to estimate the total variation of the process. Chapter 15 dealt with measurement precision, and again two sources or contributions were identified – repeatability and reproducibility – and others were hinted at (resolution, linearity, etc.).

These examples are only special cases of a more general area of statistical analysis. The definitive technique for separating and estimating the components is (as noted in section 15.5) the analysis of variance. Important and useful though it is, it lies beyond the scope of the present book, and we consider simple ways of estimating some components of variation using the statistics s_0, $s_{\bar{x}}$ and $\sqrt{(\frac{1}{2}MSSD_{\bar{x}})}$ already developed in this Chapter.

16.5.2 Let us suppose that the F-test, $ns_{\bar{x}}^2/s_0^2$, or its $\sqrt{F}$ equivalent $s_{\bar{x}}\sqrt{n}/s_0$, has indicated significant instability. If this is not the case, there is no point in proceeding with the estimation of components of variation other than s_0. The v^2 (or v') test may either indicate systematic patterns in the instability, or may suggest (where v^2 or v' are close to 1.0) that the instability is of a random nature. The magnitude of v^2, i.e. >1, $\doteq 1$ or <1, will determine the path to be followed through the flow chart of Fig. 16.4. We identify three possible contributions to the overall variation in the process, in addition to the within-sample element already considered, s_0:

- s_1 an additional component of variation arising from **random** between-sample common causes;
- s_2 a measure of **systematic** fluctuations in mean level from trends, cycles, etc. These may be either special or common causes: for example, a thermostatic cycle is part of the unavoidable system variation and would generally need to be allowed for as a common cause:
- s_3 a contribution from hunting or over-control: generally to be considered as a special cause requiring removal or investigation.

In a more complete analysis other recognisable sources could include differences between machines, operators/shifts, batches of raw material, and so on; here we limit our attention to features that are likely to occur in the routine analysis of control chart data. Because of this restriction, it is also necessary to note that s_2 and s_3, if *both* occur, may mask each other and therefore the method of analysis in this chapter may fail to detect either component.

16.5.3 Some questions that may now be addressed include:

(i) What is the overall process variation, say s_t?
(ii) What relative amounts of variation are due to the various components? and hence what are the priorities for improvement action?
(iii) If some components (e.g. the special causes) could be eliminated, what variation would remain?
(iv) Where components such as s_1 cannot be eliminated, can they be monitored?
(v) What is the correct measure to use for control limits, especially for $\bar{x}$?
(vi) How can the true process capability be measured?

In this last respect, one must caution that some features like warm-up trends, tool wear, reagent degradation, cycles, etc., may yield overall distributions (of individual values) with broader humps and shorter tails (platykurtic) than the Normal distribution. In such cases capability indices based on assuming Normality may be misleading, and are likely to be pessimistic.

16.6 Estimating the components

16.6.1 As already noted, the analysis of variance is the definitive technique for measuring components of variation. For hunting, cyclic and other serially

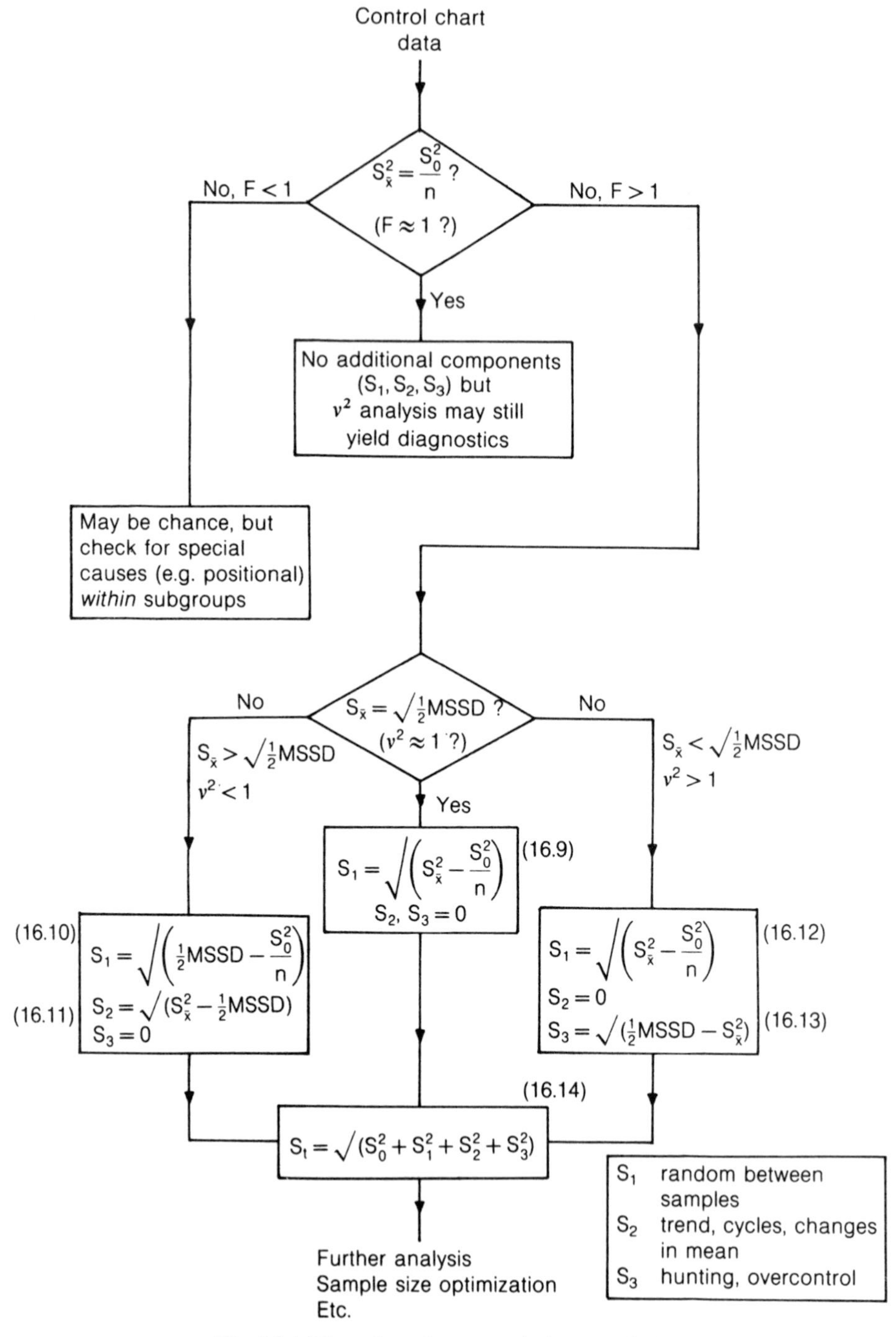

Fig. 16.4 Flow-chart for control chart analysis.

dependent situations, some time-series techniques can usefully extend the simpler analysis presented here.

Assuming we have k samples of n items each, and the estimates (to a reasonable number of significant digits, say four or more) listed below, the analysis follows the flow chart of Fig. 16.4. The required inputs are:

$$\bar{\bar{x}},\ s_0,\ s_{\bar{x}},\quad \sqrt{(\tfrac{1}{2}MSSD)}\ \text{(or } \tfrac{1}{2}MSSD \text{ before square-rooting)}$$

We note first that if $s_{\bar{x}}$ approximately equals $s_0/\sqrt{n}$, no further analysis is required, and s_1, s_2 and s_3 are assumed to be zero. A special case occurs if $s_{\bar{x}}$ is *less* than $s_0/\sqrt{n}$, i.e. the $\sqrt{F}$ of 16.4 is less than 1.0. Section 16.3.4 deals with this situation.

We move on to consider v^2 and its implications.

(i) If $s_{\bar{x}}$ exceeds $s_0/\sqrt{n}$ and $s_{\bar{x}}$ approximately equals $\sqrt{(\frac{1}{2}MSSD)}$ (i.e. $v^2 \doteq 1$), an additional common-cause element of between-sample variation is indicated. This component, s_1 is obtained as

$$s_1 = \sqrt{\left(s_{\bar{x}}^2 - \frac{s_0^2}{n}\right)} \tag{16.9}$$

Note that $s_{\bar{x}}^2$ is preferred to $\frac{1}{2}MSSD$ in this expression. Although the two estimates are roughly equal, $s_{\bar{x}}^2$ is based on $(k-1)$ degrees of freedom, and is therefore more efficient than $\frac{1}{2}MSSD$ which uses approximately $2(k-1)/3$. We note in passing that $\bar{\delta}/1.128$ uses only $3(k-1)/5$, and thus loses a further 10% of the degrees of freedom.

(ii) If $s_{\bar{x}}$ exceeds $s_0/\sqrt{n}$ and also exceeds $\sqrt{(\frac{1}{2}MSSD)}$, we have $v^2 < 1$ and an indication of both special (systematic) and random between-sample variation. The random element may now be estimated as

$$s_1 = \sqrt{\left(\tfrac{1}{2}MSSD - \frac{s_0^2}{n}\right)} \tag{16.10}$$

and the systematic element from trends, cycles or shifts in average as

$$s_2 = \sqrt{(s_{\bar{x}}^2 - \tfrac{1}{2}MSSD)} \tag{16.11}$$

The estimate s_2 is fairly crude, having very few degrees of freedom (which depend in turn on the relative magnitudes of s_0, s_1 and s_2). Nevertheless it provides some indication of the extent of the systematic element of variation.

(iii) if $s_{\bar{x}}$ exceeds $s_0/\sqrt{n}$ but is smaller than $\sqrt{(\frac{1}{2}MSSD)}$, we have $v^2 > 1$ and an indication of both hunting (or over-control) and random between-sample variation. The random element is now estimated as

$$s_1 = \sqrt{\left(s_{\bar{x}}^2 - \frac{s_0^2}{n}\right)} \tag{16.12}$$

and the hunting or over-control contribution as

$$s_3 = \sqrt{(\tfrac{1}{2}MSSD - s_{\bar{x}}^2)} \tag{16.13}$$

(iv) Any components not expressly indicated in (16.9) to (16.13) are set to zero. Thus,

in (i), s_2, s_3 are set at 0,
in (ii), s_3 is set at 0,
in (iii), s_2 is set at 0,

noting as earlier that the present method may fail to detect either s_2 or s_3 if both exist.

16.6.2 One must also observe that if the cause of s_2 or s_3 is identified and removed, other features may be revealed by later data. For example, suppose that

$s_0 = 2.5$ with $n = 5$,
$s_{\bar{x}} = 1.8$ with $k = 25$,
and $\sqrt{(\frac{1}{2}MSSD)} = 2.3$.

Then $v^2 = 1.633$, which is highly significant and indicates over-control. We estimate

$$s_1 = 1.41 \text{ and } s_3 = 1.43$$

using (16.12) and (16.13).

Now we also suppose that s_3 is found to be due to an insufficiently damped control device. This is rectified, and further data yields

$$s_0 = 2.3, \quad s_{\bar{x}} = 1.6, \quad \sqrt{(\tfrac{1}{2}MSSD)} = 1.05$$

(again with $n = 5$, $k = 25$).

We now have $v^2 = 0.431$, just significant at the 5% two-tail level, and with $\sqrt{F} = 1.56$ we suspect both a random and a systematic between-samples component. Using (16.10) and (16.11),

$$s_1 = 0.211, \quad s_2 = 1.21.$$

The random component is very small, but s_2 may represent cycles or other features that were previously masked by the hunting.

16.6.3 We now continue with the example of Table 16.2; in section 16.4.6 the value of v^2 gave a significant indication of systematic shifts in average, diagnosed as a cyclic pattern using Fig. 16.3. First, we recapitulate the statistics already derived:

$s_0 = 0.705$, $s_{\bar{x}} = 0.44208$, $\bar{\delta}/1.128 = 0.350\,56$;
$\sqrt{F} = s_{\bar{x}}\sqrt{n}/s_0 = 1.402$, $\nu_1 = 29$, $\nu_2 = 109$;
$v^2 = 0.6288$ $(u = -2.10)$.

The procedure, with $v^2 < 1.0$, uses (16.10) and (16.11) to give

$$s_1 = \sqrt{\left(0.350\,56^2 - \frac{0.705^2}{5}\right)} = 0.153\,26$$

$$s_2 = \sqrt{(0.442\,08^2 - 0.350\,56^2)} = 0.269\,34$$

with s_3 presumed zero (as it cannot be estimated in the presence of s_2 using the present methods).

Thus most of the variation appears to be contributed by the within-samples component. Of the significant, though smaller, between-samples component, rather more is contributed by the cyclic element (measured by s_2) than by between-sample common causes (measured by s_1).

16.7 Using the components of variation

16.7.1 Having estimated the components, there are several ways in which they can be used. One of the more obvious is to synthesize the overall variation in the system, and here the additive property of variances is invoked. Where sources of variation act independently in a system, their overall effect is given by the sum of the separate variances. This principle was used in section 15.4 to obtain the combined repeatability and reproducibility (σ_R) from the two components σ_r and σ_L.

Denoting the overall process variation by s_t, we may synthesize the various components via

$$s_t = \sqrt{s_0^2 + s_1^2 + s_2^2 + s_3^2} \tag{16.14}$$

It might be argued that a simpler way would be simply to combine all the $n \times k$ data values and calculate their standard deviation. There are two reasons for preferring (16.14). One is that this expression may also be used, omitting particular components (or entering alternative values) to predict the likely process variation if changes were made to the system: for example the reduction of a tool-wear effect or the elimination of a cyclic pattern.

The second point is that the standard deviation obtained by pooling all the data values depends to a small extent on the sample size structure as well as on the components themselves. This effect is rarely serious unless σ_1 is very large compared with σ_0, but the overall **apparent** standard deviation is always slightly smaller than the correct value obtained by combining components. For readers familiar with ANOVA, the point can be readily illustrated via mean squares and their expectations. Table 16.4 illustrates the effect via a hypothetical case with $\sigma_1 = 12$, $\sigma_0 = 5$, $\sigma_t = 13$, using various sample configurations all with a total of 100 individual values.

Table 16.4 Estimation of overall standard deviation (Bias in σ_t when pooling all observations)

All configurations have $k \times n = 100$; $\sigma_1 = 12$, $\sigma_0 = 5$
'True' $\sigma_t = \sqrt{12^2 + 5^2} = 13$

	Sums of squares	Degrees of freedom	Mean squares	Expected mean squares
(a) $k = 100$, $n = 1$				
Between	16731	99	169	$\sigma_0^2 + \sigma_1^2$
Within	–	0	–	σ_0^2
Total	16731	99	(169)	
(b) $k = 50$, $n = 2$				
Between	15337	49	313	$\sigma_0^2 + 2\sigma_1^2$
Within	1250	50	25	σ_0^2
Total	16587	99	$(167.\dot{5}\dot{4})$	
(c) $k = 20$, $n = 5$				
Between	14155	19	745	$\sigma_0^2 + 5\sigma_1^2$
Within	2000	80	25	σ_0^2
Total	16155	99	$(163.\dot{1}\dot{8})$	
(d) $k = 10$, $n = 10$				
Between	13185	9	1465	$\sigma_0^2 + 10\sigma_1^2$
Within	2250	90	25	σ_0^2
Total	15435	99	$(155.\dot{9}\dot{0})$	
(e) $k = 5$, $n = 20$				
Between	11620	4	2905	$\sigma_0^2 + 20\sigma_1^2$
Within	2375	95	25	σ_0^2
Total	13995	99	$(141.\dot{3}\dot{6})$	

Overall estimate of $\sigma_t = \sqrt{}$(total mean square) if observations are pooled: (a) 13.0, (b) 12.94, (c) 12.77, (d) 12.49, (e) 11.89.

16.7.2 The value of s_t is the appropriate measure for estimating capability indices in cases where the between-sample component cannot be eliminated. Before calculating indices like C_p or C_{pk}, it is wise to check on the approximate Normality of the overall pattern of variation. For the data of Table 16.2 we cannot actually check this, as only $\bar{x}$ and R values are available, but if we assume reasonable

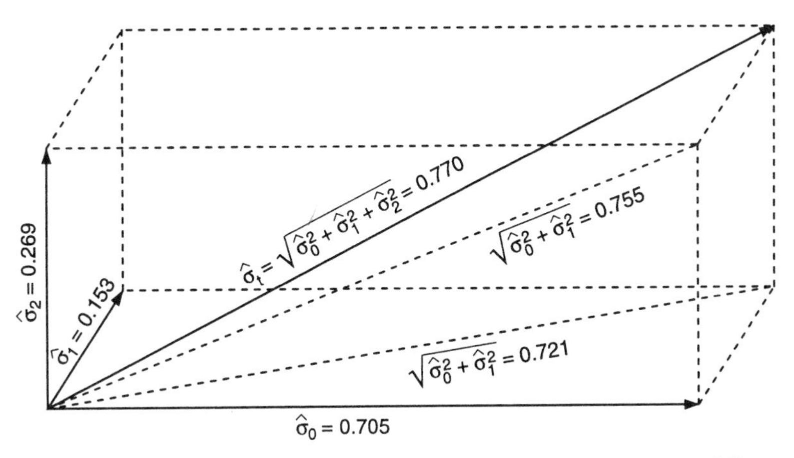

Fig. 16.5 Graphical representation of $\sigma_0, \sigma_1, \sigma_2, \sigma_t$. (Data of Table 16.2).

Normality, we have:

$$s_t = \sqrt{(0.705^2 + 0.153\,26^2 + 0.269\,34^2 + 0^2)} = 0.7701$$

For a specification of $(50.0 \pm 2.5)g$ the capability indices (with $\bar{\bar{x}} = 50.048$) are

$$C_p = 5/(6 \times 0.7701) = 1.082,$$
$$C_{pk} = (52.5 - 50.048)/(3 \times 0.7701) = 1.061.$$

The corresponding values based only on s_0 (assuming the process to be strictly in-control with $s_0 = 0.705$) are 1.182 and 1.159, respectively, some 10% greater than the true capability.

In this case, it is likely that the between-sample common causes represented by s_1 would be difficult to trace, but suppose that the cause of the cyclic component, s_2, can be identified and eliminated. We would then have the overall variation s_t reduced to

$$s_t = \sqrt{(0.705^2 + 0.15326^2 + 0^2 + 0^2)} = 0.7215$$

giving $C_p = 1.155$ and $C_{pk} = 1.118$. The slight (random) instability represented by s_1 does not have a serious effect on capability, but it could be worth trying to eliminate the cyclic effect.

Figure 16.5 illustrates the interplay of s_0, s_1 and s_2 and their effects on s_t.

16.7.3 A second important use of the components is in deriving realistic standard errors for sampling plans or control schemes. These in turn provide the basis for confidence intervals in estimating parameters, or for control limits. Consider further the preceding example, and let us assume that the cyclic contribution has been (or can be) eliminated, leaving $s_0 = 0.705$ and $s_1 = 0.153\,26$. Now if s_1 is a random between-samples element of common-cause variation, the standard error of the mean of a sample of n items is

$$\hat{\sigma}_{\bar{x}} = \sqrt{\left(s_1^2 + \frac{s_0^2}{n}\right)} \tag{16.15}$$

Note the use of $\hat{\sigma}_{\bar{x}}$ rather than $s_{\bar{x}}$ here; the latter is the observed standard error of a sequence of $\bar{x}$ values for a specific sample size, whereas $\hat{\sigma}_{\bar{x}}$ is a prediction (for any sample size n) based on the estimated components. This means that although in the example $n = 5$ was used, if a different sample size is preferred, say $n = 3$, the appropriate standard error can be predicted from (16.15).

Thus for $n = 5$, we have

$$\hat{\sigma}_{\bar{x}} = \sqrt{\left(0.15326^2 + \frac{0.705^2}{5}\right)} = 0.350\,56$$

smaller than $s_{\bar{x}}$ ($= 0.442\,08$) because of the elimination of the cyclic component s_2. If $n = 3$ were preferred, then

$$\hat{\sigma}_{\bar{x}} \rightarrow \sqrt{\left(0.15326^2 + \frac{0.705^2}{3}\right)} = 0.434\,93$$

We may look at the effect on control limits of using $\hat{\sigma}_{\bar{x}}$ rather than $D_4 \times \bar{R}$ when only within-sample variation is allowed for. With $\bar{R} = 1.64$, conventional 3-sigma limits via $\bar{R}$ would be (for $n = 5$):

$$\text{Target} \pm 0.577 \times \bar{R} = T \pm 0.946$$

These are, of course, identical to

$$T \pm 3 \times \frac{s_0}{\sqrt{n}} = T \pm 3 \times \frac{0.705}{\sqrt{5}} = T \pm 0.946$$

The limits are widened if it is necessary to allow for the between-sample common causes, and using (16.15) with $n = 5$, the 3-sigma limits are

$$T \pm 3 \times 0.35056 = T \pm 1.052$$

Evidently the same approach can be used when UK action and warning lines are required, or when the standard error is needed to set up decision criteria for cusum schemes, sampling inspection plans, or in setting confidence intervals for sample estimates of batch averages, etc.

16.7.4 The discussion of possible alternative sample sizes in the previous section raises the question of whether there is an ideal sample size/frequency configuration for any particular application. The use, especially in connection with $\bar{x}$, R-charts, of $n = 5$ is widespread, but is based more on rule-of-thumb and calculation convenience than on any optimal property. When are larger or smaller sample sizes appropriate? And if total sampling effort is a constraint, are more frequent samples of smaller size preferable to less frequent larger samples?

These questions are far-reaching and involve many statistical considerations: ideas of precision, speed of response (e.g. average run length), cost effectiveness and the nature of possible process problems (slippage, trend, erratic behaviour) may all be involved. Here we consider simply one aspect of cost effectiveness when between-sample and within-sample sources of variation are present.

As a very general rule, when s_1 is present, and especially when it is large compared with s_0, more frequent samples of small size are preferable to less frequent but larger samples.

Suppose that we have $s_1 = 1.0$, $s_0 = 2.0$, and within a specified period we can 'afford' (by cost, time, effort etc.) a total of $k \times n = 100$ items. For various combinations of k and n, the standard error of $\bar{\bar{x}}$ (i.e. the precision of the estimate of the process mean) is as follows:

$$k = 100, \quad n = 1; \quad \hat{\sigma}_{\bar{\bar{x}}} = \sqrt{\left(\frac{s_1^2}{100} + \frac{s_0^2}{100 \times 1}\right)} = 0.223\,6;$$

$$k = 20, \; n = 5; \quad \hat{\sigma}_{\bar{\bar{x}}} = \sqrt{\left(\frac{s_1^2}{20} + \frac{s_0^2}{20 \times 5}\right)} = 0.300\,0;$$

$$k = 5, \quad n = 20; \quad \hat{\sigma}_{\bar{\bar{x}}} \text{ (similarly)} = 0.489\,9;$$

$$k = 1, \quad n = 100; \quad \hat{\sigma}_{\bar{\bar{x}}} = 1.02.$$

While $k = 1$, $n = 100$ provides the most precise estimate of the overall mean, it provides no estimate of within-sample variation, so for practical reasons $k = 50$, $n = 2$, with $\hat{\sigma}_{\bar{\bar{x}}} = 0.244\,9$ would probably be preferred.

Of course, samples are not without costs; if some knowledge of cost per sampling occasion and per item sampled are available, a value-for-money plan can be devised. Let c_1 represent the cost of drawing a sample, irrespective of its size. This could include the cost of the time involved in 'visiting' the process, but excluding time to actually select the items; or the cost of setting up a test machine, preparing reagents for an analysis, interrupting a production run, etc. Let c_0 represent the additional cost per item sampled, such as its value if the test destroys it, the cost of time taken to test, measure, select it, etc. Then for a sample of size n, the total sampling cost becomes $c_1 + nc_0$.

If between-sample and within-sample components of variation, s_1 and s_0, are available, we also have the standard error of the sample mean via (16.15); the best value-for-money sample size then occurs with

$$n \risingdotseq \frac{s_0}{s_1}\sqrt{\left(\frac{c_1}{c_0}\right)} \tag{16.16}$$

Note that the ratios of s_1 to s_0 and c_1 to c_0 are the most relevant parameters. The $\risingdotseq$ symbol takes account of the fact that n must be an integer, and also that the cost function is often quite flat over several adjacent values of n. Generally, for process control purposes, one would also choose n to be at least

2 (even if $n=1$ is marginally more cost effective) so that some within-sample information is available.

We note that when the optimum n is very large (because s_0 greatly exceeds s_1, or c_1 is much larger than c_0, or both) one must consider other aspects before opting for infrequent but large samples. If there are long intervals between sampling occasions, there is the obvious risk that the process may operate out of control for some time before the control system signals its condition. There are methods for designing 'minimum cost' procedures, but they require detailed input of the likely frequency of out-of-control excursions, the costs of both unnecessary action due to false alarms and of delays in detecting slippage of a given magnitude, and the average run length performance of the control procedure for both in-control and specified out-of-control conditions. They tend to overlook the objective of continuous improvement, and the fact that 'out-of-control' may include numerous unpredictable effects as well as a simple slippage of the mean from a target level to some other value.

16.7.5 Table 16.5 illustrates the effect of sampling costs and components of variation for three widely differing situations. The earlier example with a cyclic component (assuming the cycle can be reduced or eliminated) leaves a small sample-to-sample component, relative to s_0. An example from a package filling line has s_1 of about the same size as s_0, and the example from section 11.3.1 for a forging process has s_1 (the component related to furnace variations) dominating s_0.

For the first case, the ratio of c_1 to c_0 has little overall effect unless c_1 is either negligible or dominant. In the second case, relatively small sample sizes (thereby permitting more frequent sampling) are preferable unless c_1 is dominant.

Table 16.5 Optimizing sample size: three situations

Table shows sample size for optimum cost per unit of informotion; $n \doteqdot \frac{s_0}{s_1}\sqrt{\left(\frac{c_1}{c_0}\right)}$.

	Process with small s_1/s_0	Filling tine $s_1 \doteqdot s_0$	Forge-out-furnace $s_1 > s_0$
c_1/c_0	$s_1=0.153$ $s_0=0.705$	$s_1=1.53$ $s_0=1.75$	$s_1=0.0063$ $s_0=0.0029$
0.2	2.06 → 2	0.51 → 1 or 2*	0.21 → 1 or 2
0.5	3.26 → 3	0.81 → 1 or 2*	0.33 → 1 or 2
1.0	4.61 → 5	1.14 → 2	0.46 → 1 or 2
2.0	6.52 → 6 or 7	1.62 → 2	0.65 → 1 or 2
5.0	10.30 → 10	2.56 → 2 or 3	1.03 → 1 or 2
10.0	14.6 → 15	3.62 → 3 or 4	1.46 → 2
20.0	20.1 → 20	5.12 → 5	2.06 → 2

*Sample size $n=2$ essential, at least occasionally, to monitor within-sample variation.

Table 16.6 Examples of alternative sampling plans and costs

k	n	c_1/c_0	cost $k\left(\frac{c_1}{c_0}+n\right)$	Standard error $\sqrt{\left(\frac{s_1^2}{k}+\frac{s_0^2}{k_n}\right)}$
5	7	2	45	0.1374
6	7	2	54	0.1254
3	10	5	45	0.156
2	15	10	50	0.168

($s_1 = 0.153$, $s_0 = 0.705$)

For the forge-and-furnace case, samples of more than two items are a poor return for expenditure or effort.

In cases like the first, with s_1/s_0 less than 1, the cost-efficiency curve is quite flat so that precise costs are not essential to make reasonable decisions. Suppose for example that we can afford to spend about 50 cost units within a given time period, but the ratio of c_1/c_0 is only vaguely known, say between 2 and 10. We then have the following options:

Say $c_1 = 2c_0$, then $n = 7$ and cost per sample $= 2 + 7 = 9$ units;
we can 'afford' about 5 or 6 samples.
If $c_1 = 5c_0$, $n = 10$ and cost per sample $= 15$,
we can afford three samples.
If $c_1 = 10c_0$, $n = 15$, and with cost per sample $= 25$,
we can afford two samples.

The overall costs and standard errors for the set of $k \times n$ observations are as in Table 16.6.

From Table 16.6, $k = 3$, $n = 10$ provides a 'middle of the road' sampling format with reasonable standard error (compared with others). If the value of c_1/c_0 were actually 2, the cost of this plan would fall to 36 units; if it were actually 10, the plan would cost 60 units. Thus for small s_1/s_0, convenience and other non-statistical factors can dictate the final choice of sampling plan.

16.8 Analysis for counted data

16.8.1 Similar techniques are available for counted data, and even the estimation of components of variation may occasionally be useful. We shall not venture into details for the varying sample size situations involved in p and u-charts, except to note that the methods of section 12.9, and especially (12.10) and (12.12) are applicable in these cases. We shall briefly indicate some useful methods for np and c charts, based on the dispersion tests introduced in section 6.4.3.

The statistics required from the control chart data are the sample size n (in the case of np charts), the arithmetic mean of the number of non-conforming items ($n\bar{p}$) or non-conformities ($\bar{c}$), and the standard deviations of these numbers of occurrences, say s_{np} and s_c for np and c applications respectively.

16.8.1 We deal firstly with np charts. The equivalent of the $\sqrt{F}$ test for assessing the overall stability is to test s_{np} against the theoretical variation in np-values which is used to set up the control limits. This theoretical standard error was introduced in Table 5.4, and is given by $\sqrt{[np(1-p)]}$. We therefore calculate the test statistic

$$G = \frac{s_{np}}{\sqrt{[np(1-p)]}} \tag{16.17}$$

and its significance may be evaluated via Table A.2 using $v = k-1$ degrees of freedom, where k is the number of np-values used to calculate s_{np}.

The corresponding statistic for a c-chart again compares the actual standard deviation of the c-values, s_c, with the theoretical value, $\sqrt{c}$. We thus have

$$G = \frac{s_c}{\sqrt{c}}$$

again using Table A.2 with $(k-1)$ degrees of freedom to assess the stability.

Note that Table A.2 covers both tails of the distribution $\sqrt{(\chi^2/v)}$. Where the test statistic is significantly *smaller* than 1.0, there is a too-good-to-be-true level of stability, i.e. an unusual degree of regularity in the np or c-values. This usually implies that the non-conforming items or non-conformities are not occurring randomly. There may, for example, be one of several machine positions that gives most of the faults (rather like the within-sample special cause associated with $\sqrt{F} < 1$ in section 16.3.4).

Conversely, $G > 1.0$ indicates (like $\sqrt{F} > 1$) that there are significant sample-to-sample fluctuations greater than can be attributed to the natural variation in a binomial (np) or Poisson (c) process.

16.8.2 As for measured characteristic, the v^2 test may be used to assess whether the sample-to-sample fluctuations are random or systematic. Especially for low counts, with frequent occurrences of 0 or 1, the v' alternative is not suitable. The diagnostics provided by v^2 are similar to those of section 16.4.4, noting that the cycles, trends or other systematic features apply to instability in the **proportion** of non-conforming items or the **rate** of occurrence of events. Such patterns may be due to materials, variations in process settings, differences between shifts, and so on.

It is even possible, following the lines of section 16.6 and Fig. 16.4, to estimate components of variation, but they are less likely to be useful than in the measurements situation.

16.8.3 As a first example, consider the c-chart data from the multiple-characteristics chart of Fig. 10.3. Here we have $\bar{c} = 9.85$, and one marginally out-of-control point at sample 16. We now obtain, from the completed data, the additional statistics

$$s_c = 3.61685, \quad \sqrt{(\tfrac{1}{2}MSSD)} = 3.81824$$

(excluding sample 16, $\bar{c} = 9.263$, $s_c = 2.557$, $\sqrt{(\tfrac{1}{2}MSSD)} = 2.461$).

Then for the complete data,

$$G = \frac{3.61685}{\sqrt{9.85}} = 1.152$$

with $\nu = 19$. Table A.2 indicates this to be within the 10% upper value for $\sqrt{(\chi^2/\nu)}$ and is thus not significant.

Excluding sample 16, G falls to 0.84, close to the *lower* 25% point and again non-significant.

One would not anticipate any useful indication from v^2, though unusual patterns can sometimes be detected in superficially in-control data. In this case, $v^2 = 1.1145$ for the complete data ($u = 0.537$, non-significant) and 0.9263 ($u = -0.338$, non-significant) on excluding sample 16.

Finally, we present two examples of features in counted data that may be detected using these methods.

16.8.4 Data for a np-chart with $n = 100$ are shown in Fig. 16.6. From earlier data, $n\bar{p}$ was $3.9\dot{6}$ giving $\bar{p} = 0.039\dot{6}$ and an upper control limit (US) of 9.82.

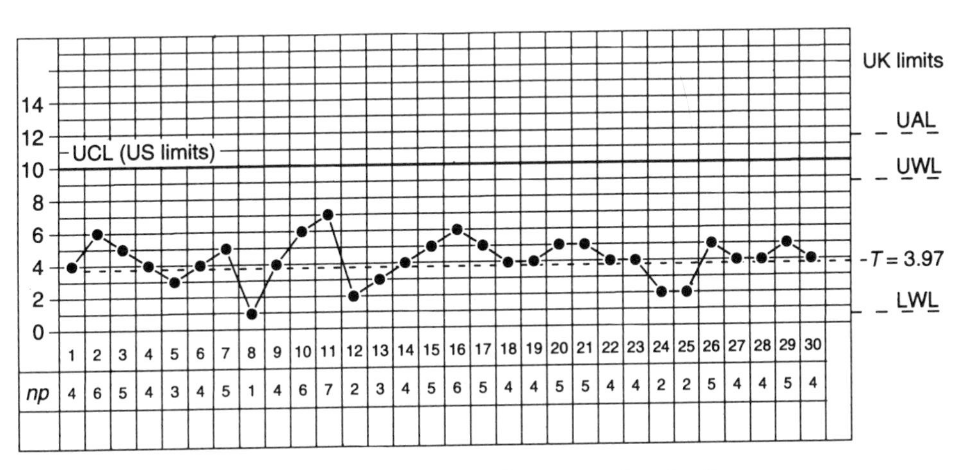

Fig. 16.6 Example of np-chart for non-conforming items.

From Fig. 10.1, UK limits would be UAL = 11.5, UWL = 8.5, LWL = 0.5, LAL not applicable (using $\alpha = 0.001$ for action lines and $\alpha = 0.025$ for warning lines). The chart shows an absence of warning values for UK limits, though only one or two would be expected, and about eight values in the outer thirds for US limits. No values exceed either UCL for the US version or UAL for the UK chart. The process would not therefore be judged out of control.

The retrospective analysis gives:

$$n\bar{p} = 4.2,\ s_{np} = 1.32353,\ \sqrt{(\tfrac{1}{2}MSSD)}\ 1.231\,76,\ \bar{p} = 0.042.$$

Then from (16.17), $G = 0.659\,82$, which lies between the lower 0.1% and 0.5% points for 29 degrees of freedom in Table A.2. There is significant indication of unexpectedly small variation in the *np* values – rather too much regularity in numbers of non-conforming items for their occurrence to be considered purely binomial.

In this case, v^2 offers no further clues, giving a value of 0.866 1 with $u = -0.757$ (not significantly below 1.0).

An example of a *c*-chart appears in Fig. 16.7. The previous value of $\bar{c}$ had been close to 6, giving UCL = 13.35 for US procedure, and UAL = 15.5, UWL = 11.5, LWL = 1.5, LAL not applicable for a UK chart. For this chart, signals would have been noted at samples 22 and 30, and the UK chart would give warning indications on samples 17, 14 and 20. The latter half of the data also has twelve points above the target line and only three below, so that there are ample on-line prompts to corrective action. How do the G and v^2 tests react to this situation – in retrospect, of course?

From the data at the foot of Fig. 16.7:

$$\bar{c} = 7.4, \quad s_c = 3.410, \quad \sqrt{(\tfrac{1}{2}MSSD)} = 2.487\,9.$$

then $G = 1.2534$, close to the upper 2.5% point for 29 degrees of freedom,

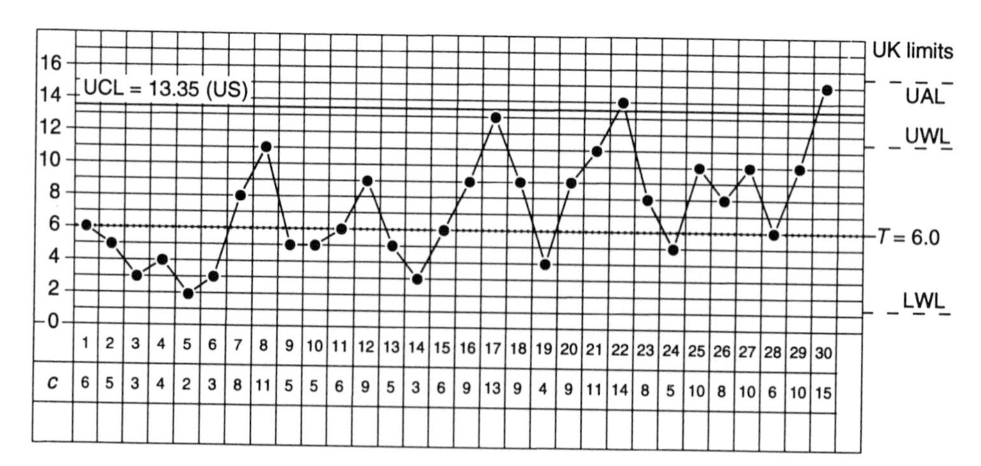

Fig. 16.7 Example of *c*-chart for non-conformities.

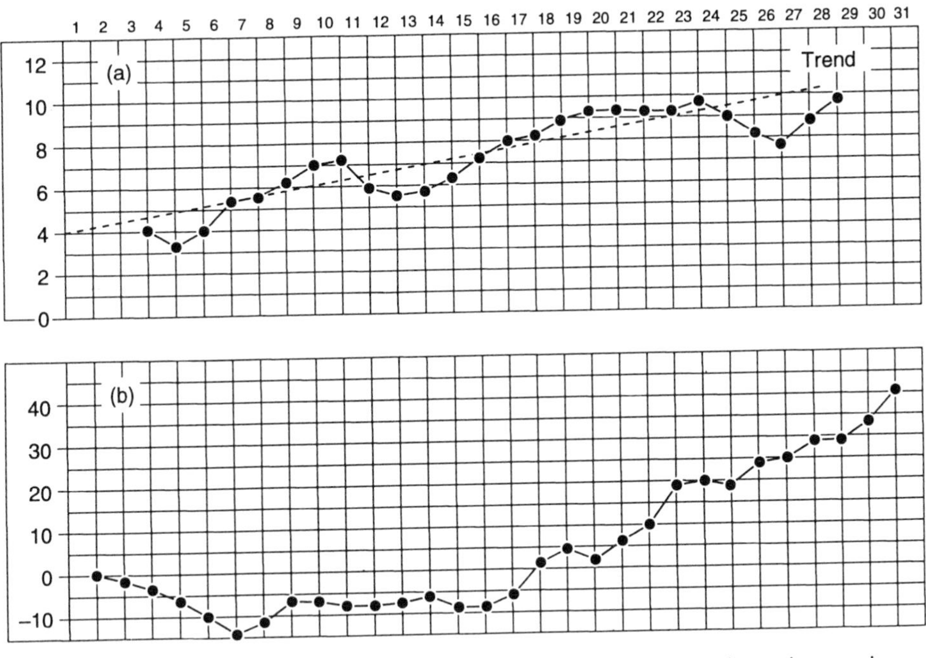

Fig. 16.8 Moving average and cusum of data from Fig. 16.6. (a) Five-point moving average (centred on middle value); (b) cusum, $T = 6.0$.

indicating greater sample-to-sample variation than expected from a Poisson variable with mean of 7.4. We also have $v^2 = 0.532\,3$, and $u(v^2) = -2.646$, which suggests trend, cyclic behaviour or other shifts in average. Note also that $\sqrt{(\frac{1}{2}MSSD)}$ is close to $\sqrt{7.4}$ $(=2.72)$, so that after allowing for the systematic shifts in average the remaining between-sample variation is consistent with the Poisson model.

Inspection of the control chart suggests a change in mean between the first and second halves of the data, although there is a vague possibility of a four-to-five sample cycle as well. A five-point moving average, shown in Fig. 16.7(a), smooths this cycle but shows a fairly steady upward trend, and the steadily increasing slope of the cusum in Fig. 16.7(b) gives a similar indication, with little evidence of cycling.

The example shows that even when series data is not treated pro-actively via a continuously updated control chart, the G and v^2 statistics as routine summaries can signal (retrospectively) features worth investigation. The same is true of the F and v^2 (or v') ratios for measurements.

17

SPC in 'non-product' applications

17.1 Introduction

17.1.1 The title of this Chapter must not be taken to imply a separation between the activities of manufacture and those of the many supporting functions essential to production. Rather, it indicates that there often needs to be change of emphasis in the techniques used for monitoring and improvement of service, clerical and administrative operations as compared with those for products where often a customer specification provides a guide to the quality required.

17.1.2 Statistical process control is generally introduced to an organization at the output end of an industrial process or production line. The techniques adopted need, initially at least, to be simple. Thus they are likely to include Pareto analysis, brain-storming and cause-effect diagrams and control charts such as $\bar{x}$, R, np and c (or perhaps p and u). From the output of such charts, measures of capability in terms of specification limits are developed (e.g. C-indices).

Before long, attention shifts to the process itself. This is essential if problems are to be solved rather than shelved, and if reduction of variation is to be achieved. A knowledge of the sources of this variation, and an improved understanding of how the process works and how its subdivisions interact, are necessary. One of the main objects of this book is to provide a range of useful techniques for furthering this understanding.

17.2 SPC in the process

17.2.1 SPC within the process, rather than on the product, needs careful consideration of the type of measurement that may be useful. Three of the early 'discoveries' are frequently that

(i) in-process measurements are often obtained singly, and subgrouping to force them into the $\bar{x}$, R format is impracticable or counter-productive;

(ii) emphasis is often on the detection of trends or medium-term shifts in performance rather than occasional large and individual-value violations;
(iii) process measurements, such as temperatures, flow rates, properties of materials, often give non-Normal distribution patterns.

It is also realized that except for simple engineering processes, so-called 'control adjustment' is often a complicated business. Changing one variable in a process to rectify a particular output characteristic can have adverse effects on another characteristic, or may itself affect other process variables. This is among the reasons for requiring improved **understanding** of the whole system: equipment, machinery, technology, environmental and human aspects.

17.2.2 At this stage, manufacturing support functions must be included within the SPC scene: maintenance, engineering, materials supply and handling, laboratories and of course suppliers of both materials and equipment. While the basic philosophy of SPC is essentially applicable, a broader base of methods and careful choice of techniques are essential.

For simple processes, there is a direct feedback loop from product to process. If machined components are undersize, a tool setting can be adjusted; if bottles are overfilled, a timer or cut-off setting can be adjusted. Where there are more complex relationships, it may even be necessary to incorporate regression techniques or the design and analysis of appropriate experiments in order to pursue improvement. There are numerous sources of information on these methods – some are listed in the bibliography.

17.2.3 When dealing with process, rather than product, characteristics, the term 'specification' may take on a rather different meaning than its familiar use as a basis for capability measurement. The process specification limits are often technical 'state of the art' bands or zones, usually combined with a target value or aim at which the process is believed to perform well. There is rarely any formal interpretation in terms of failing to satisfy a customer or violating a product standard if values fall outside these zones. They serve to indicate when attention to some aspect of the process deserves attention or investigation in order to forestall difficulties. It is worth noting some of the differences between such process guidelines and product specifications.

(i) Initially the limits are likely to be based on technical rather than statistical considerations, and from the strict control chart viewpoint, an out-of-control state probably exists due to (as yet) unmeasured sources of variation.
(ii) The calculation of capability indices against the limits is rarely relevant. For example, occasional values outside a process guideline do not imply that the process output is unsatisfactory or will fail a customer specification.
(iii) In many cases involving process measurements, there will be control (in the technological sense) via sensors, thermostats, pressure valves, etc. These often result in grossly non-Normal distributions from their mode of opera-

tion. Thus a pure sine-wave thermostat will generate an arc sine distribution (Fig. 14.5), a simple on/off controller may yield a uniform distribution. In general, these pure forms will be modified by other sources of variation, but the essential point is that non-Normality may, in effect, be perfectly normal (small n) in the conventional sense. Capability indices can then present a very distorted impression: in the two cases above, C_p of 0.5 could mean that the process **never** goes outside the desired range.

17.2.4 The evolution of SPC for in-process measurements may follow the steps below, though not always in the same order: for example (i) and (iv) may be interchanged if a reasonable understanding of the process already exists.

(i) Data is collected on carefully selected process characteristics to measure the extent and identify the sources of existing variation;
(ii) The process measurements are, where possible related to product or output characteristics to develop an improved understanding of the process;
(iii) Using cause/effect analysis, process flow sheets, brain storming or other means, possible means of reducing variation and removing cases of disturbance are developed;
(iv) Process targets or aim zones are set to achieve satisfactory performance and encourage improvement;
(v) A system to react to out-of-zone occurrences must be instituted and maintained;
(vi) When sufficient experience is gained, statistically-based control limits or other decision criteria can be progressively introduced. Attempting this step too early can produce a 'crying wolf' effect if the system is not yet fairly stable and understood;
(vii) The system is regularly reviewed and priorities re-assessed for technical effort, noting that when real stability is achieved occasional auditing may replace more intense monitoring. This step requires that process parameters themselves are now well controlled.

Attention to these principles can promote an atmosphere in which management, technical and operational workforce resources are combined to improve all aspects of the system. Stereotyped application of techniques must be avoided, and adaptation of methods to suit circumstances is necessary for greatest effect.

17.3 SPC in business areas

17.3.1 Because all business and commercial systems have the INPUT→ ACTIVITY→OUTPUT features that characterize a process, the concepts of SPC are applicable to business areas that have little or no immediate connection with production. All these systems have customer-satisfaction implications, especially when the many supplier/customer relationships within the system are properly recognized.

It is widely recognized that the use of SPC methods combined with dynamic problem-solving can considerably reduce costs and increase productivity. There are many areas, not necessarily related to external customer satisfaction or to product performance, where similar savings can be achieved. The potential for improving the effective use of materials, fuel, utilities and human resources are enormous, but surprisingly little statistical effort seems to be applied in these areas. Efficient handling of stocks, transport and maintenance activities are further areas for cost saving, and statistical methods can provide a useful contribution.

17.3.2 From off-line production-related activities it is but a short step to business management functions, involving literally all departments. Delays, errors and re-work occur just as often, and sometimes with even more serious results, in administration as in production. Customers are frustrated, even lost, by poor systems of correspondence, order handling , invoicing and delivery as much as by lapses in product quality standards.

Many SPC techniques, with suitable modification, are appropriate to financial, budgetary and sales monitoring. In fact moving average methods were used in such applications long before their value in industrial quality control was realized. Attribute or event charts are as applicable to monitoring trends in absence, sickness, epidemiology or accidents as they are to product defects.

17.4 Tools for business-area SPC

17.4.1 Recognizing that the problems to be tackled in business areas may be different in nature from those in production and quality, we consider the role of various SPC techniques.

There is, in whatever application one takes, a need for well-planned data collection and presentation. Data may come from surveys, questionnaires, routine records, meter readings and many other sources. It may not always be in the most convenient format for statistical analysis, and computer or manual summaries may be needed as a first step.

As in the process area, applications will often involve relationships between variables. Apart from more specialized methods, scatter plots are invaluable for investigating patterns. Time plots superimposed or arranged one below another can also be revealing, often showing sympathetic or antithetic changes, although one must be careful not to assume cause-and-effect mechanisms when in fact both variables may be affected by some other factor.

17.4.2 In control chart activities, there will not usually be a specification (a performance target does not have the same implications), so that capability indices will rarely be required. Run charts form a useful preliminary to more formal methods such as $\bar{x}$, R; indeed, plots of individual values, moving averages, business ratios and activity indices are more likely to occur, and cusum techniques will prove useful for detecting sustained changes. Methods based on the Normal distribution are often less relevant than in product quality SPC. There is a wealth of data exploration tools available, and some (like box-and-whisker plots, blob diagrams and the many non-parametric methods based on ranked data) make no assumptions about Normality or will clearly indicate when it is unsafe to assume it.

17.4.3 In the modern office, personal computers or networks with access to extensive software are increasingly available, and more and more staff become proficient in their routine use. Their speed, power and graphic potential may be exploited to make SPC and TQ (total quality) more effective, but the emphasis must be to assist people in their activities, and not simply to pass the problem on to a machine. The computer is a powerful data-handling tool, and whilst it will not solve problems it can greatly assist those who are concerned with problem-solving.

17.4.4 The need for full commitment by senior management is even more apparent when tackling 'quality' in non-production areas than in production-line SPC. There is a well-known barrier, sometimes termed 'administrative arrogance' that results in an attitude of 'SPC is all right for the works, but we don't need it in the office'. Because the vast majority of problems arise from the **system** rather than from its operation, people working in an established system frequently fail to see the problems it causes, or are afraid to rock the boat if they do see them. Unless the boat is rocked, the system will stagnate, even

Table 17.1 Areas of potential for SPC application

Accounts	General services	Reception (and telephones)
Administration	Invoicing	Research/Development
Advertising	Laboratory	Safety
Catering	Legal	Salaries/wages
Cleaning	Library	Secretarial
Computing	Maintenance	Security
Costing	Marketing	Standards
Customer relations	Method study	Stores
Data processing	Order handling	Technical services
Design	Personnel	Training
Despatch/packing	Planning	Transport
Drawing	Progress	Travel
Engineering	Public relations	Welfare/medical
Environmental	Purchasing	Word processing/typing
Fire prevention	Quality Assurance	

creating an environment where improvement becomes impossible. Table 17.1 lists numerous areas where systems operate, data is collected or is available, and where improvement can result in smoother operation, cost savings, better human relations, environmental enhancement and in general, greater efficiency and competitiveness.

Bibliography and references

Standards, manuals and tables

British Standards Institution, London

BS 600 (1935) The application of statistical methods to industrial standardization and quality control.
BS 600R (1942) Quality control charts (now withdrawn).
BS 2564 (1955) Control chart technique when manufacturing to a specification.
BS 2846* (1975–84) Statistical interpretation of quantitative data.

(1975) Part 1: Statistical analysis of quantitative data.
(1975) Part 2: Estimation of the mean-confidence interval.
(1975) Part 3: Determination of a statistical tolerance interval.
(1976) Part 4: Techniques of estimation and tests relating to means and variances.
(1977) Part 5: Power of tests relating to means and variances.
(1976) Part 6: Comparison of two means in the case of paired observations.
(1984) Part 7: Tests for departure from normality.
(1984) Part 8: Comparison of a proportion with a given value.

BS 5700 (1984) Process control using quality control chart methods and cusum techniques.
BS 5701 (1980) Number defective charts for quality control.
BS 5703 (1980–82) Data analysis and quality control using cusum techniques.

(1980) Part 1: Introduction to cusum charting.
(1980) Part 2: Decision rules and statistical tests for cusum charts and tabulations.
(1981) Part 3: Cusum methods for process/quality control by measurement.
(1982) Part 4: Cusum methods for counted/attributes data.

BS 5497† (1987) Precision of test methods. Part 1: Guide to the determination of repeatability and reproducibility.

Miscellaneous

Department of Trade and Industry (1979) *Code of Practical Guidance for Packers and Importers.* [Weights and Measures Act (1979)], HMSO, London.

*BS 2846 (1975 onwards) is continuously being updated: a selection only is presented above.
†BS 5497 includes other Parts (some in preparation) as UK equivalents of International Standard 5725.

Department of Trade and Industry (1979) *Manual of Practical Guidance for Inspectors.* [Weights and Measures Act (1979)], HMSO, London.

Ford, Chrysler, General Motors and American Society for Quality control (1990) *Measurement Systems Analysis Reference Manual*, Automotive Industry Action Group on behalf of the Companies, Troy, MI.

Ford Motor Company (1986) *Statistical Process Control for Dimensionless Materials*, EU 881A, Brentwood, Essex.

Ford Motor Company (1987) *Statistical Process Control Instruction Guide*, EU 880a, Brentwood, Essex.

Murdoch, J. and Barnes, J. A. (1986) *Statistical Tables for Science, Engineering, Management and Business Students*, Macmillan Educational, London.

National Bureau of Standards (February 1954) *Statistical Theory of Extreme Values and Some Practical Applications*, Applied Mathematics Series, 33, US Department of Commerce, Washington DC.

National Bureau of Standards (July 1953) *Probability Tables for the Analysis of Extreme Value Data*, Applied Mathematics Series, 22, US Department of Commerce, Washington DC.

Pearson, E. S. and Hartley, H. O. (Eds) (1954, 1972) *Biometrika Tables for Statisticians*, 2 vols, Cambridge University Press, Cambridge.

Society of Motor Manufacturers and Traders (SMMT), *Guidelines to Statistical Process Control*, London.

Textbooks

General statistical methods

Chatfield, C. (1984) *Statistics for Technology*, Chapman & Hall, London.

Ehrenberg, A. S. C. (1982) *Data Reduction*, John Wiley, Chichester.

Goldsmith, P. L. (Ed.) (1972) *Statistical Methods in Research and Production*, Oliver & Boyd, Edinburgh.

Moroney, M. J. (1951) *Facts from Figures*, Penguin, Harmondsworth.

Wetherill, G. B. (1972) *Elementary Statistical Methods*, Chapman & Hall, London.

Process control, quality control, and total quality management

Duncan, A. J. (1974) *Quality Control and Industrial Statistics*, Richard D. Irwin, Homewood, Illinois.

Deming, W. Edwards (1982) *Quality, Productivity and Competitive Position*, American Society for Quality Control, Milwaukee.

Grant, E. L. and Leavenworth, R. S. (1980) *Statistical Quality Control*, McGraw-Hill, New York.

Hahn, G. J. and Shapiro, S. S. (1967) *Statistical Models in Engineering*, John Wiley, New York.

Hald, A. (1952) *Statistical Theory with Engineering Applications*, John, Wiley, New York.

Ishikawa, K. (1982) *Guide to Quality Control*, Asian Productivity Organization, Tokyo.

Juran, J. K. (1979) *Quality Control Handbook*, McGraw-Hill, New York.

Oakland, J. S. (1986) *Statistical Process Control*, Heinemann, London.

Owen, Mal (1989) *SPC and Continuous Improvement*, IFS Publications, Kempston (UK) and Springer Verlag, Berlin.

Price, F. (1984) *Right First Time*, Gower, London.

Shewart, W. A. (1931) *Economic Control of Quality of Manufactured Product*, Van Nostrand, New Jersey. (Reprinted by the American Society for Quality Control, Milwaukee, 1980).

Wetherill, G. B. and Brown, D. W. (1991) *Statistical Process Control*, Chapman & Hall, London.

Experimental design and related topics

Box, G. E. P., Hunter, W. G. and Hunter, J. S. (1978) *Statistics for Experimenters*, John Wiley, New York.

Fisher, R. A. (1990) *Statistical Methods, Experimental Design and Scientific Inference*, Oxford University Press, Oxford.

John, J. A. and Quenouille, M. H. (1977) *Experiments: Design and Analysis*, Charles Griffin, London.

Ross, P. J. (1988) *Taguchi Techniques for Quality Engineering*, McGraw-Hill, New York.

Technical papers and monographs

Bissell, A. F. (1988) Control Chart limits for attributes and events. *Journal of Applied Statistics*, **15**, 97–103.

Bissell, A. F. (1990) How reliable is your capability index? *Journal of the Royal Statistcal Society C (Applied Statistics)*, **39**, 331–340.

Bissell, A. F. (1990) Control charts and cusums for high precision processes. *Total Quality Management*, **1**, 221–228.

Bissell, A. F. (1990) Weighted cusums – method and applications. *Total Quality Management*, **1**, 391–402.

Bissell, A. F. (1990) Estimating variation from data with varying element sizes. *Journal of Applied Statistics*, **18**, 287–295.

Bissell, A. F. and Williamson, R. J. (1988) Successive difference tests, theory and practice. *Journal of Applied Statistics*, **15**, 305–323.

Champ, G. W. and Woodall, W. H. (1987) Exact results for Shewhart control charts with supplementary runs rules. *Technometrics*, **29**, 393–399.

Clements, J. A. (1989) Process capability calculations for non-normal distributions. *Quality Progress*, **22**, 95–100.

Grubbs, F. E. (1969) Procedures for detecting outlying observations in samples. *Technometrics*, **11**, 1–21.

Kolmogorov, A. N. (1933) Sulla determinazione empirica di'una legge di distribuzione. *Gionarli Istituto Italiano degli Attuari*, **4**, 83–91.

Shapiro, S. S. and Wilk, M. B. (1965) An analysis of variance test for normality. *Biometrika*, **52**, 191.

Smirnov, N. V. (1948) Table for estimating the goodness of fit of empirical distributions. *Annals of Mathematical Statistics*, **19**, 279–281.

Appendix: means and standard errors of functions of variables

Table A.1 The Normal distribution

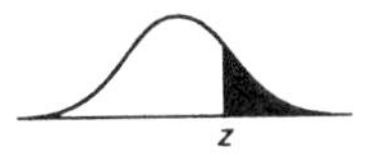

(i) *Percentage points.* Values of z (or u) for stated one-tail proportions (in %, or parts per million for $z \geqslant 3$).

%	z	%	z	p.p.m.	z	p.p.m	z
50	0.000	10	1.282	500	3.291	5	4.42
45	0.126	5	1.645	250	3.481	2.5	4.56
40	0.253	2.5	1.960	100	3.719	1	4.75
35	0.385	1	2.326	50	3.891	0.5	4.89
30	0.524	0.5	2.576	25	4.06	0.25	5.03
25	0.674	0.25	2.807	20	4.11	0.1	5.20
20	0.842	0.1	3.090	15	4.17		
15	1.036	0.05	3.291	10	4.26		

(ii) *Normal probability scale.* Read % or p.p.m. for given value of z (or u), or vice versa.

z or u 0 0.1 0.2 0.3 0.4 0.5 0.6 0.7 0.8 0.9 1.0 1.1

P(z), (%) 50 48 46 44 42 40 38 36 34 32 30 29 28 27 26 25 24 23 22 21 20 19 18 17 16 15 14

z 1.0 1.1 1.2 1.3 1.4 1.5 1.6 1.7 1.8 1.9 2.0 2.1

15 14 13 12 11 10 9 8 7 6 5 4.5 4 3.5 3 2.5 2

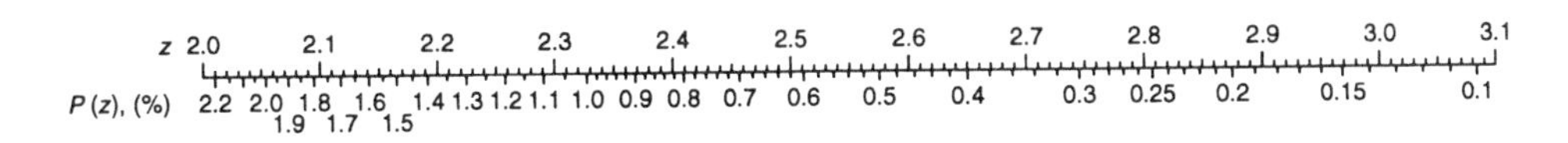

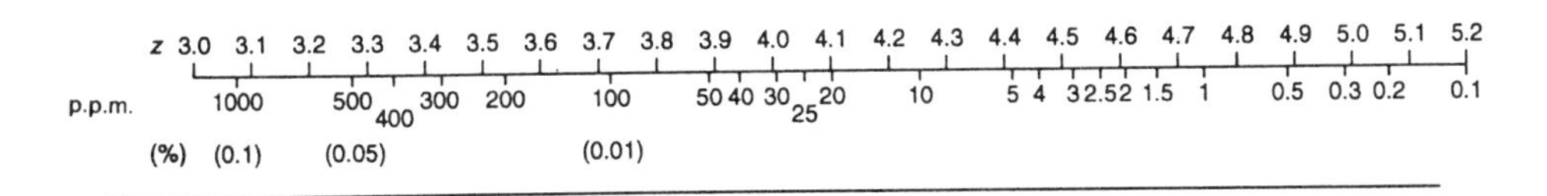

Table A.2 Percentage points of $\sqrt{(x^2/v)}$ for testing s against σ_0

n	v $(=n-1)$	Lower % points							(median) 50%	Upper % points						
		0.1	0.5	1.0	2.5	5	10	25		25	10	5	2.5	1.0	0.5	0.1
2	1	0.000013	0.0063	0.0125	0.031	0.063	0.126	0.319	0.675	1.150	1.645	1.960	2.241	2.576	2.807	3.290
3	2	0.0316	0.071	0.100	0.159	0.227	0.325	0.536	0.832	1.117	1.517	1.731	1.921	2.146	2.302	2.628
4	3	0.090	0.155	0.196	0.268	0.343	0.441	0.636	0.888	1.170	1.443	1.614	1.765	1.945	2.068	2.329
5	4	0.151	0.277	0.272	0.348	0.422	0.512	0.693	0.916	1.160	1.395	1.540	1.669	1.822	1.927	2.149
6	5	0.205	0.287	0.333	0.408	0.479	0.567	0.731	0.933	1.151	1.359	1.488	1.602	1.737	1.830	2.026
8	7	0.292	0.376	0.421	0.491	0.556	0.636	0.777	0.952	1.136	1.310	1.418	1.512	1.625	1.702	1.864
10	9	0.358	0.439	0.482	0.548	0.608	0.681	0.810	0.963	1.125	1.277	1.371	1.454	1.522	1.619	1.760
12	11	0.408	0.486	0.527	0.589	0.645	0.712	0.830	0.970	1.116	1.253	1.337	1.412	1.499	1.560	1.686
15	14	0.466	0.540	0.577	0.634	0.685	0.746	0.852	0.976	1.106	1.227	1.301	1.366	1.443	1.496	1.606
20	19	0.533	0.600	0.634	0.685	0.730	0.783	0.875	0.982	1.093	1.197	1.260	1.315	1.380	1.425	1.519
25	24	0.580	0.642	0.673	0.719	0.760	0.833	0.891	0.986	1.085	1.176	1.232	1.281	1.338	1.378	1.460
30	29	0.615	0.673	0.701	0.744	0.781	0.826	0.901	0.988	1.078	1.161	1.211	1.256	1.308	1.343	1.418
40	39	0.666	0.716	0.742	0.779	0.812	0.850	0.916	0.991	1.069	1.140	1.183	1.231	1.265	1.296	1.359
50	49	0.700	0.746	0.769	0.803	0.832	0.867	0.926	0.993	1.062	1.125	1.163	1.197	1.236	1.263	1.320
60	59	0.725	0.768	0.789	0.820	0.847	0.879	0.933	0.994	1.057	1.114	1.149	1.180	1.215	1.240	1.291
80	79	0.761	0.799	0.817	0.844	0.868	0.895	0.943	0.996	1.050	1.099	1.129	1.155	1.186	1.207	1.251
100	99	0.785	0.820	0.836	0.861	0.882	0.907	0.949	0.997	1.045	1.089	1.116	1.139	1.166	1.185	1.224
>100	z	−3.09	−2.58	−2.33	−1.96	−1.64	−1.28	−0.67	0	0.67	1.28	1.64	1.96	2.33	2.58	3.09

For $n > 100$, use $1 \pm z\sqrt{(1/2v)}$, where z is Normal variate with required tail %.

Table A.3 Percentage points of $\sqrt{F} = s_1/s_2$

(s_i, n_1, v_1 refer to sample with larger standard deviation)

$n_2(v_2)$	Percentage points for $\sqrt{F}$: 50	25	10	5	2.5	1.0	0.5	0.25	0.1
					$n_1 = 2(v_1 = 1)$				
2(1)	1.00	2.41	6.31	12.7	25.5	63.1	127	255	637
3(2)	0.82	1.60	2.92	4.30	6.21	9.92	14.1	20.0	31.6
4(3)	0.76	1.42	2.35	3.18	4.18	5.84	7.45	9.46	12.9
5(4)	0.74	1.34	2.13	2.78	3.50	4.60	5.60	6.76	8.61
6(5)	0.73	1.30	2.02	2.57	3.16	4.03	4.77	5.60	6.87
8(7)	0.71	1.25	1.89	2.36	2.84	3.50	4.03	4.59	5.41
10(9)	0.70	1.23	1.83	2.26	2.69	3.25	3.69	4.15	4.78
12(11)	0.70	1.21	1.80	2.20	2.59	3.11	3.50	3.89	4.44
15(14)	0.69	1.20	1.76	2.14	2.51	2.98	3.33	3.67	4.14
20(19)	0.69	1.19	1.73	2.09	2.43	2.86	3.17	3.48	3.88
25(24)	0.68	1.18	1.71	2.06	2.39	2.80	3.09	3.38	3.75
30(29)	0.68	1.17	1.70	2.05	2.36	2.76	3.04	3.31	3.66
50(49)	0.68	1.16	1.68	2.01	2.31	2.68	2.94	3.19	3.50
100(99)	0.68	1.16	1.66	1.98	2.28	2.63	2.87	3.10	3.39
∞	0.67	1.15	1.64	1.96	2.24	2.58	2.81	3.02	3.29
					$n_1 = 3\ (v_1 = 2)$				
2(1)	1.22	2.74	7.04	14.1	28.3	70.7	141	283	707
3(2)	1.00	1.73	3.00	4.36	6.24	9.95	14.1	20.0	31.6
4(3)	0.94	1.51	2.34	3.09	4.01	5.55	7.06	8.94	12.2
5(4)	0.91	1.41	2.08	2.64	3.26	4.24	5.13	6.16	7.83
6(5)	0.89	1.36	1.94	2.41	2.90	3.64	4.28	5.00	6.09
8(7)	0.88	1.30	1.80	2.18	2.56	3.01	3.52	3.99	4.66
10(9)	0.87	1.27	1.73	2.06	2.39	2.83	3.18	3.54	4.05
12(11)	0.86	1.26	1.69	2.00	2.30	2.68	2.99	3.29	3.72
15(14)	0.85	1.24	1.65	1.93	2.20	2.55	2.81	3.08	3.43
20(19)	0.85	1.22	1.61	1.88	2.12	2.43	2.66	2.89	3.19
25(24)	0.84	1.21	1.59	1.84	2.08	2.37	2.58	2.79	3.06
30(29)	0.84	1.21	1.58	1.82	2.05	2.33	2.53	2.72	2.97
50(49)	0.84	1.19	1.55	1.79	2.00	2.25	2.43	2.61	2.83
100(99)	0.84	1.19	1.54	1.76	1.96	2.20	2.37	2.52	2.72
∞	0.83	1.18	1.52	1.73	1.92	2.15	2.30	2.45	2.63
					$n_1 = 4\ (v_1 = 3)$				
2(1)	1.34	2.86	7.32	14.7	29.4	73.5	147	294	735
3(2)	1.07	1.78	3.03	4.38	6.26	9.96	14.1	20.0	31.6
4(3)	1.00	1.53	2.32	3.05	3.93	5.43	6.89	8.72	11.9

(Continued)

Table A.3 (*Contd.*)

$n_2(\nu_2)$	Percentage points for $\sqrt{F}$ 50	25	10	5	2.5	1.0	0.5	0.25	0.1
5(4)	0.97	1.43	2.05	2.57	3.16	4.09	4.93	5.91	7.50
6(5)	0.95	1.37	1.90	2.33	2.79	3.47	4.07	4.74	5.76
8(7)	0.93	1.31	1.75	2.08	2.43	2.91	3.30	3.72	4.33
10(9)	0.92	1.28	1.68	1.97	2.25	2.64	2.95	3.28	3.73
12(11)	0.92	1.26	1.63	1.89	2.15	2.49	2.76	3.03	3.40
15(14)	0.91	1.24	1.59	1.83	2.06	2.36	2.58	2.81	3.12
20(19)	0.90	1.22	1.55	1.77	1.98	2.24	2.43	2.62	2.88
25(24)	0.90	1.21	1.53	1.73	1.93	2.17	2.35	2.52	2.75
30(29)	0.90	1.20	1.51	1.71	1.90	2.13	2.30	2.46	2.67
50(49)	0.89	1.19	1.48	1.67	1.84	2.05	2.20	2.34	2.52
100(99)	0.89	1.18	1.46	1.64	1.80	2.00	2.13	2.26	2.42
∞	0.89	1.17	1.44	1.61	1.77	1.94	2.07	2.18	2.33
				$n_1 = 5\ (\nu_1 = 4)$					
2(1)	1.35	2.93	7.47	15.0	30.0	75.0	150	300	750
3(2)	1.10	1.80	3.04	4.39	6.26	9.96	14.1	20.0	31.6
4(3)	1.03	1.55	2.31	3.02	3.89	5.36	6.80	8.60	11.7
5(4)	1.00	1.44	2.03	2.53	3.10	4.00	4.81	5.77	7.31
6(5)	0.98	1.38	1.88	2.28	2.72	3.38	3.94	4.59	5.58
8(7)	0.96	1.31	1.72	2.03	2.35	2.80	3.17	3.57	4.15
10(9)	0.95	1.27	1.64	1.91	2.17	2.58	2.82	3.12	3.54
12(11)	0.95	1.25	1.59	1.83	2.07	2.38	2.62	2.87	3.22
15(14)	0.94	1.23	1.55	1.76	1.97	2.24	2.45	2.66	2.94
20(19)	0.93	1.21	1.51	1.70	1.89	2.12	2.30	2.47	2.70
25(24)	0.93	1.20	1.48	1.67	1.84	2.05	2.21	2.37	2.57
30(29)	0.93	1.19	1.47	1.64	1.81	2.01	2.16	2.30	2.49
50(49)	0.92	1.18	1.44	1.60	1.75	1.93	2.06	2.18	2.34
100(99)	0.92	1.17	1.42	1.57	1.71	1.87	1.99	2.10	2.24
∞	0.92	1.16	1.40	1.54	1.67	1.82	1.93	2.03	2.15
				$n_1 = 6\ (\nu_1 = 5)$					
2(1)	1.38	2.97	7.57	15.2	30.4	75.9	152	304	759
3(2)	1.12	1.81	3.05	4.39	6.27	9.96	14.1	20.0	31.6
4(3)	1.05	1.55	2.30	3.00	3.86	5.31	6.74	8.52	11.6
5(4)	1.02	1.44	2.01	2.50	3.06	3.94	4.74	5.68	7.19
6(5)	1.00	1.38	1.86	2.25	2.67	3.31	3.87	4.49	5.45
8(7)	0.98	1.31	1.70	1.99	2.30	2.73	3.09	3.47	4.03
10(9)	0.97	1.27	1.62	1.87	2.12	2.46	2.73	3.02	3.42
12(11)	0.96	1.25	1.57	1.79	2.01	2.31	2.53	2.77	3.09
15(14)	0.96	1.23	1.52	1.72	1.91	2.17	2.36	2.55	2.81
20(19)	0.95	1.21	1.48	1.66	1.83	2.04	2.20	2.36	2.57

Table A.3 (*Contd.*)

$n_2(\nu_2)$	50	25	10	5	2.5	1.0	0.5	0.25	0.1
				Percentage points for $\sqrt{F}$					
25(24)	0.95	1.20	1.45	1.62	1.78	1.97	2.12	2.26	2.44
30(29)	0.94	1.19	1.43	1.60	1.74	1.93	2.06	2.20	2.36
50(49)	0.94	1.17	1.40	1.55	1.68	1.85	1.97	2.08	2.22
100(99)	0.94	1.16	1.38	1.52	1.64	1.79	1.90	1.99	2.12
∞	0.93	1.15	1.36	1.49	1.60	1.74	1.83	1.92	2.03
				$n_1 = 8\ (\nu_1 = 7)$					
2(1)	1.41	3.02	7.68	15.4	30.8	77.0	154	308	770
3(2)	1.14	1.83	3.06	4.40	6.27	9.97	14.1	20.0	31.6
4(3)	1.07	1.56	2.29	2.98	3.82	5.26	6.67	8.43	11.5
5(4)	1.04	1.44	1.99	2.47	3.01	3.87	4.65	5.57	7.05
6(5)	1.02	1.38	1.84	2.21	2.62	3.23	3.77	4.37	5.31
8(7)	1.00	1.30	1.67	1.95	2.23	2.64	2.98	3.34	3.88
10(9)	0.99	1.27	1.58	1.81	2.05	2.37	2.62	2.89	3.27
12(11)	0.98	1.24	1.53	1.74	1.94	2.21	2.42	2.64	2.94
15(14)	0.98	1.22	1.48	1.66	1.84	2.07	2.24	2.42	2.66
20(19)	0.97	1.20	1.43	1.59	1.75	1.94	2.08	2.23	2.42
25(24)	0.97	1.18	1.41	1.56	1.70	1.87	2.00	2.12	2.29
30(29)	0.96	1.18	1.39	1.53	1.66	1.82	1.94	2.06	2.21
50(49)	0.96	1.16	1.36	1.48	1.60	1.74	1.84	1.94	2.06
100(99)	0.96	1.15	1.33	1.45	1.56	1.68	1.77	1.85	1.96
∞	0.95	1.14	1.31	1.42	1.51	1.62	1.70	1.77	1.86
				$n_1 = 10\ (\nu_1 = 9)$					
2(1)	1.42	3:04	7.74	15.5	31.0	77.6	155	310	776
3(2)	1.16	1.83	3.06	4.40	6.28	9.97	14.1	20.0	31.6
4(3)	1.08	1.56	2.29	2.97	3.80	5.23	6.62	8.37	11.4
5(4)	1.05	1.44	1.98	2.45	2.98	3.83	4.60	5.50	6.96
6(5)	1.03	1.38	1.82	2.18	2.58	3.19	3.71	4.31	5.22
8(7)	1.01	1.30	1.65	1.92	2.20	2.59	2.92	3.27	3.79
10(9)	1.00	1.27	1.56	1.78	2.01	2.31	2.56	2.82	3.18
12(11)	0.99	1.24	1.51	1.70	1.89	2.15	2.35	2.56	2.85
15(14)	0.99	1.23	1.46	1.63	1.79	2.01	2.17	2.34	2.57
20(19)	0.98	1.21	1.41	1.56	1.70	1.88	2.01	2.14	2.32
25(24)	0.98	1.20	1.38	1.52	1.64	1.80	1.92	2.04	2.19
30(29)	0.97	1.19	1.36	1.49	1.61	1.76	1.87	1.97	2.11
50(49)	0.97	1.18	1.33	1.43	1.55	1.67	1.76	1.85	1.96
100(99)	0.97	1.17	1.30	1.41	1.50	1.61	1.69	1.76	1.86
∞	0.96	1.16	1.28	1.37	1.45	1.55	1.62	1.68	1.76

(Continued)

Table A.3 (*Contd.*)

$n_2(\nu_2)$	Percentage points for $\sqrt{F}$: 50	25	10	5	2.5	1.0	0.5	0.25	0.1
				$n_1 = 12$ ($\nu_1 = 11$)					
2(1)	1.44	3.06	7.78	15.6	31.2	78.0	156	312	780
3(2)	1.17	1.84	3.07	4.41	6.28	9.97	14.1	20.0	31.6
4(3)	1.09	1.56	2.29	2.96	3.80	5.21	6.60	8.34	11.3
5(4)	1.06	1.44	1.98	2.44	2.97	3.80	4.56	5.46	6.91
6(5)	1.04	1.37	1.81	2.17	2.57	3.16	3.67	4.26	5.16
8(7)	1.02	1.30	1.64	1.90	2.18	2.57	2.88	3.22	3.73
10(9)	1.01	1.26	1.55	1.76	1.99	2.28	2.51	2.76	3.12
12(11)	1.00	1.23	1.49	1.68	1.88	2.11	2.31	2.51	2.79
15(14)	1.00	1.21	1.44	1.60	1.77	1.97	2.12	2.28	2.50
20(19)	0.99	1.18	1.39	1.53	1.68	1.83	1.96	2.09	2.25
25(24)	0.99	1.17	1.36	1.49	1.62	1.76	1.87	1.98	2.12
30(29)	0.98	1.16	1.34	1.46	1.59	1.71	1.81	1.91	2.04
50(49)	0.98	1.14	1.31	1.41	1.52	1.62	1.71	1.79	1.89
100(99)	0.98	1.13	1.28	1.37	1.48	1.56	1.63	1.70	1.78
∞	0.97	1.12	1.25	1.34	1.43	1.50	1.56	1.62	1.69
				$n = 15$ ($\nu_1 = 14$)					
2(1)	1.44	3.08	7.81	15.7	31.3	78.4	157	314	784
3(2)	1.17	1.85	3.07	4.41	6.28	9.97	14.1	20.0	31.6
4(3)	1.10	1.57	2.28	2.95	3.78	5.19	6.57	8.30	11.3
5(4)	1.07	1.44	1.97	2.42	2.95	3.77	4.53	5.42	6.85
6(5)	1.05	1.37	1.80	2.15	2.54	3.13	3.64	4.21	5.10
8(7)	1.03	1.30	1.63	1.88	2.14	2.52	2.83	3.17	3.67
10(9)	1.01	1.26	1.53	1.74	1.95	2.24	2.47	2.71	3.05
12(11)	1.01	1.23	1.48	1.65	1.83	2.07	2.26	2.45	2.72
15(14)	1.00	1.21	1.42	1.58	1.73	1.92	2.07	2.23	2.38
20(19)	0.99	1.19	1.37	1.50	1.63	1.79	1.91	2.03	2.19
25(24)	0.99	1.18	1.34	1.46	1.57	1.71	1.82	1.92	2.05
30(29)	0.99	1.17	1.33	1.43	1.53	1.66	1.76	1.85	1.97
50(49)	0.98	1.15	1.28	1.38	1.46	1.57	1.65	1.72	1.81
100(99)	0.98	1.12	1.25	1.33	1.41	1.51	1.57	1.63	1.71
∞	0.98	1.11	1.23	1.30	1.37	1.44	1.50	1.54	1.61
				$n_1 = 20$ ($\nu_1 = 19$)					
2(1)	1.45	3.09	7.85	15.7	31.5	78.7	157	315	787
3(2)	1.18	1.85	3.07	4.41	6.28	9.97	14.1	20.0	31.6
4(3)	1.11	1.57	2.28	2.94	3.77	5.17	6.54	8.27	11.3
5(4)	1.07	1.44	1.96	2.41	2.93	3.75	4.50	5.38	6.80
6(5)	1.05	1.37	1.79	2.14	2.52	3.10	3.60	4.17	5.05
8(7)	1.02	1.30	1.61	1.86	2.12	2.49	2.79	3.12	3.60
10(9)	1.01	1.25	1.52	1.72	1.92	2.20	2.42	2.66	2.99
12(11)	1.01	1.23	1.46	1.63	1.80	2.03	2.21	2.40	2.66

Table A.3 (*Contd.*)

$n_2(v_2)$	Percentage points for $\sqrt{F}$: 50	25	10	5	2.5	1.0	0.5	0.25	0.1
15(14)	1.00	1.20	1.40	1.56	1.69	1.88	2.02	2.17	2.37
20(19)	1.00	1.18	1.35	1.47	1.59	1.74	1.85	1.96	2.11
25(24)	1.00	1.16	1.32	1.43	1.53	1.66	1.76	1.85	1.98
30(29)	0.99	1.14	1.30	1.40	1.49	1.61	1.70	1.78	1.89
50(49)	0.99	1.13	1.26	1.34	1.42	1.52	1.58	1.65	1.78
100(99)	0.99	1.11	1.23	1.30	1.37	1.45	1.50	1.56	1.62
∞	0.98	1.09	1.20	1.26	1.31	1.38	1.42	1.47	1.52
				$n_1 = 25$ ($v_1 = 24$)					
2(1)	1.46	3.10	7.87	15.8	31.6	79.0	158	316	790
3(2)	1.18	1.85	3.07	4.41	6.28	9.97	14.1	20.0	31.6
4(3)	1.11	1.57	2.28	2.94	3.76	5.16	6.53	8.25	11.2
5(4)	1.08	1.44	1.96	2.40	2.92	3.73	4.48	5.35	6.77
6(5)	1.06	1.37	1.79	2.13	2.51	3.08	3.57	4.14	5.01
8(7)	1.04	1.29	1.60	1.85	2.10	2.46	2.76	3.09	3.57
10(9)	1.02	1.25	1.51	1.70	1.90	2.17	2.39	2.63	2.95
12(11)	1.02	1.22	1.45	1.62	1.78	2.01	2.18	2.36	2.62
15(14)	1.01	1.19	1.39	1.54	1.67	1.85	1.99	2.13	2.33
20(19)	1.00	1.17	1.34	1.45	1.57	1.71	1.82	1.93	2.07
25(24)	1.00	1.15	1.30	1.41	1.51	1.63	1.72	1.81	1.93
30(29)	1.00	1.14	1.28	1.38	1.47	1.58	1.66	1.74	1.85
50(49)	0.99	1.12	1.24	1.32	1.39	1.48	1.54	1.61	1.68
100(99)	0.99	1.10	1.21	1.28	1.34	1.41	1.46	1.51	1.57
∞	0.99	1.08	1.18	1.23	1.28	1.34	1.38	1.41	1.46
				$n_1 = 30$ ($v_1 = 29$)					
2(1)	1.46	3.11	7.89	15.8	31.6	79.1	158	317	791
3(2)	1.19	1.86	3.08	4.41	6.28	9.97	14.1	20.0	31.6
4(3)	1.11	1.59	2.27	2.94	3.75	5.15	6.52	8.22	11.2
5(4)	1.08	1.44	1.95	2.40	2.91	3.72	4.46	5.33	6.74
6(5)	1.06	1.37	1.78	2.12	2.50	3.06	3.56	4.12	4.99
8(7)	1.04	1.29	1.60	1.84	2.09	2.45	2.75	3.07	3.54
10(9)	1.03	1.25	1.50	1.69	1.89	2.16	2.37	2.60	2.93
12(11)	1.02	1.22	1.44	1.60	1.77	1.99	2.16	2.34	2.59
15(14)	1.01	1.19	1.38	1.52	1.66	1.83	1.97	2.10	2.30
20(19)	1.01	1.16	1.33	1.44	1.55	1.69	1.79	1.90	2.04
25(24)	1.00	1.15	1.29	1.39	1.49	1.61	1.70	1.78	1.90
30(29)	1.00	1.13	1.27	1.36	1.45	1.56	1.64	1.71	1.81
50(49)	1.00	1.11	1.23	1.30	1.37	1.46	1.52	1.57	1.65
100(99)	0.99	1.10	1.20	1.26	1.31	1.38	1.43	1.47	1.53
∞	0.99	1.08	1.16	1.21	1.26	1.31	1.34	1.37	1.42

(Continued)

Table A.3 (*Contd.*)

$n_2(\nu_2)$	50	25	10	5	2.5	1.0	0.5	0.25	0.1
				Percentage points for $\sqrt{F}$					
				$n_1 = 50$ ($\nu_1 = 49$)					
2(1)	1.47	3.12	7.92	15.9	31.7	79.4	159	318	794
3(2)	1.19	1.86	3.08	4.41	6.28	9.97	14.1	20.0	31.6
4(3)	1.12	1.57	2.27	2.93	3.74	5.13	6.50	8.21	11.2
5(4)	1.08	1.44	1.95	2.39	2.90	3.70	4.44	5.30	6.70
6(5)	1.06	1.37	1.77	2.11	2.48	3.04	3.53	4.09	4.95
8(7)	1.04	1.29	1.59	1.82	2.07	2.42	2.71	3.03	3.49
10(9)	1.03	1.24	1.49	1.67	1.86	2.13	2.34	2.56	2.88
12(11)	1.02	1.21	1.43	1.58	1.74	1.95	2.12	2.29	2.50
15(14)	1.02	1.18	1.37	1.50	1.63	1.79	1.92	2.06	2.24
20(19)	1.01	1.15	1.31	1.41	1.52	1.65	1.75	1.84	1.98
25(24)	1.01	1.14	1.27	1.37	1.45	1.56	1.65	1.73	1.83
30(29)	1.00	1.13	1.25	1.33	1.41	1.51	1.58	1.65	1.74
50(49)	1.00	1.10	1.20	1.27	1.33	1.40	1.45	1.50	1.57
100(99)	1.00	1.08	1.16	1.22	1.26	1.32	1.36	1.40	1.44
∞	0.99	1.06	1.12	1.16	1.20	1.24	1.26	1.29	1.32
				$n_1 = 100$ ($\nu_1 = 99$)					
2(1)	1.48	3.13	7.94	15.9	31.8	79.6	159	318	796
3(2)	1.20	1.86	3.08	4.41	6.28	9.97	14.1	20.0	31.6
4(3)	1.12	1.57	2.27	2.92	3.74	5.12	6.48	8.19	11.1
5(4)	1.09	1.44	1.94	2.38	2.88	3.68	4.42	5.28	6.67
6(5)	1.07	1.37	1.77	2.10	2.47	3.02	3.51	4.06	4.91
8(7)	1.05	1.29	1.58	1.81	2.05	2.40	2.69	3.00	3.47
10(9)	1.04	1.24	1.48	1.66	1.84	2.10	2.31	2.53	2.84
12(11)	1.03	1.21	1.42	1.57	1.72	1.93	2.09	2.26	2.49
15(14)	1.02	1.18	1.35	1.48	1.60	1.76	1.89	2.02	2.19
20(19)	1.01	1.15	1.29	1.39	1.49	1.61	1.71	1.80	1.93
25(24)	1.01	1.13	1.26	1.34	1.42	1.53	1.60	1.68	1.78
30(29)	1.01	1.12	1.23	1.31	1.38	1.47	1.54	1.60	1.69
50(49)	1.00	1.09	1.18	1.24	1.29	1.35	1.40	1.45	1.50
100(99)	1.00	1.07	1.14	1.18	1.22	1.26	1.30	1.33	1.37
∞	1.00	1.04	1.09	1.11	1.14	1.16	1.18	1.20	1.22
				$n_1 = \nu_1 = \infty$					
2(1)	1.48	3.14	7.96	15.9	31.9	79.8	160	319	798
3(2)	1.20	1.86	3.08	4.42	6.28	9.97	14.1	20.0	31.6
4(3)	1.13	1.57	2.27	2.92	3.73	5.11	6.47	8.17	11.1
5(4)	1.09	1.44	1.94	2.37	2.87	3.67	4.40	5.25	6.64
6(5)	1.07	1.37	1.76	2.09	2.45	3.00	3.48	4.03	4.88
8(7)	1.05	1.28	1.57	1.80	2.04	2.38	2.66	2.97	3.42
10(9)	1.04	1.24	1.47	1.65	1.83	2.08	2.28	2.49	2.80

Table A.3 (*Contd.*)

$n_2(\nu_2)$	Percentage points for $\sqrt{F}$ 50	25	10	5	2.5	1.0	0.5	0.25	0.1
12(11)	1.03	1.20	1.40	1.55	1.70	1.90	2.06	2.22	2.45
15(14)	1.02	1.17	1.34	1.46	1.58	1.73	1.85	1.98	2.15
20(19)	1.02	1.14	1.28	1.37	1.46	1.58	1.67	1.76	1.87
25(24)	1.01	1.12	1.24	1.32	1.39	1.49	1.56	1.63	1.72
30(29)	1.01	1.11	1.21	1.28	1.34	1.43	1.49	1.55	1.62
50(49)	1.01	1.08	1.15	1.20	1.25	1.30	1.34	1.38	1.43
100(99)	1.00	1.05	1.10	1.13	1.16	1.19	1.22	1.24	1.27
∞	1.00	1.00	1.00	1.00	1.00	1.00	1.00	1.00	1.00

Table A.4 Percentage points of 'Student's t'

See also Table 5.6 for table of t for use with confidence interval calculations. For one-tail value α in this table, and with ν degrees of freedom, the corresponding values in Table 5.6 appear with sample size $n = \nu + 1$ and confidence level $100(1 - 2\alpha)\%$

One-tail proportion, at or beyond t

ν	25%	10%	5%	2.5%	1%	0.5%	0.1%	0.05%
1	1.00	3.08	6.31	12.71	31.8	63.7	318.3	636.6
2	0.82	1.89	2.92	4.30	6.96	9.92	22.3	31.6
3	0.77	1.64	2.35	3.18	4.54	5.84	10.2	12.9
4	0.74	1.53	2.13	2.78	3.75	4.60	7.17	8.61
5	0.73	1.48	2.02	2.57	3.36	4.03	5.89	6.87
6	0.72	1.44	1.94	2.45	3.14	3.71	5.21	5.96
7	0.71	1.41	1.89	2.37	3.00	3.50	4.79	5.41
8	0.71	1.40	1.86	2.31	2.90	3.36	4.50	5.04
9	0.70	1.38	1.83	2.26	2.82	3.25	4.30	4.78
10	0.70	1.37	1.81	2.23	2.76	3.17	4.14	4.59
12	0.70	1.36	1.78	2.18	2.68	3.05	3.93	4.32
14	0.69	1.35	1.76	2.14	2.62	2.98	3.79	4.14
16	0.69	1.34	1.75	2.12	2.58	2.92	3.69	4.01
18	0.69	1.33	1.73	2.10	2.55	2.88	3.61	3.92
20	0.69	1.33	1.72	2.09	2.53	2.85	3.55	3.85
25	0.68	1.32	1.71	2.06	2.49	2.79	3.45	3.73
30	0.68	1.31	1.70	2.04	2.46	2.75	3.39	3.65
40	0.68	1.30	1.68	2.02	2.42	2.70	3.31	3.55
50	0.68	1.30	1.68	2.01	2.40	2.68	3.26	3.50
60	0.68	1.30	1.67	2.00	2.39	2.66	3.23	3.46
80	0.68	1.29	1.66	1.99	2.37	2.64	3.20	3.42
100	0.68	1.29	1.66	1.98	2.36	2.63	3.17	3.39
∞	0.67	1.28	1.64	1.96	2.33	2.58	3.09	3.29

Values for $\nu = \infty$ are indentical to z.

A.1 Introduction

At several points in this book, reference has been made (directly or indirectly) to variables derived from others in various ways. For example, the arithmetic mean is a simple function ($\Sigma x/n$) of the n values that contribute to it. Transformations, such as $\sqrt{x}$, $1/x$ or $\ln x$, have also been mentioned.

In many cases, where the means and standard deviations of the primary or contributory variables are known, the corresponding statistics of the derived variable can be obtained exactly, or in other cases, estimated approximately. This Appendix describes some of the more useful relationships, and examples are given for each broad type of function.

A.2 Linear combinations of independent variables

A.2.1 A formal definition of independence involves the distribution functions of the variables, but implies that the probability of occurrence of any value of y is unaffected by the value of another variable, x. There is a further implication that y and x are uncorrelated, i.e. have a zero coefficient of correlation. Although absence of correlation is not a sufficient definition of independence, it is generally a satisfactory guide to the use of the methods here – cases of dependent but uncorrelated variables are unusual, such as inverted u-shaped relationship between y and x.

For the most general form of linear function, we may write the output variable in terms of one or more contributory variables, x_1, x_2, x_3 etc. Then

$$y = a + bx_1 + cx_2 + dx_3 \text{ etc.} \tag{A.1}$$

For independence between x_1 and x_2, and all other pairs, we have

$$\mu_y = a + b\mu_1 + c\mu_2 + d\mu_3 \text{ etc.} \tag{A.2}$$

and

$$\sigma_y^2 = b^2\sigma_1^2 + c^2\sigma_2^2 + d^2\sigma_3^2 \text{ etc.} \tag{A.3}$$

Some examples will illustrate the importance of these results.

A.2.2 Standard Error of $\bar{x}$.

Let us write $y = \sum_1^n x_i$. Then,

$$y = x_1 + x_2 + x_3 \ldots x_n$$

i.e. a linear function of the x_i with $a = 0$ and all other coefficients equal to one. Further, all the x_i are from the same population with mean μ_x and standard deviation σ_x. Thus,

$$\mu_y = \mu_x + \mu_x + \mu_x \ldots n \text{ times}$$
$$= n\mu_x$$

and

$$\sigma_y^2 = \sigma_x^2 + \sigma_x^2 + \sigma_x^2 \ldots n \text{ times}$$
$$= n\sigma_x^2$$

Now set $z = (1/n)y$, another linear function but with $a = 0$, $b = 1/n$ and all other coefficients zero. We now have

$$\mu_2 = \frac{1}{n}\mu_y = \frac{1}{n}n\mu_x = \mu_x$$

and

$$\sigma_z^2 = \left(\frac{1}{n}\right)^2 \sigma_y^2 = \left(\frac{1}{n}\right)^2 n\sigma_x^2 = \sigma_x^2/n$$

This is the well-known result that the distribution of $\bar{x}$ has the same mean as that of x, and the standard error of $\bar{x}$ (the square root of its variance) is $\sigma/\sqrt{n}$.

A.2.3 Sums and differences of random variables

These are special cases of the linear function, and for $y = x_1 + x_2$ we have $a = 0$, $b = c = 1$ and all other coefficients zero. Then,

$$\mu_y = \mu_1 + \mu_2 \quad \text{and} \quad \sigma_y^2 = \sigma_1^2 + \sigma_2^2$$

an important result used, for example, in combining variance components from repeatability and reproducibility, and components arising from various parts or stages of a process.

Similarly for $y = x_1 - x_2$, with $a = 0$, $b = 1$, $c = -1$

$$\mu_y = \mu_1 - \mu_2$$

but because of the squaring of the coefficients, $\sigma_y^2 = \sigma_1^2 + \sigma_2^2$. Another important result, with uses in testing differences between means, and providing the basis for the $\sqrt{\tfrac{1}{2}(MSSD)}$ measure of variation in a sequence of values.

A.2.4 Numerical examples

(i) Suppose that a material is prepared by an automatic weighing and dispensing machine. For each of the ingredients dispensed by different stations, the means and standard deviations are:

Resin	$\mu_1 = 50$ kg,	$\sigma_1 = 0.5$
Catalyst	$\mu_2 = 2$ kg,	$\sigma_2 = 0.15$
Reinforcement	$\mu_3 = 30$ kg,	$\sigma_3 = 1.6$
Filler	$\mu_4 = 18$ kg,	$\sigma_4 = 0.8$

The total target 'shot' is thus $50+2+30+18=100$ kg. What is the variation in shot weight?
If the stations dispense independently,

$$\sigma_y^2=\sigma_1^2+\sigma_2^2+\sigma_3^2+\sigma_4^2$$

so

$$\sigma_{\text{total}}^2=0.5^2+0.15^2+1.6^2+0.8^2=3.4725$$

whence $\sigma_{\text{total}}=1.8635$.
It may thus be reasonable to assume that about 95% of shot weights will lie between 96.35 and 103.65 kg, and almost all (99.9%) between 93.87 and 106.13 kg, using $z=1.96$ and $z=3.29$ in these estimates.

(ii) A sample of packaged goods is found to have to an average gross weight of 522 g and standard deviation 5.6. The containers average 18 g with standard deviation 0.8. What proportion of net weights are likely to be less than 500 grams?

Here, the combination of net and container weights gives the gross weight distribution. Thus the linear model is

$$\text{Gross} = \text{Net} + \text{Container}$$

so that y represents gross, x_1 is the (unknown) net weight and x_2 the container weights.
Then,

$$522=\mu_{\text{net}}+18, \quad \text{whence } \mu_{\text{net}}=504,$$
$$5.6^2=\sigma_{\text{net}}^2+0.8^2, \text{ so } \sigma_{\text{net}}=\sqrt{30.72}=5.54.$$

At 500 g, $z=(500-504)/5.54=-0.72$, giving about 24% of contents at or below 500 g.

The expressions for linear combinations find many applications in measurement precision, tolerancing and estimating the total effect of several contributions to process variation.

A.3 Linear combination of correlated variables

A.3.1 Although correlation has not been covered in this book, the correlation coefficient is likely to be familiar to many readers, and this section is therefore added for completeness. For the same linear model (A.1) we require the pair-wise correlations between x_1 and x_2, x_1 and x_3, x_2 and x_3 etc.

Then,

$$\mu_y=a+b\mu_1+c\mu_2+d\mu_3 \text{ etc.} \quad \text{(as for (A.2))}$$

but

$$\left.\begin{aligned}\sigma_y^2&=b^2\sigma_1^2+c^2\sigma_2^2+d^2\sigma_3^2\\&\quad+2bc\rho_{12}\sigma_1\sigma_2+2bd\rho_{13}\sigma_1\sigma_3+2cd\rho_{23}\sigma_3 \text{ etc.}\end{aligned}\right\} \quad \text{(A.4)}$$

Note that the signs of the correlation coefficients (ρ) may have the effect of either increasing σ_y^2 if they are positive or of reducing it if they are negative.

A.3.2 The variation in a series of individual values has been estimated using $\sqrt{\frac{1}{2}(MSSD)}$. An investigation reveals that because of a semi-cyclic pattern, there is a correlation of 0.4 between successive values. If the overall standard deviation is 1.7, what is the effect of the correlation on $\sqrt{(\frac{1}{2}MSSD)}$?

Here we are only concerned with σ_y^2 and not with μ_y; in any case, for the model $y = x_1 - x_2$ with x_1 and x_2 from the same distribution, $\mu_y = 0$. We have

$$\sigma_y^2 = \sigma_1^2 + \sigma_2^2 + \rho_{12}\sigma_1\sigma_2$$

with $\sigma_1 = \sigma_2 = 1.7$, $\rho_{12} = 0.4$. Using (A.4)

$$\sigma_y^2 = 2.89 + 2.89 - 2 \times 0.4 \times 1.7 \times 1.7 = 3.468$$

This would be the quantity estimated by $MSSD$, so that $\sqrt{(\frac{1}{2}MSSD)}$ would become 1.317, appreciably underestimating the real variation in the system. This would result in control limits too close together, giving rise to over-control. If the correlation had been negative, $\sqrt{(\frac{1}{2}MSSD)}$ would over-estimate σ, giving a value of about 2.01 instead of the true 1.7.

The example illustrates the importance of the phrase 'for independent observations' that often precedes statistical formulae.

A.4 Products and quotients

A.4.1 We state these in general forms for situations where the coefficients of variation are less than about 0.3 for both variables concerned. It is also simplest to use these coefficients of variation rather than the variances of the variables, so we have

$$C_1 = \sigma_1/\mu_1, \quad C_2 = \sigma_2/\mu_2 \quad \text{and} \quad C_y = \sigma_y/\mu_y$$

A.4.2 For the product $y = x_1 \cdot x_2$ with x_1, x_2 independent,

$$\mu_y = \mu_1\mu_2, \tag{A.5}$$

$$\sigma_y^2 = \mu_2^2\sigma_1^2 + \mu_1^2\sigma_2^2 \tag{A.6}$$

but using coefficients of variation by dividing (A.6) throughout by $\mu_1^2\mu_2^2$,

$$C_y^2 = C_1^2 + C_2^2 \tag{A.7}$$

A.4.3 For the quotient $y = x_1/x_2$, the expressions are approximations, but are excellent when C_2 is less than 0.3.

$$\mu_y \doteqdot \frac{\mu_1}{\mu_2}\left(1 + \frac{\sigma_2^2}{\mu_2^2}\right), \quad \text{i.e.} \ \frac{\mu_1}{\mu_2}(1 + C_2^2) \tag{A.8}$$

$$C_y^2 \doteqdot C_1^2 + C_2^2$$

as for (A.7) except that for the quotient it is an approximate rather than an exact relationship.

A slightly better expression for σ_y^2 is

$$\sigma_y^2 \doteqdot \mu_y^2(C_1^2 + C_2^2 + 3C_1^2C_2^2) \quad \text{(A.9)}$$

the $3C_1^2C_2^2$ term becoming unimportant when either or both of C_1, C_2 are very small. Taking the square root of (A.9) gives

$$\sigma_y = \mu_y\sqrt{(C_1^2 + C_2^2 + 3C_1^2C_2^2)}$$

A.4.4 When x_1 and x_2 are correlated, some additional terms need to be incorporated. We have:

(i) $y = x_1 \cdot x_2$, correlation ρ_{12};

$$\mu_y \doteqdot \mu_1\mu_2 + \rho_{12}\sigma_1\sigma_2 \quad \text{(A.10)}$$

$$C_y^2 \doteqdot C_1^2 + C_2^2 + 2\rho_{12}C_1C_2 \quad \text{(A.11)}$$

(ii) $y = x_1/x_2$, correlation ρ_{12};

$$\mu_y \doteqdot \frac{\mu_1}{\mu_2}(1 + C_2^2 - \rho_{12}C_1C_2) \quad \text{(A.12)}$$

$$C_y^2 \doteqdot C_1^2 + C_2^2 - 2\rho_{12}C_1C_2 \quad \text{(A.13)}$$

Note that in (A.10) to (A.13), the sign of ρ_{12} must also be taken into account.

A.4.5

(i) Suppose we have measurements of the length and width of some moulded components. For reasons connected with post-shrinkage, there is some correlation between the length and width dimensions, say $\rho = 0.3$. We wish to obtain the mean and standard deviation of the areas of the mouldings. Say for length, $\mu_1 = 4$ cm and $\sigma_1 = 0.05$, for width $\mu_2 = 2.5$ cm and $\sigma_2 = 0.05$ Then

$$\mu_y(\text{area}) \doteqdot 4 \times 2.5 + (0.3 \times 0.05 \times 0.05) = 10.000\,75\ \text{cm}^2,$$

$$C_y \doteqdot \sqrt{\left[\left(\frac{0.05}{4}\right)^2 + \left(\frac{0.05}{2.5}\right)^2 + 2 \times 0.3\left(\frac{0.05}{4}\right)\left(\frac{0.05}{2.5}\right)\right]}$$

$$= 0.02658.$$

Then with $\mu_y = 10.000\,75$, $\sigma_y \doteqdot 10.000\,75 \times 0.026\,58 = 0.265\,8$ cm.

The effect of the correlation on μ_y is negligible, but σ_y is some 12.5% greater than the value would be for uncorrelated length and width.

(ii) As an example for a quotient, this time for uncorrelated variables, the volume of a liquid product is estimated indirectly by dividing the weight (kilograms) by the density (grams per ml). Setting x_1 = weight, with $\mu_1 = 5.5$ and $\sigma_1 = 0.23$, and x_2 = density with $\mu_2 = 0.903$ and $\sigma_2 = 0.007$, we have from (A.8) and (A.9)

$$\mu_y \doteqdot \frac{5.5}{0.903}\left[1+\left(\frac{0.007}{0.903}\right)^2\right]=6.091$$

$$\sigma_y \doteqdot 6.091\sqrt{\left[\left(\frac{0.23}{5.5}\right)^2+\left(\frac{0.007}{0.903}\right)^2+3\left(\frac{0.23}{5.5}\right)^2\left(\frac{0.007}{0.903}\right)^2\right]}$$

$$=0.2591$$

(the omission of the final term in the brackets would make no difference to the first four significant figures).

A.5 Non-linear functions

A.5.1 Non-linear functions of one random variable occur in numerous applications of statistics. In Chapter 14 we have noted the logarithmic, reciprocal and $\sqrt{x}$ transformations, among others, as a means of converting data to a form more likely to have an approximately Normal distribution. Taguchi uses signal-to-noise ratios based on logarithmic transformations, such as $-10\log_{10}(s/\bar{x})$. In chemistry, physics and engineering, it may be useful to predict from $\bar{x}$ and s_x what the corresponding measures may be for $y=x^3$ or $y=\sin x$.

Setting $y=f(x)$ for the general case, some useful approximate relationships exist for μ_y and σ_y in terms of μ_x and σ_x when the first and second derivations of $f(x)$ are obtainable. The value of C_x should not exceed 0.3, and the function must be monotonic over the region of practical interest. This means that either y continually increases with x, or continually decreases as x increases. Using $f'(x)$ to denote d_y/d_x and $f''(x)$ for d_y^2/dx^2 the expressions, via Taylor series expansions, are

$$\mu_y \doteqdot f(\mu_x)+\tfrac{1}{2}\sigma_x^2 f''(\mu_x) \tag{A.14}$$

$$\sigma_y \doteqdot \sigma_x \cdot f'(\mu_x) \tag{A.15}$$

Some examples will clarify these expressions. The terms $f(\mu_x), f'(\mu_x)$ and $f''(\mu_x)$ indicate the algebraic forms for $f(x)$, d_y/d_x and d_y^2/dx^2 with the value of μ substituted for x. Thus if $y=\ln x$, we have

$$\frac{dy}{dx}=\frac{1}{x}, \qquad \frac{d^2y}{dx^2}=-\frac{1}{x^2}$$

Then,

$$f(\mu_x)=\ln(\mu_x), \qquad f'(\mu_x)=\frac{1}{\mu_x}, \qquad f''(\mu_x)=-\frac{1}{\mu_x^2}$$

The approximations for μ_y and σ_y become

$$\mu_y \doteqdot \ln(\mu_x)-\frac{1}{2}\sigma_x^2\left(\frac{1}{\mu_x^2}\right)=\ln(\mu_x)-\frac{1}{2}C_x^2$$

$$\sigma_y \doteqdot \sigma_x\left(\frac{1}{\mu_x}\right)=C_x$$

Another interesting case is that for $y = x^2$. Here we have

$$\frac{dy}{dx} = 2x, \qquad \frac{d^2y}{dx^2} = 2$$

noting that the latter may be written as $2x^0$.
Then,

$$\mu_y \fallingdotseq \mu_2^2 + \tfrac{1}{2} \times 2 \times \sigma_x^2 = \mu_x^2 + \sigma_x^2$$

and

$$\sigma_y \fallingdotseq 2\sigma_x \times \mu_x$$

A.5.2 A widely used tranformation is $\sqrt{x}$, or $y = x^{1/2}$. One of its many applications is in the analysis of Poisson counts of occurrences, the square-root transformation tending to Normalize the distribution (though it remains discrete, of course). Another useful property is that it stabilizes the variance, i.e. the variance of y does not depend on μ_y, whereas for the Poisson distribution itself $\sigma_x^2 = \mu_x$. The method of (A.14) and (A.15) demonstrates how this stabilization occurs.

For $y = x^{1/2}$,

$$\frac{dy}{dx} = \frac{1}{2}x^{-1/2} \quad \text{and} \quad \frac{d^2y}{dx^2} = -\frac{1}{4}x^{-3/2}$$

These may be written in the forms

$$f(x) = \sqrt{x}, \qquad f'(x) = 1/(2\sqrt{x}), \qquad f''(x) = -1/(4x\sqrt{x})$$

Now with $\sigma_x = \sqrt{\mu_x}$, and applying (A.14), (A.15)

$$\mu_y \fallingdotseq \sqrt{\mu_x} + \frac{1}{2}\left(\frac{-\sigma_x^2}{4\mu_x\sqrt{\mu_x}}\right) = \sqrt{\mu_x} - \left(\frac{1}{8\sqrt{\mu_x}}\right)$$

$$\sigma_y \fallingdotseq \sigma_x \times \left(\frac{1}{2\sqrt{\mu_x}}\right) = \frac{1}{2} \quad \text{(i.e. a constant value for any } \mu_x\text{)}$$

A.5.3 In some cases functions may not be differentiable (or one may not remember how to differentiate them!). The following empirical method can then give useful results. From the function $y = f(x)$ calculate two dummy y-values using

$$\left.\begin{aligned} y_1 &= f(\mu_x + a\sigma_x) \\ y_2 &= f(\mu_x - a\sigma_x) \end{aligned}\right\} \tag{A.16}$$

The value of a should be such that $\mu_x \pm 2a\sigma_x$ does not yield invalid values of x, e.g. negatives or other banned regions, such as proportions exceeding 1, etc.

Then,

$$\mu_y \doteq \tfrac{1}{2}(y_1 + y_2) \tag{A.17}$$

$$\sigma_y = \tfrac{1}{2a}(y_1 - y_2) \tag{A.18}$$

In many cases, $a = 1.0$ will be convenient, but one should check that $\mu_x \pm 2\sigma_x$ yields valid values.

A.5.4 Two numerical examples will serve to illustrate these methods.

(i) Results for the residual precious metal level in an effluent, measured in parts per million, are approximately log Normally distributed. A typical set of 50 results gave $\bar{x} = 1.4514$, $s_x = 7075$. Thus $C_x = 0.4875$. Applying (A.14) and (A.15) by substituting $\bar{x}, s_x$ for μ_x, σ_x yields, for $y = \ln(x)$,

$$\hat{\mu}_y \doteq \ln(1.4514) - \tfrac{1}{2} \times 0.4875^2 = 0.2537,$$
$$\hat{\sigma}_y \doteq 0.4875$$

In fact by taking $\ln(x)$ for each of the 50 values, it was found that $\bar{y} = 0.2652$, $s_y = 0.4733$. The match is very good, especially as C_x is well above the value of 0.3 suggested as a limit in A.5.1.

To use the empirical method of A.5.3, we must note that $(1.4514 - 0)/0.7075 = 2.0514$. In other words, lower-tail deviations of more than two standard deviations give x-values in the impossible negative region. We therefore choose $a = 1.0$.

Then,

$$y_1 = \ln(1.4514 + 0.7075) = 0.7696,$$
$$y_2 = \ln(1.4514 - 0.7075) = -0.2958.$$

(negative logarithms are not invalid; they represent very small p.p.m.).

$$\hat{\mu}_y \doteq \tfrac{1}{2}(0.7696 - 0.2958) = 0.2369,$$
$$\hat{\sigma}_y \doteq \tfrac{1}{2}(0.7696 + 0.2958) = 0.5327.$$

The approximate values are reasonable, but not as good as those from the differentiated functions.

(ii) The average number of calls on an engineering service facility is 8.4 per hour, with a standard deviation of 2.0. For planning purposes, a manager enquires about the corresponding statistics for the interval between calls, assuming the calls are mutually independent.

Here with $\mu_x = 8.4$, $\sigma_x = 2.0$, we are interested in $y = 1/x$ or x^{-1}. Then via (A.14), (A.15),

$$f'(x) = -\frac{1}{x^2}, \qquad f''(x) = \frac{2}{x^3}$$

This gives

$$\mu_y \doteq \frac{1}{8.4} + \frac{1}{2} \times 2^2 \times \frac{1}{8.4^3} = 0.1224,$$

and

$$\sigma_y \doteqdot 2 \times \frac{1}{8.4^2} = 0.028\,34$$

These could be expressed as a mean of 7.34 minutes and standard deviation 1.70 minutes.

If the empirical method is used, we note that $\mu_x \pm 2\sigma_x$ stays clear of negative (and therefore invalid) numbers of calls, so that $a = 1.0$ is reasonable in (A.16) to (A.18).

Then

$$y_1 = 1/(8.4 + 2.0) = 0.961\,5,$$
$$y_2 = 1/(8.4 - 2.0) = 0.156\,25,$$
$$\mu_y \doteqdot 0.1262, \quad \sigma_y \doteqdot 0.0300,$$

quite close to the more rigorous method.

A.6 Dealing with composite functions

A.6.1 For more complex cases, functions may be broken down into two or more stages. Thus suppose we are interested in the function $z = \sin^2(x + y)$. We must first ensure that the sum $(x + y)$ remains within one quadrant (0–90°, 90–180°, etc.) otherwise the sine will be an oscillating rather than monotonic function. If this condition is satisfied, and given the means and standard deviations of x and y, we can proceed thus:

(i) Set $v = x + y$ and deduce μ_v and σ_v;
(ii) Set $w = \sin v$, and hence find μ_w and σ_w;
(iii) Finally set $z = w^2$ to obtain μ_z and σ_z.

When dealing with composite cases, it may be necessary to tighten the requirement for coefficients of variation (CV), for example to ensure that for the separate variables none has CV greater than 0.2.

A.6.2 A practical example concerned the estimation of capability for output current from an electrical circuit. The circuit included five resistors in parallel, all drawn from a production stream with mean = 10, standard deviation = 0.2. The electrical supply across the circuit was a nominal 12 volts, with standard deviation 0.25.

Let r denote individual resistance, R the resistance of the five components in parallel, v the supply voltage and a the output current. Then, from electrical theory,

$$\frac{1}{R} = \frac{1}{r_1} + \frac{1}{r_2} + \frac{1}{r_3} + \frac{1}{r_4} + \frac{1}{r_5} = \sum_1^5 \frac{1}{r}$$

and

$$A = V \div R.$$

We shall thus require intermediate variables say $x = 1/r$ and $y = 1/R$ so that $R = 1/y$.

Dealing first with $x = 1/r$, we have

$$f(r) = \frac{1}{r}, \quad f'(r) = -\frac{1}{r^2}, \quad f''(r) = \frac{2}{r^3}, \text{ giving}$$

$$\mu_x \doteqdot \frac{1}{10} + \frac{1}{2}.0.2^2 \times \frac{2}{1000} = 0.100\,04,$$

$$\sigma_x \doteqdot 0.2 \times \frac{1}{100} = 0.002$$

Now with $y = x_1 + x_2 + x_3 + x_4 + x_5$, with common μ_x, σ_x for all x's, and using (A.2), (A.3) with b, c, d, e, f all equal to 1,

$$\mu_y = 5\mu_x \doteqdot 0.500\,2,$$
$$\sigma_y^2 = 5\sigma_x^2 \quad \text{whence } \sigma_y \doteqdot 0.004\,472$$

Next, for $R = 1/y$ we again have the reciprocal transformation leading to

$$\mu_R \doteqdot \frac{1}{0.500\,2} + \frac{1}{2} \times 0.004\,472^2 \times \frac{2}{0.5002^3} = 1.999\,36,$$

$$\sigma_R \doteqdot 0.004\,472 \times \frac{1}{0.500\,2^2} = 0.017\,87$$

Finally we have $A = V \div R$, with $\mu_v = 12$, $\sigma_v = 0.25$ and μ_R, σ_R as developed above. Then, using (A.8) and (A.9),

$$\mu_A \doteqdot \frac{12}{1.999\,36}\left[1 + \left(\frac{0.017\,87}{1.999\,36}\right)^2\right] = 6.002\,4$$

or approximately 6 amp,
and

$$\sigma_A \doteqdot 6.002\,4\sqrt{\left[\left(\frac{0.25}{12}\right)^2 + \left(\frac{0.017\,87}{1.999\,36}\right)^2 + 3\left(\frac{0.25}{12}\right)^2\left(\frac{0.017\,87}{1.999\,36}\right)^2\right]}$$

Again the final term has little effect, and we find

$$\sigma_A \doteqdot 0.136\,1$$

If the specification is for 6.0 ± 0.5 amp, we have

$$C_p = \frac{6.5 - 5.5}{6 \times 0.1361} = 1.225$$

The values of C_{pU}, C_{pL} and C_{pk} will, of course, depend on the actual mean resistance and voltage achieved in manufacture or operation.

A.7 Other useful standard errors

A.7.1 Standard errors of several statistical functions have been noted elsewhere in this book. Table 5.4, for example, in addition to the standard deviations of the binomial (np) and Poisson (c) variables, lists the standard errors of p and u. The standard error of $\bar{x}$ has occurred not only in Table 5.4 but by implication in all sections dealing with tests or control charts for sample means.

Table 16.1 listed standard errors for R and s, but in the case of s there are algebraic functions available for μ_s and σ_s, though these involve gamma (Γ) functions which are only available on mathematical software or some scientific calculators. For the interested reader, they are listed below.

Other functions that may be of interest include the sample median, M and the variance, s^2. These are given below for samples from a Normal distribution, but may not apply to other distribution forms, not only because of their shape but because of relationships between μ and σ. For the Normal distribution, sample values of $\bar{x}$ and s are statistically independent, but this can be quite untrue for other forms.

A.7.2 Some of the cases noted above are as follows:

(i) Mean and standard error for s in samples of size n; σ is the true standard deviation of the population.

$$\mu_s = \sigma\sqrt{\left(\frac{2}{n-1}\right)} \times \left(\frac{\Gamma(\frac{1}{2}n)}{\Gamma(\frac{1}{2}(n-1))}\right) \quad \text{(A.19)}$$

$$\sigma_s = \sigma\sqrt{\left[1 - \left(\frac{2}{n-1}\right)\left(\frac{\Gamma(\frac{1}{2}n)}{\Gamma(\frac{1}{2}(n-1))}\right)^2\right]} \quad \text{(A.20)}$$

These values are those listed in Table 16.1, except that σ_s is divided by μ_s to express it in terms of the estimate $\bar{s}$.

For large n, the expressions are approximated by

$$\mu_s \doteqdot \sigma\sqrt{\left[1 - \frac{1}{2(n-1)}\right]}$$

$$\sigma_s \doteqdot \sigma/\sqrt{[2(n-1)]}$$

For example, with $n = 10$, these approximations give $\mu_s \doteqdot 0.9718$, $\sigma_s \doteqdot 0.2357$; the true values are 0.972 66 and 0.232 24, respectively.

(ii) For the variance, the mean of s^2 equals σ^2 (i.e. s^2 is the unbiassed estimator of σ^2). The standard error of s^2 also has the simple form **for a Normal variable,**

$$\sigma(s^2) = \sigma^2\sqrt{\left(\frac{2}{n-1}\right)} \quad \text{(A.21)}$$

(iii) The situation for the median is, perhaps surprisingly, more complicated. This is because for odd-sized samples the median is estimated from the central (ordered) value alone, whereas for even-sized samples it is taken as the mean of the two central values, and is thus slightly less variable.

For large samples, from a Normal distribution,

$$\sigma(M) = \sigma\sqrt{\left(\frac{\pi}{2n}\right)} = 1.2533\frac{\sigma}{\sqrt{n}} \tag{A.22}$$

For finite sample sizes, especially those in the range $n = 2$ to 25 as used in SPC applications, the factor 1.2533 can be replaced by

$1.2533\left(1 - \frac{2}{9n}\right)$ for odd values of n

or

$1.2533\left(1 - \frac{\sqrt{2}}{2n+3}\right)$ for even values of n

These approximations are generally correct to three decimal places, and the error is never great enough to affect the second decimal place.

Thus for $n = 7$,

$$1.2533\left(1 - \frac{2}{9 \times 7}\right) = 1.2135$$

compared with the true value of 1.214 (to three decimal places).

Similarly for $n = 10$,

$$1.2533\left(1 - \frac{\sqrt{2}}{2 \times 10 + 3}\right) = 1.1762$$

compared with a true value of 1.177.

(iv) Finally, the use of log-transformations is often recommended for analysis or charting of standard deviations or variances. Applying (A.14), (A.15) to $\ln(s)$ or $\ln(s^2)$, using the standard errors for s and s^2 given in (i) and (ii) above:

$$\text{For } \ln(s),\ \mu \doteqdot \ln\left[\sigma\sqrt{\left(\frac{2n-3}{2n-2}\right)}\right] - \frac{1}{4n-6}$$

$$\sigma \doteqdot \sqrt{\left(\frac{1}{2n-3}\right)}$$

$$\text{For } \ln(s^2),\ \mu \doteqdot \ln\sigma^2 - \left(\frac{1}{n-1}\right)$$

$$\sigma \doteqdot \sqrt{\left(\frac{2}{n-1}\right)}$$

Thus for, say, $n = 5$, and setting $\sigma = 1$,
for $\ln(s)$, mean $\doteqdot -0.1382$, standard deviation $\doteqdot 0.3780$;
and for $\ln(s^2)$, mean $\doteqdot -0.25$, standard deviation $\doteqdot 0.7071$.

Index

Page numbers appearing in **bold** refer to figures and page numbers appearing in *italic* refer to tables.